SUSTAINABLE MEMBRANE TECHNOLOGY FOR ENERGY, WATER, AND ENVIRONMENT

SUSTAINABLE MEMBRANE TECHNOLOGY FOR ENERGY, WATER, AND ENVIRONMENT

Edited by

AHMAD FAUZI ISMAIL
TAKESHI MATSUURA

WILEY

A JOHN WILEY & SONS, INC., PUBLICATION

Published by John Wiley & Sons, Inc., Hoboken, New Jersey.
Published simultaneously in Canada.

Library of Congress Cataloging-in-Publication Data:

International Conference on Membrane Science and Technology (7th : Kuala Lumpur, Malaysia : 2009)
Sustainable membrane technology for energy, water, and environment / edited by Ahmad Fauzi Ismail and Takeshi Matsuura.
p. cm.
Papers presented at the 7th International Conference on Membrane Science and Technology held in Kuala Lumpur, Malaysia on May 13–15, 2009.
Includes index.
ISBN 978-1-118-02459-1 (cloth)
1. Membranes (Technology)–Congresses. I. Ismail, A. F., 1966– II. Matsuura, Takeshi, 1936– III. Title.
TP159.M4I5676 2009
660'.283–dc23

2011035193

Printed in the United States of America.

ISBN: 9781118024591

10 9 8 7 6 5 4 3 2 1

CONTENTS

FOREWORD

> Sustainable development is development that meets the needs of the present without compromising the ability of future generations to meet their own needs.
> —*The Report of the U.N. Brundtland Commission, Our Common Future, 1987*

To fulfill the mission of sustainable development, it is necessary to achieve economic growth without excessive demand on natural resources while little stress is imposed upon the environment. A significant technology shift is necessary to achieve the goal. Fortunately, it is also a fact that the technologies to be used for this purpose are already available. Membrane separation technology is nowadays recognized as one of such key technologies that can contribute to sustainable development by enhancing energy efficiency, producing clean water, and alleviating environmental contamination.

In the spirit of enhancing and promoting the roles of membrane, Advanced Membrane Technology Research Center (AMTEC), Universiti Teknologi Malaysia, took the initiative to organize the 7th International Conference on Membrane Science and Technology (MST 2009) at Kuala Lumpur, Malaysia, with the main theme of "Sustainable Energy, Water, and Environment."

To highlight advances and new findings in membrane science and technology and their impacts on technology development, the main aims were:

- To promote interdisciplinary collaboration for the common goal to achieve sustainable development,
- To foster global collaborations and joint ventures,
- To promote and facilitate growth of scientific and technical development in the field of membrane technology development in Malaysia and the ASEAN region.

With a large population of the ASEAN nations (about 300 million) and rich natural resources, for example, agricultural products in Thailand, and petroleum, natural gas, and water in Indonesia and Malaysia, sustainable growth of the region based on sensible utilization of the valuable resources will have a strong impact on the growth of the entire global economy.

This book is unique in providing information to assess the quantity and quality of current research and development activities in the ASEAN region, especially on membrane science and technology. The book also reflects upon those in the Middle East region, where the sustainable development is equally important. It is believed that the book will open up a new avenue for the establishment of global collaborations to achieve our common goal of welfare of the human society.

Prof. Dato' Ir. Dr. Zaini Ujang

Vice Chancellor
Universiti Teknologi Malaysia

PREFACE

This book consists of the papers selected from the 7th International Conference on Membrane Science and Technology 2009 (MST 2009) held on the theme of "Sustainable Technology for Energy, Water, and Environment" on May 13 to 15, 2009. MST has been held every year in different countries of the ASEAN region, such as Thailand (2003, 2008), Malaysia (2004), Indonesia (2005), and Singapore (2006). In the year 2009, MST was organized by the Advanced Membrane Technology Research Center (AMTEC), Faculty of Chemical and Natural Resources of Universiti Teknologi Malaysia and held in Kuala Lumpur (Malaysia) with the support of the Ministry of Science, Technology and Innovation Malaysia, British Council and Universiti Teknologi Malaysia. The conference enjoyed truly international flavor with more than 200 participants from 23 different countries. In particular, 13 eminent plenary and keynote speakers from all over the world delivered lectures on their fields of expertise. Papers of high quality were presented in 5 technical sessions and among those, 27 papers were specially chosen to be included in this book. In addition, focus was placed on the papers of the host country Malaysia, where strong membrane R&D activities are currently being conducted.

In particular, the book includes the following parts.

Part I covers new membrane material and membrane development: The topics cover hollow fiber spinning (Chapter 1), mixed matrix membrane for ultrafiltration (Chapter 2), composite membrane based on bacterial cellulose membrane as the substrate (Chapter 3), posttreatment of nanofiltration membrane (Chapter 4), and formation and characterization of electro-spun nanofiber membranes (Chapter 5).

Part II covers membrane applications for gas and vapor separation: The topics cover the through discussions on CO_2 permeation-structure relationship for 6FDA-based polyimide membranes (Chapters 6 and 7), mixed matrix and nanocomposite membrane for gas separation (Chapter 8), and inorganic membranes for gas separation (Chapter 9).

Part III covers membrane applications in water treatment: Water treatment processes such as skim natural rubber latex concentration, removal of diethanolamine (DEA) from wastewater, removal of copper ions, microfiltration of nonalcoholic beer and wastewater reclamation are included in this chapter. In particular, membrane fouling is discussed in Chapters 10 and 13. Both commercial reverse osmosis (RO), nanofiltration (NF), and ultrafiltration (UF) tubular membranes (Chapter 11) and laboratory made membranes (chitosan [Chapter 12]) were used for water treatment. And, finally, membrane bioreactor was used in Chapter 14 for wastewater reclamation.

Part IV covers environmental applications: Three chapters in this part address CO_2 capture, accomplished either by membrane contactor (Chapters 16 and 17) or by W/O emulsion liquid membrane (Chapter 18). Furthermore, this chapter covers topics such as surface treatment of polypropylene hollow fibers for liquid phase microextraction (Chapter 15) and the removal of dye from of liquid waste solutions (Chapter 19).

Part V covers energy applications: The topics cover combined solar energy—seawater desalination process (Chapter 20), development of polymeric membranes for fuel cell applications (Chapter 21), nanocomposite membranes for direct methanol fuel cell (DMFC) (Chapter 22), membrane development for biogas purification (Chapter 23), and for biodiesel separation (Chapter 24).

Finally, Part VI covers other industrial applications, including methyl tertiary butyl ether (MTBE) separation from methanol mixture by pervaporation (Chapter 25), fractionation of ternary xylene mixtures (Chapter 26), and ammonia removal from salty water by direct contact membrane distillation (DCMD) (Chapter 27).

The editors believe that many innovative ideas on the development of membrane science and technology are assembled in this book. The readers may also have a glimpse into the rapidly growing R&D activities in the ASEAN and the Middle East regions that are emerging as the next-generation R&D centers of membrane technologies, especially owing to their need of water and natural gas production technology. As the table of contents indicates, the book addresses applications of membrane science and technology for energy, water, and environment. Hence, this book will be useful not only for the engineers, scientists, professors, and graduate students who are engaged in the R&D activities of the field, but also for those who are interested in the sustainable development of these particular regions.

The editors would like to express their sincere thanks to the staff of the Advanced Membrane Technology Research Center (AMTEC), Universiti-Teknologi Malaysia, Johor Bahru, Johor, Malaysia for their support, encouragement, and their understanding during the period of the book's writing. The relationship and friendship between authors and AMTEC members have created a very conducive environment which motivated us and ensured the completion of this book despite many hardships. Among them are

Dr. Muhammad Noorul Anam, Mohd. Suhaimi Abdullah, Dr. Lau Woei Jye, Ng Be Cheer, Goh Pei Sean, Dr. Suhaila Sanip, Dayang Salyani, Dr. Nurmin Bolong, Dr. Hatijah Basri, Erna Yuliwati, Dr. Azeman Mustafa, Dr. Abdel Latif Hashemifard, Agung Mataram, Dr. Juhana Jaafar, Norhaniza Yusof, Farhana Aziz, Mohd Razis, Dr. Amir Mansourizadeh, Dr. Mohd Ali Aroon, and Dr. Gholamreza Bakeri just to name a few. T. Matsuura would like to thank Universiti Teknologi Malaysia for his appointment to Distinguished Visiting Professor during the years 2009 and 2010, which enabled him to dedicate his time to the edition of the book.

Ahmad Fauzi Ismail
Takeshi Matsuura

CONTRIBUTORS

Jamal Aalaei (Dr.), Polymer Science and Technology Division, Research Institute of Petroleum Industry, West Blvd., Azadi Sport Complex, Tehran, Iran

Nurbaiti Abdul Hanid (Ms.), Department of Engineering Science, Faculty of Science and Technology, Universiti Malaysia Terengganu, Mengabang Telipot, 21030, Kuala Terengganu, Malaysia

Ahmad Zuhairi Abdullah (Dr.), School of Chemical Engineering, Engineering Campus, Universiti Sains Malaysia, Seri Ampangan, 14300 Nibong Tebal, Seberang Perai Selatan, Pulau Pinang, Malaysia

Sani N.A. Abdullah, School of Chemical Engineering, Engineering Campus, Universiti Sains Malaysia, Seri Ampangan, 14300 Nibong Tebal, S.P.S, Penang, Malaysia

Ahmedy Abu Naim (Assoc. Prof. Dr.), Separation Science and Technology Group, IbnuSina Institute for Fundamental Science Studies, Universiti Teknologi Malaysia, 81310 UTM Johor Bahru, Malaysia; Department of Chemistry, Faculty of Science, Universiti Teknologi Malaysia, 81310 UTM Johor Bahru, Malaysia

Abdul Latif Ahmad (Prof. Dr.), School of Chemical Engineering, Engineering Campus, Universiti Sains Malaysia, Seri Ampangan, 14300 Nibong Tebal, S.P.S, Penang, Malaysia

Asmadi Ali (Mr.), Department of Engineering Science, Faculty of Science and Technology, Universiti Malaysia Terengganu, Mengabang Telipot, 21030, Kuala Terengganu, Malaysia

Nora'aini Ali (Dr.), Department of Engineering Science, Faculty of Science and Technology, Universiti Malaysia Terengganu, Mengabang Telipot, 21030, Kuala Terengganu, Malaysia

Mohammed Al-Jarrah (Eng.), Department of Chemical Engineering, Jordan University of Science and Technology, PO Box 3030, Irbid-Jordan

Marinah Ariffin (Dr.), Department of Chemistry, Faculty of Science and Technology, Universiti Malaysia Terengganu, Mengabang Telipot, 21030 Kuala Terengganu, Malaysia

Mohammad Ali Aroon (Dr.), Advanced Membrane Technology Research Centre (AMTEC), Universiti Teknologi Malaysia, 81310 Skudai, Johor, Malaysia

Gangasalam Arthanareeswaran (Dr.), Membrane Research Laboratory, Department of Chemical Engineering, National Institute of Technology Tiruchirappalli 620015, India

Chulsung Bae (Prof. Dr.), Department of Chemistry, University of Nevada Las Vegas, 4505 Maryland Parkway, Box 454003, Las Vegas, Nevada 89154-4003, USA

Fawzi Banat (Prof.), Department of Chemical Engineering, Jordan University of Science and Technology, PO Box 3030, Irbid-Jordan

Jalal Barzin (Dr.), Department of Biomaterials, Iran Polymer and Petrochemical Institute, P.O. Box 14965/115, Tehran, Iran

Subhash Bhatia (Prof. Dr.), School of Chemical Engineering, Engineering Campus, Universiti Sains Malaysia, Seri Ampangan, 14300 Nibong Tebal, Seberang Perai Selatan, Pulau Pinang, Malaysia

Inamullah Bhatti, Faculty of Petroleum and Renewable Energy Engineering, University Teknologi Malaysia, 81310 UTM Skudai, Johor, Malaysia

Nurmin Bolong (Dr.), Civil Engineering Programme, School of Engineering & IT, Universiti Malaysia Sabah (UMS), 88999 Kota Kinabalu, Sabah, Malaysia

Budiyono (Assoc. Prof. Dr.), Faculty of Engineering, University of Diponegoro, Jl. Prof. Sudharto Tembalang, Semarang, Indonesia

Abdullah Busairi (Prof. Dr.), Faculty of Engineering, University of Diponegoro, Jl. Prof. Sudharto Tembalang, Semarang, Indonesia

Ying Chang (Dr.), Department of Chemistry, University of Nevada Las Vegas, 4505 Maryland Parkway, Box 454003, Las Vegas, Nevada 89154-4003, USA

Chel-Ken Chiam (Mr., M.Sc.), Membrane Technology Research Group, Centre of Materials and Minerals, School of Engineering and Information

Technology, Universiti Malaysia Sabah, Jalan UMS, 88400, Kota Kinabalu, Sabah, Malaysia

W. James Noel Fernando (Prof. Assoc.), School of Chemical Engineering, Engineering Campus, Universiti Sains Malaysia, 14300 Nibong Tebal, Pulau Pinang, Malaysia

Luc Fillaudeau (Dr.), Laboratoire d'Ingénierie des Systèmes Biologiques et des Procédés, CNRS UMR5504, INRA UMR792, INSA, 135, Avenue de Rangueil F-31077 Toulouse, France

Azadeh Ghaee (Mrs.), Faculty of New Science and Technology, University of Tehran, PO Box 14399-55941, Tehran, Iran

Gagik Badalians Gholikandi, Faculty of Water Engineering, Power and Water University of Technology (PWUT) and Water Research Institute (WRI), Tehran, Iran

Ani Idris (Prof. Dr.), Department of Bioprocess Engineering, Faculty of Chemical Engineering, Universiti Teknologi Malaysia, 81310 UTM Skudai, Johor

Ahmad Fauzi Ismail (Prof. Dr.), Advanced Membrane Technology Research Centre (AMTEC), Universiti Teknologi Malaysia (UTM), 81310 Skudai, Johor, Malaysia

Manal Ismail (Dr.), Department of Chemical and Process Engineering, Faculty of Engineering and Built Environment, Universiti Kebangsaan Malaysia, 43600 UKM Bangi, Selangor, Malaysia

Juhana Jaafar (Ms.),Advanced Membrane Technology Research Centre, Universiti Teknologi Malaysia (UTM), 81310 Skudai, Johor Malaysia

Ahmad Jusoh (Prof. Ir. Dr.), Department of Engineering Science, Faculty of Science and Technology, Universiti Malaysia Terengganu, Mengabang Telipot, 21030 Kuala Terengganu, Terengganu

Khairul Sozana Nor Kamarudin, Faculty of Petroleum and Renewable Energy Engineering, University Teknologi Malaysia, 81310 UTM Skudai, Johor, Malaysia

Shinji Kanehashi (Mr.), Department of Applied Chemistry, Meiji University, Higashi-mita, Tama-ku, Kawasaki, 214-8571, Japan

Ghader Khanbabaei (Dr.), Polymer Science and Technology Division, Research Institute of Petroleum Industry, West Blvd., Azadi Sport Complex, Tehran, Iran

Mahdi Khosravi, Faculty of Water Engineering, Power and Water University of Technology (PWUT), Hakimieh, Tehranpars, Shahid Abbaspour Blvd., PO Code 16589-54381, PO Box 16765-313, Tehran, Iran

Tutuk Joko Kusworo (Assoc. Prof. Dr.), Faculty of Engineering, University of Diponegoro, Jl. Prof. Sudharto Tembalang, Semarang, Indonesia

Kok Keong Lau (Dr.), Department of Chemical Engineering, Universiti Teknologi PETRONAS, Bandar Seri Iskandar, 31750 Tronoh, Perak, Malaysia

Amir Mansourizadeh (Dr.), Advanced Membrane Technology Research Centre (AMTEC), Universiti Teknologi Malaysia, 81310 Skudai, Johor, Malaysia

Agung Mataram (Mr.), Advanced Membrane Technology Research Centre (AMTEC), Universiti Technologi Malaysia, 81310 Skudai, Johor, Malaysia, and Department of Mechanical Engineering, Sriwijaya University, Sumatera Selatan, Indonesia

Takeshi Matsuura, Industrial Membrane Research Laboratory, Department of Chemical and Biological Engineering, University of Ottawa, Ottawa, Ontario, K1N 6N5 Canada

Norlisa Mili (Ms.), Department of Chemical Engineering, Faculty of Chemical Engineering, Universiti Teknologi Malaysia, 81310 UTM Skudai, Johor

Mazidatulakmam Miskam (Ms.), Department of Chemistry, Faculty of Science, Universiti Teknologi Malaysia, 81310 UTM Johor Bahru, Malaysia

Sou Miyata (Dr.), Department of Applied Chemistry, Meiji University, 1-1-1 Higashi-mita, Tama-ku, Kawasaki, 214-8571, Japan

Siti Nur Hidayah Mohamad (Ms.), Food and Biomaterial Eng. Group, Bioprocess Eng. Department, Faculty of Chemical Engineering, Universiti Teknologi Malaysia, 81310 Skudai, Johor. Malaysia

Abdul Wahab Mohammad (Prof. Dr. Ir.), Department of Chemical and Process Engineering, Faculty of Engineering and Built Environment, Universiti Kebangsaan Malaysia, 43600 UKM Bangi, Selangor, Malaysia

Ida Idayu Muhamad (Assoc. Prof. Dr.), Food and Biomaterial Eng. Group, Bioprocess Engineering Department, Faculty of Chemical Engineering, Universiti Teknologi Malaysia, 81310 Skudai, Johor, Malaysia

Hilmi Mukhtar (Assoc. Prof. Dr.), Department of Chemical Engineering, Universiti Teknologi PETRONAS, Bandar Seri Iskandar, 31750 Tronoh, Perak, Malaysia

Azeman Mustafa (Assoc. Prof. Dr.), Advanced Membrane Technology Research Center (AMTEC), Universiti Teknologi Malaysia, 81310 Skudai, Johor,M alaysia

Kazukiyo Nagai (Prof.), Department of Applied Chemistry, Meiji University 1-1-1 Higashi-mita, Tama-ku, Kawasaki, 214-8571, Japan

Mohd Ghazali Mohd Nawawi (Assoc. Prof. Dr.), Department of Chemical Engineering, Faculty of Chemical Engineering, University Teknologi Malaysia, 81310 UTM Skudai, Johor Bahru, Malaysia

Zairossani Mohd Nor (Dr.), Technology and Engineering Division, Rubber Research Institute of Malaysia, Malaysian Rubber Board, 47000 Sungai Buloh, Selangor, Malaysia

Norasikin Othman (Assoc. Prof. Dr.), Department of Chemical Engineering, Faculty of Chemical Engineering, Universiti Teknologi Malaysia, 81310 UTM Skudai, Johor, Malaysia

Rahimah Othman (Ms.), Department of Chemical and Process Engineering, Faculty of Engineering and Built Environment Universiti Kebangsaan Malaysia, 43600 UKM Bangi, Selangor, Malaysia

Norhayati Pa'e (Ms.), Food and Biomaterial Eng. Group, Bioprocess Eng. Department, Faculty of Chemical Engineering, Universiti Teknologi Malaysia 81310 Skudai, Johor. Malaysia

Yanuardi Raharjo (Mr.), Department of Chemistry, Faculty of Science,Universiti Teknologi Malaysia, 81310 UTM Johor Bahru, Malaysia

Sunarti Abd Rahman (Mrs.), Faculty of Chemical and Natural Resources Engineering, Universiti Malaysia Pahang, 25000 Kuantan, Pahang, Malaysia

Ali Rahmatpour (Dr.), Polymer Science and Technology Division, Research Institute of Petroleum Industry, West Blvd., Azadi Sport Complex, Tehran, Iran

Nur Azrini Ramlee (Ms.), Department of Chemical Engineering, Faculty of Chemical Engineering, University Teknologi Malaysia, 81310 UTM Skudai, Johor Bahru, Malaysia

Rosalam Sarbatly (Assoc. Prof. Dr.), Membrane Technology Research Group, Centre of Materials and Minerals, School of Engineering and Information Technology, Universiti Malaysia Sabah, Jalan UMS, 88400, Kota Kinabalu, Sabah, Malaysia

Mohd Razman Salim (Prof. Dr.), Advanced Membrane Technology Research Centre (AMTEC), Universiti Teknologi Malaysia (UTM), 81310 Skudai, Johor Malaysia

Jumat Salimon (Prof. Dr.), Department of Chemical Science and Food Technology, Faculty of Science and Technology, Universiti Kebangsaan Malaysia, 43600 UKM Bangi, Selangor, Malaysia

Syairah Salleh (Ms.), Department of Chemistry, Faculty of Science, Universiti Teknologi Malaysia, 81310 UTM Johor Bahru, Malaysia

Mohd Marsin Sanagi (Prof. Dr.), Separation Science and Technology Group, IbnuSina Institute for Fundamental Science Studies, Universiti Teknologi

Malaysia, 81310 UTM JohorBahru, Malaysia, and Department of Chemistry, Faculty of Science,Universiti Teknologi Malaysia, 81310 UTM Johor Bahru, Malaysia

Shuichi Sato (Mr.), Department of Applied Chemistry, Meiji University, Higashi-mita, Tama-ku, Kawasaki, 214-8571, Japan

Binyam Seyoum (Mr.), Department of Chemical Engineering, Universiti Teknologi PETRONAS, Bandar Seri Iskandar, 31750 Tronoh, Perak, Malaysia

Mojtaba Shariaty-Niassar (Prof.), Department of Chemical Engineering, University College of Engineering, Tehran University, PO Box 11365-4563, Tehran, Iran

Mohammad Soltanieh (Prof.), Department of Chemical and Petroleum Engineering, Sharif University of Technology, Tehran, Iran

Seyyed Ali Reza Tabatabaei Nejad (Dr.), Department of Chemical Engineering, Sahand University of Technology, Tabriz, Iran

Hamid Reza Tashaouei, National Water and Wastewater Company (NWWC), Iran

Radhe Shyam Thakur (Mr.), Membrane Research Laboratory, Department of Chemical Engineering, National Institute of Technology, Tiruchirappalli 620015, India

Wan Aini Wan Ibrahim (Prof. Dr.), Separation Science and Technology Group, IbnuSina Institute for Fundamental Science Studies, Universiti Teknologi Malaysia, 81310 UTM Johor Bahru, Malaysia, and Department of Chemistry, Faculty of Science, Universiti Teknologi Malaysia, 81310 UTM Johor Bahru, Malaysia

Devaraj Veerasamy (Dr.), Technology and Engineering Division, Rubber Research Institute of Malaysia, Malaysian Rubber Board, 47000 Sungai Buloh, Selangor, Malaysia, and Advanced Membrane Technology Research Centre (AMTEC), Universiti Teknologi Malaysia (UTM), 81310 Skudai, Johor Bahru, Johor, Malaysia

Mehdi Yazdanshenas (Dr.), Department of Chemical Engineering, Sahand University of Technology, Tabriz, Iran

Yin Fong Yeong (Dr.), School of Chemical Engineering, Engineering Campus, Universiti Sains Malaysia, Seri Ampangan, 14300 Nibong Tebal, Seberang Perai Selatan, Pulau Pinang, Malaysia

Abdul Halim Mohammad Yusof (Dr.), Food and Biomaterial Eng. Group, Bioprocess Eng. Department, Faculty of Chemical Engineering, Universiti Teknologi Malaysia, 81310 Skudai, Johor. Malaysia

Siti Nazrah Zailani (Ms.), Department of Chemical Engineering, Faculty of Chemical Engineering, Universiti Teknologi Malaysia, 81310 UTM Skudai, Johor, Malaysia

Fathie Ahmad Zakil (Ms.), Faculty of Chemical & Natural Resources Engineering, Universiti Malaysia Pahang, LebuhrayaTunRazak, 26300 Gambang, Kuantan, Pahang, Malaysia

Sharif Hussein Sharif Zein (Prof. Assoc.), School of Chemical Engineering, Engineering Campus, Universiti Sains Malaysia, Seri Ampangan, 14300 Nibong Tebal, S.P.S, Penang, Malaysia

PART I

MEMBRANE MATERIAL AND MEMBRANE DEVELOPMENT

1

SPINNING EFFECT OF POLYETHERSULFONE HOLLOW FIBER MEMBRANE PREPARED BY WATER OR POLYVINYLPYRROLIDONE IN TERNARY FORMULATION

Nurmin Bolong, Ahmad Fauzi Ismail, and Mohd Razman Salim

1.1 INTRODUCTION

The art in membrane making has been found to be influenced by many factors starting from the dope contents and formulation up to the fabrication process, including posttreatment and module preparation of the membranes. Since Leob and Sourirajan invented the first flat anisotropic cellulose acetate membranes by the phase inversion process in the late 1950s [1], the synthetic polymeric membrane technology has obtained much attention from both academia and industry. Compared to the flat sheet, hollow fiber membranes possess the following advantages: (1) much larger membrane area per unit volume of membrane module which results in higher productivity, (2) large packing density that offers the lowest cost per unit membrane area (3), self-mechanical support which enables back-flushing to recover the membrane flux, (4) good flexibility and easy handling during module fabrication and in the operation [2–4].

Sustainable Membrane Technology for Energy, Water, and Environment, First Edition.
Edited by Ahmad Fauzi Ismail and Takeshi Matsuura.

The ability of producing self-made membranes allows a greater chance to produce and achieve a successful treatment system since during the production process of membrane it can be tailored to suit the requirements. However, many important factors need to be taken into account and properly controlled in order to allow production of membranes that meet the desired characteristics. In general, membrane fabrication involves trial-and-error method which covers optimal material formulation, selection of critical operating variables, and proper conditions during fiber spinning. Although a great deal of works have been carried out in the past, fabrication of membrane for specified application is still extremely difficult and becomes an interesting subject of study.

One of the common polymer materials used for membrane formation is polyethersulfone (PES). PES is principally favorable due to the characteristics of wide temperature limits, wide pH tolerances, fairly good chlorine resistance, easiness to fabricate membranes in a wide variety of configurations and modules, wide range of pore sizes ranging from 10 Å to 0.2 μm, and good chemical resistance to aliphatic hydrocarbons, alcohols, and acids [5, 6].

In the immersion process, choice of the additive or the nonsolvent system has strong influence in membrane fabrication. In particular, to increase hydrophilicity of the resultant membrane, a hydrophilic polymer, polyvinylpyrrolidone (PVP), is often blended into relatively hydrophobic PES [7, 8]. The addition of a second polymer such as PVP in the casting solution may produce highly porous membranes with well-connected pores and surface properties [6]. Water is also frequently used as a nonsolvent [9], and can form a complex multiple hydrogen bonding with solvents such as *N,N*-disubstituted amides (e.g., *N*-methyl pyrrolidone, NMP) [10]. Furthermore, water can be used to control the porosity of the membranes by inducing macrovoids [11].

In this work, an attempt is made to study and compare the effect of water and PVP as additives in the polymer solution toward PES as a base polymer in fabricating hollow fiber membranes by the phase inversion technique. The hollow fiber membranes are characterized by pure water permeation (PWP) and sodium chloride rejection. The effect of spinning conditions, particularly that of the dope extrusion rate (DER), is investigated. Furthermore, dry-wet or wet-wet spinning influences are also compared.

1.2 EXPERIMENTAL

Hollow fiber membranes were fabricated in the laboratory using a simple phase inversion technique [12]. The homemade hollow fiber membranes were spun at spinning parameters given in Table 1.1, using water as coagulant both in the coagulation tank and as the bore fluid. Prior to spinning, a spinning dope of an exclusive formulation consisting of PES (Radel A-300 with a molecular weight of about 15,000 Dalton (Da), purchased from Amoco Performance Products Inc., Atlanta, GA), water (deionized) or PVP (Kollidon® 15, or PVP K-15) with a weight average molecular weight (M_w) of 10,000 Da (E. Merck

TABLE 1.1. Hollow Fiber Spinning Parameters

Hollow Fiber Composition (wt %)	Spinning Variation		Membrane ID
	Air Gap Distance (cm)	Dope Extrusion Rate (cm^3/min)	
PES/NMP/Water (22/73/5)	5	1.0	W1.0-5
	5	2.5	W2.5-5
	5	4.0	W4.0-5
PES/NMP/Water (22/73/5)	0	1.0	W1.0-0
	0	2.5	W2.5-0
	0	4.0	W4.0-0
PES/NMP/PVP (22/73/5)	5	1.0	P1.0-5
	5	2.5	P2.5-5
	5	4.0	P4.0-5
PES/NMP/Water (22/73/5)	0	1.0	P1.0-0
	0	2.5	P2.5-0
	0	4.0	P4.0-0

KGaA, Darmstadt, Germany), and *N*-methyl-2-pyrrolidone (NMP, >99%, Aldrich Chemical Comp. Inc., Milwaukee, WI) was prepared to produce an asymmetric hollow fiber membrane.

The fabricated hollow fiber membranes were characterized by filtration tests using a cross-flow filtration setup as detailed elsewhere [13]. Prior to the filtration test, 30 hollow fibers (approximately of 10×10^{-3} m^3 surface area) were potted into a bundle, which was slowly inserted into a stainless steel 30 mm housing and assembled with the permeation test system. After 0.5 hour of pressurization at 90 psig, the PWP flux was first measured at an operating pressure of 80 psig and at ambient temperature. Then the separation experiments were performed using 0.01 M sodium chloride. The sodium chloride solute concentrations in the feed and the permeate were determined by a conductivity meter (WTW handheld meter model LF330).

1.3 RESULTS AND DISCUSSION

The results of filtration experiments are illustrated in Figure 1.1. The pressure normalized pure water flux (also termed as PWP) was calculated for all the membranes and divided into the two main groups (i.e., PES-water and PES-PVP), which then were plotted against the sodium chloride rejection. Regardless of spinning conditions, the results typically reveal that PVP addition could reach an optimum NaCl rejection of ~40%. The highest PWP was ~15 L/m^2.h.bar. Whereas, use of water as nonsolvent has shown a larger range of separation performance with up to ~45% NaCl rejection and the highest PWP of ~28 L/m^2.h.bar.

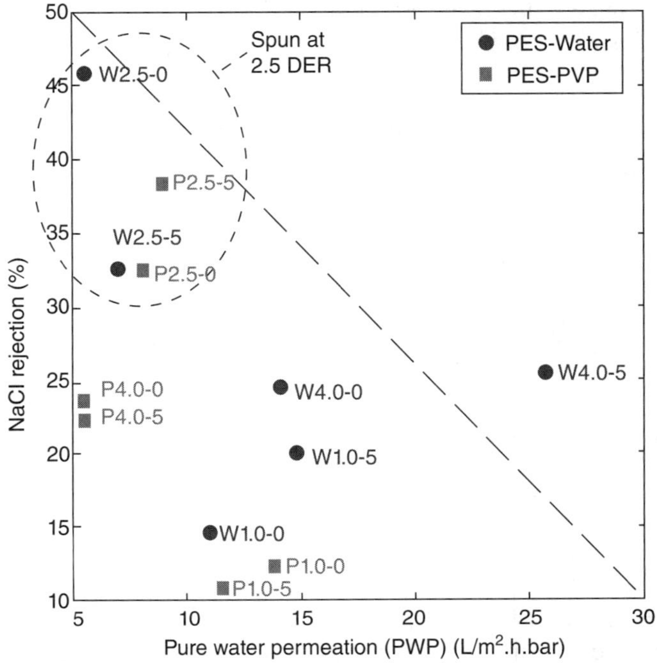

Figure 1.1. Influence of the membrane composition on the water separation performance.

Water or PVP certainly had an effect on the thermodynamic interaction properties of the casting dopes and the process of demixing and thus on membrane structures and performances. The existence of ternary component in the dope formulation has caused a shift in the composition path toward binodal or liquid–liquid demixing. This thermodynamically indicates that the addition of water or a second polymer such as PVP to the casting solution enhances coagulation [6] and induces occurrence of instantaneous demixing [9]. Addition of both water and PVP generally leads to an increment of PWP compared to pure PES. For example, PES membrane fabricated under the condition 2.5 DER at 5 cm air gap without additive showed PWP = 2.18 L/m^2.h.bar and ~47% NaCl separation. PVP is known to hydrophilize the membrane [13] and influence membrane surface porosity [6]. On the other hand, water is a strong nonsolvent for PES, and primarily affects the phase inversion rate when diffused into the polymer solution.

From the data shown in Figure 1.1, the addition of PVP could enhance sodium chloride rejection but not necessarily contribute to the enhancement of PWP, whereas the addition of water affected PWP and NaCl rejection in a much wider range. This might be related to the overall morphological properties of the membranes and due to the stronger effect of water on demixing rate than PVP.

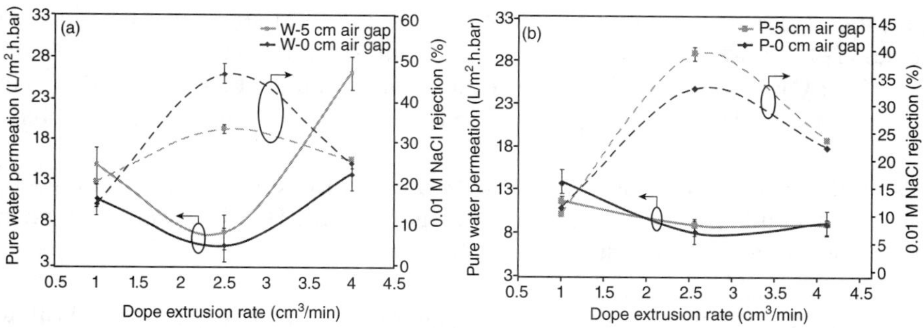

Figure 1.2. Effect of spinning condition (DER and air gap) on pure water permeation and NaCl separation for (a) PES-Water and (b) PES-PVPmembranes.

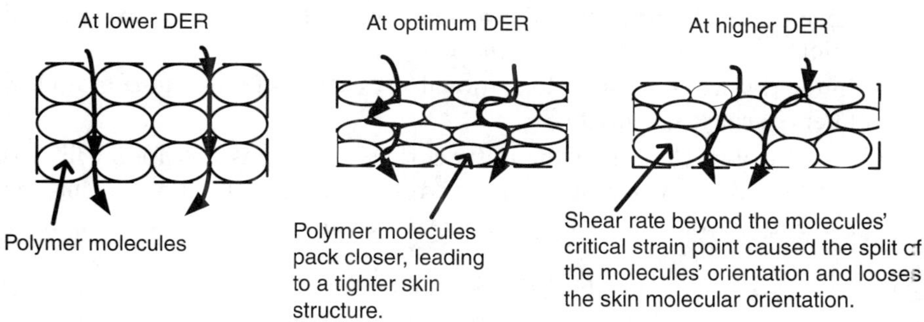

Figure 1.3. Illustration of macromolecule arrangement in the hollow fiber at different DERs.

Figure 1.1 further indicates that all the high NaCl rejection data are found in a region surrounded by the broken line where hollow fibers were spun at DER of 2.5 cm^3/min. This trend can also be found in Figure 1.2 where PWP and NaCl rejection are plotted versus DER for the addition of (1) water and (2) PVP, respectively. Figure 1.2 shows the effect of DER and air gap on PWP and NaCl rejection. The PWP of both PES/water and PES/PVP membranes were found to have a minimum at 2.5 DER. Conversely, NaCl rejection had a maximum at this DER. This is due to the effect of the shear rate on the macromolecular packing and alignment in the hollow fiber.

Figure 1.3 explains schematically the effect of DER on the macromolecular arrangement. At the optimum DER, the shear force working on the outer surface of the nascent hollow fiber causes polymer molecules to be packed closer, leading to a tighter skin structure with enhanced molecular orientation. As a result, hollow fiber membranes of denser outer skin layers with small pore sizes are formed [14, 15]. Hence, PWP decreases and the solute separation increases. However, further increase in DER beyond a critical point disrupts the molecular orientation and loosens the outer skin structure. This in turn enlarges the pore sizes and thus reduces solute separation and increases water flux. The DER effect on PWP was found more significant for PES-water than PES-PVP formulation, indicating that PES-PVP has stronger molecular chain interaction.

In Figure 1.2, comparison can also be made between the wet-spinning (air gap; 0 cm) and the dry-wet spinning (air gap; 5 cm). The differences between these two spinning procedures are as follows:

1. In the wet spinning, the die swell is immediately frozen when the hollow fibers enter the coagulation bath.
2. In the wet spinning, the molecular orientation induced by the shear stress on the outer surface is also frozen.
3. In the dry-wet spinning, the die swell disappears while the nascent hollow fiber is traveling through the air gap, and relaxation of shear-induced orientation of polymer takes place.
4. In the dry-wet spinning, elongational stress works on the nascent hollow fibers because of gravity.
5. In the dry-wet spinning, the outer surface of the nascent hollow fiber is not in contact with water while traveling through the air gap, and the solvent/nonsolvent exchange is delayed at the outer surface.

All these have effects on the hollow fiber morphology and performance as observed in other studies [4, 6]. In Figure 1.4, the air gap region, a phenomenon called die swell experienced when exiting from the spinneret is illustrated.

Looking into Figure 1.2, PES/water hollow fiber membranes fabricated by dry-wet spinning (with 5 cm air gap) showed higher permeation rate compared to wet-wet spinning (with 0 cm air gap). Increase in pore size with an increase

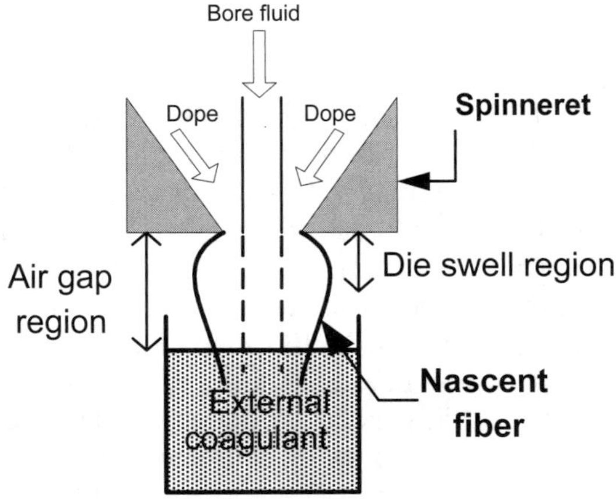

Figure 1.4. Illustration of die swell in hollow fiber formation at spinneret.

in the air gap distance of the same range for the spinning of PES/PVP dope was reported earlier by other researchers [16]. The sodium chloride rejection, on the other hand, decreased for the dry-wet spinning, showing typical trade-off effect between flux and rejection. The difference between the wet spinning and dry-wet spinning was less significant for the flux of PES/PVP hollow fibers, as shown in Figure 1.2b. Furthermore, in contrast to PES/water formulation, dry-wet spinning showed higher NaCl rejection than wet spinning for the PES/PVP formulation. The presence of PVP has significantly provoked changes on membrane separation properties, which were related to variation of demixing rates between PVP-PES and water-PES and diffusion velocity of solvent and nonsolvent (water in the bore fluid and coagulation tank).

1.4 CONCLUSION

Polymeric hollow fiber membranes were fabricated by the phase inversion technique. Two types of additives, water and PVP, were added as tertiary additives to the spinning dope of PES. In addition, wet-spinning and dry-wet spinning were conducted, and results were compared. The experimental data showed that the addition of both water and PVP could increase water flux but simultaneously give a trade-off effect to the sodium chloride rejection. Spinning conditions including the DER have significant effects on membrane performance. Pure water permeation showed minimum effects while the solute separation showed maximum effects when the dope shear rate was increased.

ACKNOWLEDGMENT

The authors would like to acknowledge the financial support from of the Ministry of Higher Education (MOHE) grant and Universiti Malaysia Sabah (UMS). The author (Nurmin Bolong) is thankful to the Universiti Teknologi Malaysia (UTM), especially Advanced Membrane Technology Research Centre (AMTEC), for providing excellent research environment to complete this work.

REFERENCES

[1] Leob S. and S. Sourirajan. Sea water demineralization by means of an osmosis membrane. *Adv. Chem. Ser.* ACS38 (1963) 117.

[2] Ho W. S. W. and K. K. Sirkar (eds.), *Membrane Handbook*. Van Nostrand Reinhold, New York, 1992.

[3] Lesjean B., S. Rosenberger, J. S. Schrotter, and A. Recherche. Membrane-aided biological wastewater treatment—an overview of applied system. *Membr. Tech.* 8 (2004) 5.

[4] Chung T.-S., J.-J. Qin, and J. Gu. Effect of shear rate within the spinneret on morphology, separation performance and mechanical properties of ultrafiltration polyethersulfone hollow fiber membranes. *Chem. Eng. Sci.* 56 (2001) 5869.

[5] Cheryan M. *Ultrafiltration & Microfiltration Handbook*. Technomic Publishing Co., Lancaster, PA, 1998.

[6] Xu Z. L. and F. AlsalhyQusay. Polyethersulfone (PES) hollow fibre ultrafiltration membranes prepared by PES/non-solvent/NMP solution. *J. Memb. Sci.* 233 (2004) 101.

[7] Lafreniére L. Y., D. F. Talbot, T. Matsuura, and S. Sourirajan. Effect of polyvinylpyrrolidone additive on the performance of polyethersulfone ultrafiltration membranes. *Ind. Eng. Chem. Res.* 26 (1987) 2385.

[8] Qin J. J., F. S. Wong, Y. Li, and Y. T. A. Liu. A high flux ultrafiltration membrane spun from PSU/PVP(K90)/DMF/1,2-propanediol. *J. Memb. Sci.* 211 (2003) 139.

[9] Mulder M. *Basic Principles of Membrane Technology*. Kluwer Academic Publishers, Dordrecht, The Netherlands, 2002.

[10] Symons M. C. R., J. M. Harvey, and S. E. Jackson. Spectroscopic studies of water–aprotic-solvent interactions in the water-rich region. *J.C.S. Faraday I* 76 (1980) 256.

[11] Schäfer A. I., A. G. Fane, and T. D. Waite (eds.), *Nanofiltration: Principles and Applications*. Elsevier, Oxford, UK, 2006.

[12] Ismail A. F., M. I. Mustaffar, R. M. Illias, and M. S. Abdullah. Effect of dope extrusion rate on morphology and performance of hollow fibers membrane for ultrafiltration. *Sep. Purif. Tech.* 49 (2006) 10.

[13] Ulbricht M. Advanced functional polymer membranes. *Polymer* 47 (2006) 2217.

[14] Wang K. Y., T. Matsuura, T.-S. Chung, and W. F. Guo. The effects of flow angle and shear rate within the spinneret on the separation performance of poly(ethersulfone) (PES) ultrafiltration hollow fiber membranes. *J. Memb. Sci.* 240 (2004) 67.

[15] Ismail A. F. *Novel Studies of Molecular Orientation in Synthetic Polymeric Membranes for Gas Separation*. Doctor of Philosophy, University of Strathclyde, Glasgow, UK, 1997.

[16] Khulbe K. C., C. Y. Feng, T. Matsuura, D. C. Mosqueada-Jimenaez, M. Rafat, D. Kingston, R. M. Narbaitz, and M. Khayet. Characterization of surface-modified hollow fiber polyethersulfone membranes prepared at different air gaps. *J. Appl. Polym. Sci.* 104 (2007) 710.

2

EFFECT OF INORGANIC PARTICLE ON THE PERFORMANCE OF POLYETHERSULFONE-CELLULOSE ACETATE ULTRAFILTRATION MEMBRANES

Gangasalam Arthanareeswaran and Radhe Shyam Thakur

2.1 INTRODUCTION

Membrane separation processes offer a number of advantages in terms of low energy use and capital investments. Membrane-based separation processes, such as reverse osmosis (RO), nanofiltration (NF), ultrafiltration (UF), microfiltration (MF), electrodialysis (ED), pervaporation (PV), and gas separation have been developed for various applications [1]. UF, an intermediate between NF and MF, is a pressure-driven process used for removing solutes, such as oils, particulate matters, bacteria, suspended solids, large macromolecules, and proteins, and UF membranes have pore sizes ranging from 0.005 to 0.1 μm [2]. Both polymeric and inorganic materials have been used for the preparation of UF membranes. A familiar method for the preparation of polymeric membranes is the phase inversion process [1, 2]. Since Loeb and Sourirajan first introduced the phase inversion method [3], much investigation has been made to know the mechanism of membrane formation with asymmetric structure using polymers. In a phase inversion process, the cast solution film on a substrate is immersed and is precipitated in a nonsolvent coagulation bath. During

Sustainable Membrane Technology for Energy, Water, and Environment, First Edition.
Edited by Ahmad Fauzi Ismail and Takeshi Matsuura.

the process, the solvent in the casting solution film is exchanged with a nonsolvent, and phase separation occurs in the film. Generally, polymeric membranes are ineffective to their physical integrity in solvents because of their tendency to swell or dissolve. Polymer blending is a possibility of modifying and improving membrane material properties. This modification holds for membrane preparation to tailor a specific separation performance [4]. Blending of two polymeric materials is also intrinsically limited by a trade-off between their compatibility and their heterogeneity, they have been difficult to form homogenous membranes for high-performance separation applications. According to the literature [5–9], cellulose acetate (CA) has been blended with additional polymers such as polyethyleneglycol (PEG), polyethyleneimine, polysulfone (PSf), sulfonated PSf, and epoxy resin for their improved performance. Chen et al. [6] reported that CA as a matrix polymer was blended with polyethyleneimine in a mixture of solvents to prepare a modified MF membrane. Synthesis of a polymer blend membrane is motivated by the necessity to superimpose requisite properties upon the basic transport properties of base polymer. Thus, the hydrophilic–hydrophobic balance as well as properties of a membrane system can be easily altered if the membrane is prepared from multicomponent polymer blends [8]. Malaisamy et al. [8] reported that the beyond 25% of sulfonated PSf content in CA resulted in phase separation of the blend, due to incompatibility between CA and sulfonated PSf. Furthermore, according to Mahendran et al. [9], at a higher blend ratio of CA and epoxy resin, it was unable to form membrane because of incompatibility. Although there have been works reported [10–14] using PEG and polyvinylpyrrolidone (PVP) as additives, those works involved mostly polymeric blend systems such as PSf/sulfonated polyetheretherketone (SPEEK), CA/polyethersulfone (CA/PES), CA/polycarbonate, CA/sulfonated poly(ether imide), and CA/PSf except for the use of inorganic particles as modifier. There is yet no report regarding the effect of inorganic particles on the performance of polymer blend UF membranes. In view of this, an attempt is made to investigate the effect of Al_2O_3 particles as modifier for PES/CA blend UF membranes. PES has repeating phenylene rings which enhance physical and mechanical stability. However, the use of PES for aqueous phase is restricted due to its hydrophobicity and may be improved by modification of PES, through blending.

In recent years, organic–inorganic composite membranes where inorganic particles are dispersed inside organic (polymer) membrane matrix have been of great interest due to their intrinsic advantages over pure polymeric–inorganic membranes [15]. These membranes are often called mixed-matrix membranes. In these mixed-matrix membranes, membrane materials involve various fundamental tasks, such as polymer-chain rigidity, free volume, and the altered interface which influences transport of solutes through the membrane [16]. In studies of organic–inorganic membranes, the mineral fillers used are mainly silica, zirconia, titanium (IV) oxide, bentonite, aerosil, carbon, clay, zeolite, and alumina [17–23].

Bottino et al. [17] prepared and characterized the novel organic–inorganic composite membranes by adding fine silica particles in the porous matrix of poly(vinylidene fluoride) (PVDF). According to Bottino et al. [18], incorporation of inorganic particles into polymer materials leads to increased membrane permeability and control of membrane-surface properties. Genne et al. [19] prepared a PSf/ZrO_2 composite membrane and reported that the permeability increases and the rejection of dextran decreases upon addition of increasing amounts of ZrO_2 to the membrane casting solution. Oh et al. [20] modified the surfaces of UF membranes by dispersing nano-sized titanium (IV) oxide (TiO_2) particles in a PVDF solution. Finken [21] reported the incorporation of modified bentonite into casting solutions of cellulose di acetate/cellulose tri acetate blend membranes. Adoor et al. [22] fabricated poly(vinyl alcohol)-based mixed-matrix membranes loaded with 5 and 10 wt % of sodium montmorillonite (Na + MMT) clay particles by solvent casting method. Qiu et al. [23] developed UF membrane by blending PSf and functionalized multi-walled carbon nanotubes (MWNTs) with different compositions dissolved in N,N-dimethylformamide (DMF) using a classical phase inversion method. It was found that the MWNT's concentration was an important factor influencing the morphology and permeation properties of the blend membranes.

Fe_3O_4 as a mineral filler was recently investigated by Jian et al. [24], to improve the UF performance of PSf membrane. Yan et al. [16] prepared PVDF/nano-sized alumina membranes by phase inversion process and characterized the membranes by UF experiments in terms of water flux and molecular weight cutoff, and reported that water flux and surface hydrophilicity increased by adding the hydrophilic inorganic nano-sized Al_2O_3 particles. Further, Maximous et al. [25] investigated the effect of Al_2O_3 nanoparticles on the membrane fouling characteristics by preparing membranes from casting dopes with different weight ratios (0.01, 0.03, 0.05, 0.1, and 0.2) of Al_2O_3 to PES. In this investigation, an attempt has been made to prepare PES/CA blend UF membranes composed of Al_2O_3 particles using phase inversion process. The effects of the addition of Al_2O_3 particles to PES/CA blend polymer on the membrane properties were investigated by examining the UF performances such as pure water flux, pore size distribution, membrane resistance, and morphological structure.

2.2 EXPERIMENTAL

2.2.1 Materials

Commercial-grade PES (Gafone 3300) was obtained as a gift sample from Gharda Chemicals Pvt., Ltd (India) and was used as supplied. Commercial grade MYCEL Cellulose Acetate CDA 5770 (acetyl content 39.99 wt %) was procured from Mysore Acetate and Chemicals Company Ltd., India. DMF from Qualigens Fine Chemicals, Glaxo India Ltd. was sieved through molecular sieves (Type-4 Å) to remove moisture and stored in dry conditions prior to use. Sodium lauryl sulfate (SLS) of analar grade was obtained from Qualigens Fine Chemicals Ltd., India and was used as surfactant. The commercial Al_2O_3 particle in the average particle size of 100 μm was obtained from SRL India Ltd. Proteins, viz, bovine serum albumin (69 kDa), pepsin (35 kDa), and trypsin (20 kDa) were purchased for the measurement of the pore size distribution from SRL Pvt. Ltd., India. Egg albumin (45 kDa) was obtained from CDH Ltd., India.

2.2.2 Preparation of Solution Blend of Polymers

The PES and CA polymers were first dissolved in DMF at a total polymer concentration of 20 wt % and then thoroughly mixed with the Al_2O_3 particles under constant mechanical stirring at a moderate speed of rotation in a round bottom flask for 3–4 hours at 40°C. Special care was taken to ensure homogeneous dispersion of the Al_2O_3 particles. The homogeneous solution obtained was allowed to stand for at least 3 hours in an airtight condition to get rid of air bubbles. A series of such polymer solutions were prepared by varying the composition of PES/CA with an increment of Al_2O_3 particles, as shown in Table 2.1. It has been shown that the introduction of CA beyond 25% in the total polymer composition (20 wt %) of PES/CA resulted in heterogeneous solution leading to phase separation during membrane formation. Hence, we selected an optimum of 25% of CA in the PES/CA blend composition and introduced 10, 20, and 30% of Al_2O_3 particles into PES/CA blend by reducing the composition of PES as 70, 60, and 50%, respectively.

Membrane casting conditions: casting solution temperature 42 ± 2°C; casting temperature 25 ± 1°C; casting relative humidity 50 ± 2%; solvent evaporation time 30 seconds.

2.2.3 Membrane Preparation

UF membranes were prepared by dry-wet phase inversion technique from PES/CA/DMF solution blended with Al_2O_3 particles. The polymer solution

TABLE 2.1. Compositions of PES/CA Membrane with Different Concentrations of Al_2O_3 Particles

Total Blend Composition 20 wt %			
PES (%)	CA (%)	Al_2O_3 (%)	Solvent DMF (wt %)
100	0	0	80
0	100	0	80
75	25	0	80
70	20	10	80
60	20	20	80
50	20	30	80

was poured over a smooth glass plate and spread with a film applicator with side runners to produce 0.22 mm thickness of the membrane. The casting environment, namely, relative humidity (50%) and temperature (25°C), were standardized and maintained for the preparation of membranes with better physical properties such as homogeneity, thickness, and morphology. The nonsolvent of DMF/SLS was prepared from 2 L water consisting of 2.5% (v/v) DMF and 0.2 wt % SLS in distilled water and was maintained to 4°C. The solvent was evaporated at 30 seconds in the casting chamber and then the glass plate along with the polymer film was immersed in the nonsolvent bath. The wet phase inversion was carried out by immersing the polymer film into a nonsolvent bath containing water and SLS/DMF mixture as nonsolvents. The formed membranes were eventually peeled from the glass plate and after the gelation, membranes were stored at 5°C in distilled water until they were tested. Every membrane sheet was inspected with a lightbox to detect defects, such as thin spots and pinholes, and defective membranes were discarded. Prepared membranes (76 mm diameter) were cut from the flat sheets for the UF experiments.

2.2.4 Experimental Setup

The UF experiments were carried out in a 400 mL batch type stirred cell (UF cell XFUF 076 01, Millipore, India) fitted with a Teflon-coated magnetic paddle. The effective membrane area available for UF was 38.5 cm^2. The feed solution or pure water filled in the cell was stirred at 400 rpm using a magnetic stirrer. All the experiments were carried out at 30 ± 2°C and compaction at 414 kPa transmembrane pressure while other permeation studies were at 345 kPa transmembrane pressure. The pure water was collected from the bottom of the cell.

2.3 MEMBRANE CHARACTERIZATION

2.3.1 Pure Water Flux

Membranes were subjected to pure water flux estimation at a transmembrane pressure of 414 kPa. The permeability was measured under steady-state flow. The pure water flux was calculated using Equation 2.1:

$$J_w = \frac{Q}{A \cdot \Delta T}, \tag{2.1}$$

where Q is the quantity of permeate collected (L), J_w is water flux (Lm^{-2}/h), ΔT is the sampling time (h), and A is the membrane area (m^2).

2.3.2 Water Content

The membranes were soaked in water for 24 hours and weighed after mopping with blotting paper. These wet membranes were placed in a vacuum oven at 75°C for 48 hours, and the dry weights were determined. From these two values, the percent water content was derived using Equation 2.2:

$$\%\,Water\ Content = \frac{Wet\ sample\ weight - Dry\ sample\ weight}{Wet\ sample\ weight} \times 100. \quad (2.2)$$

2.3.3 Membrane Resistance (R_m)

To determine membrane resistance (R_m), the pure water flux of membranes was measured at different transmembrane pressures (ΔP) namely at 69, 138, 207, 276, 345, and 414 kPa. The resistance of the membrane was evaluated from the slope of water flux versus transmembrane pressure difference (ΔP) plot, using Equation 2.3:

$$R_m = \frac{\Delta P}{J_w}. \quad (2.3)$$

2.3.4 Pore Size Distribution

The pore size distribution was determined using the protein solutions of different molecular weights. From the feed and permeate concentrations, the percentage rejection was calculated using the equation reported in the literature [8, 10]. The molecular weight of the solute that has a solute rejection (SR) above 80% was used to evaluate the average pore size, $\bar{R}$, of the membranes by the following equation [26]:

$$\bar{R} = 100\left(\frac{\alpha}{\%SR}\right), \quad (2.4)$$

where $\bar{R}$ is the average pore size (radius) of the membrane (Å), and α is the average solute radius (Å). The average solute radii, also known as the Stoke radii, were obtained from the plot of solute molecular weight versus solute radius in aqueous solution, which was developed by Sarbolouki [26].

The surface porosity, ε, of the membrane was calculated by the orifice model given below, assuming that only the skin layer of the membrane is effective in separation [27]:

$$\varepsilon = \frac{3\pi\eta_w J_w}{\bar{R} \times \Delta P}, \quad (2.5)$$

where ε is the surface porosity; η_w is the viscosity of the deionized water (g/cm s); J_w is the pure water flux (cm/s), and ΔP is the applied pressure (dyn/cm^2). From the values of ε and $\bar{R}$ (cm), the pore density in the membrane surface was calculated using Equation 2.6:

$$n = \frac{\varepsilon}{\pi \times \bar{R}^2}, \quad (2.6)$$

where n is number of pores/cm^2.

2.3.5 Morphological Studies

The top surface and cross section of membranes with 75/25% of PES/CA blend and Al_2O_3 in 10 and 30% were studied with scanning electron microscopy (SEM) (Hitachi S-3000H, Japan). The membranes were cut into pieces of various sizes and mopped with filter paper. These samples were gold sputtered for producing electric conductivity, and photomicrographs were taken in very high vacuum conditions operating between 15 and 25 kV.

2.4 RESULTS AND DISCUSSION

2.4.1 Effect of Al_2O_3 Concentration on

Pure water permeation experiments are one of the most important methods for finding the permeability of the membranes. Pure water flux is affected by membrane formation variables such as nonsolvent condition, gelation bath temperature, and solvent evaporation time. After compaction of the membranes for 4–5 hours at 414 kPa, the membranes were studied their water permeability under 345 kPa. The pure water flux of all the membranes was measured after an initial stabilization period of 30–60 minutes, and the results are presented in Table 2.2. The pure PES membrane showed a low flux of 5.98 L/m^2/h. The pure CA membrane exhibited a pure water flux value of 29.87 L/m^2/h. It can be clearly observed that the hydrophilic membrane (CA)

TABLE 2.2. Pure Water Flux of PES/CA Membrane with Different Concentrations of Al_2O_3 Particles at 414 kPa

Total Blend Composition (20 wt %)			Pure Water Flux (L/m^2/h) Nonsolvent	
PES (%)	CA (%)	Al_2O_3 (%)	DMF/SLS Mixture	Water
100	0	0	14.12	5.98
0	100	0	29.87	13.78
75	25	0	33.87	16.93
70	20	10	58.23	37.58
60	20	20	89.98	64.84
50	20	30	103.22	92.63

experienced a higher flux compared to the hydrophobic membrane (PES). This observation indicates that membrane properties, especially hydrophilicity, play an important role in the UF process. The continuation of sulfonyl groups with ring structure provides stiffness as well as dense structure, relative hydrophobicity, and high thermal stability [28], which led to a low flux of the pure water. The high hydrophobicity of membranes could also lead to adsorption of molecules on the membrane surface [29, 30]. When CA is blended with PES, the pure water flux increases to 33.87 L/m^2/h. On the other hand, the CA membrane has carbonyl groups (C=O), stretching, which appear as hydrophilic branches in the structure. In contrast, the CA membranes have two or more hydroxyl groups. In the present work, Al_2O_3 particle was added into PES/CA blend casting solution playing the role of membrane surface modifier. PES/CA/Al_2O_3 blend membranes prepared from 30% of Al_2O_3 have considerably higher flux compared to all the other membranes studied. The results indicate that the pure water flux of PES/CA (75/25%) blend membranes can be increased by the addition of increasinvg Al_2O_3 concentration from 10% to 30% in place of CA. Further, its hydrophilicity can be improved due to the addition of Al_2O_3 particles, which have some characteristics such as hydrophilic and bigger surface area [31]. Thus, water fluxes were increased. It was reported that the membrane performance was influenced by the formation of pores on the surface, and the pore formation was affected by the interfacial intension between solvent/nonsolvent mixtures and the polymer matrix [32]. When the coagulation bath was used as SLS/DMF mixture, the pure water flux of all the prepared membranes was improved substantially. The explanation for this improving trend could be ascribed to the diffusion rate of the solvent–nonsolvent (here the effect of DMF/SLS) and the increase of the Al_2O_3 particle concentration in the casting solution. Genne et al. [19] reported that on adding small amount of inorganic ZrO_2 particles into PSf polymer, the permeability was increased. Sikder et al. [33] studied the effect of pure water flux on CA/PSf blend MF membranes by using water and glycerin–water mixture as a nonsolvent in the membrane formation.

2.4.2 Effect of Al_2O_3 Concentration on Water Content

Water content is a vital parameter in membrane characterization as it is closely related to porosity of membranes. Water content of the formed membranes was calculated as shown in Table 2.3. It is seen that the water contents of the PES/CA membranes with addition of Al_2O_3 particles are higher than that of the neat PES membrane. Further, the water content was increased with an increase of Al_2O_3 amount in the PES/CA blend. When concentration of Al_2O_3 was increased from 10% to 30% in the PES/CA blend, water content increased from 82.1% to 86%. This increasing trend confirms the presence of increasing number of pores in the membrane with the increase of Al_2O_3 amount. The hydrophilic and porous properties of Al_2O_3 attract more water molecules inside the membrane and thus increase the void size [34]. The water content

TABLE 2.3. Water Content of PES/CA Membrane with Different Contents of Al_2O_3 Particles

Total Blend Composition (20 wt %)			Water Content (%)	
			Nonsolvent	
PES (%)	CA (%)	Al_2O_3 (%)	DMF/SLS Mixture	Water
100	0	0	80.2	73.4
0	100	0	76.4	70.1
75	25	0	82.1	75.7
70	20	10	84.5	77.9
60	20	20	85.9	78.6
50	20	30	86.0	78.9

TABLE 2.4. Membrane Resistance of PES/CA Membrane with Different Concentrations of Al_2O_3 Particles

Total Blend Composition (20 wt %)			Membrane Resistance (kPa/L/m^2/h)	
			Nonsolvent	
PES (%)	CA (%)	Al_2O_3 (%)	DMF/SLS Mixture	Water
100	0	0	14.08	30.34
0	100	0	29.41	68.92
75	25	0	12.92	24.72
70	20	10	6.83	11.66
60	20	20	5.97	6.54
50	20	30	4.01	4.45

of neat PES membrane was found to have 80.2% in presence of DMF/SLS mixture as nonsolvent in gelation bath. This may be attributed to the controlled exchange rate between DMF in polymer and SLS/DMF mixture in the nonsolvent gelation bath. Further, a slight increase in water content was observed for 25% CA in 75% PES. The reason was that the addition of CA increased the wettability of the PES membranes, and more pores in the PES membranes with CA would be wetted and filled with water. The similar results have also been observed by Mu et al. [35] in their study of the water content of PVDF/CA blend membrane.

2.4.3 Effect of Al_2O_3 Concentration on Membrane Resistance

The membrane resistance has reflects the difference of surface structure and morphology of membranes. As evident from Table 2.4 and Figure 2.1, the membrane resistance of pure PES exhibited a high value of 14.08 kPa m^2 h/L.

This may be due to its relatively low porosity and hydrophobicity. The membrane resistance was reduced to 12.92 kPa/m^2/h for 75/25% PES/CA

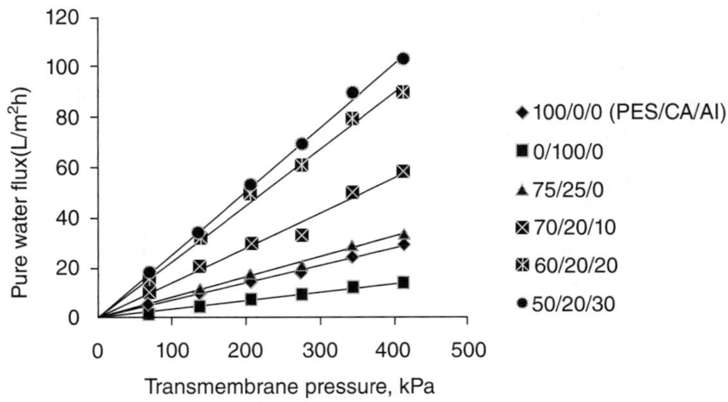

Figure 2.1. Effect of transmembrane pressure on pure water flux for DMF/SLS mixture as nonsolvent.

blend membrane due to the addition of more hydrophilic CA. The membrane hydrophilicity can also be considered as a parameter responsible for membrane resistance against fouling. The membrane resistance of PES/CA blend membrane was decreased significantly (4.0 kPa/m^2/h) when concentration of Al_2O_3 was increased to 30%. This may be due to the formation of larger pores caused by the addition of Al_2O_3 particles. The decrease in membrane resistance may be due to the incompatibility of Al_2O_3 and PES/blend chains; the polymer chains did not attach to the Al_2O_3 wall tightly, forming narrow gaps surrounding the Al_2O_3. According to Mu et al. [35], a decrease in membrane resistance allows some solutes to permeate through the membranes, resulting in lower rejection coefficient. The presence of Al_2O_3 particles might have caused looser association between polymer segments, resulting in more porous membrane. Consequently, the molecules are able to permeate more easily through the membrane. As for the effect of coagulation bath, it can be observed from Tables 2.2 and 2.3 that pure water flux and water content were high for the membranes formed by nonsolvent of SLS/DMF mixture than nonsolvent of water. In accordance with those data, the membrane resistance was low when DMF/SLS/mixture was used as a nonsolvent. This low membrane resistance may be due to the commencement of surface tension gradient at the DMF/SLS mixture for the convective mass transfer phenomena of nonsolvent to solvent or vice versa.

2.4.4 Effect of Al_2O_3 Concentration on Pore Size Distribution

The permeability of UF membranes is mainly governed by pore size distribution. The larger pore size and higher pore density result in enhanced permeation of water through membrane [36]. It is also seen in our experimental data when DMF/SLS mixture was added to the coagulation bath. The pore size,

TABLE 2.5. Pore Size Distribution of the PES/CA Blend Membranes with Different Concentrations of Al_2O_3 Particle

Total Blend Composition (20 wt %)			Average Pore Radius, R (Å)		Surface Porosity, $\varepsilon \times 10^{-2}$		Pore Density, $\eta \times 10^9$, cm^{-2}	
			Nonsolvent		Nonsolvent		Nonsolvent	
PES (%)	CA (%)	Al_2O_3 (%)	DMF/SLS Mixture	Water	DMF/SLS Mixture	Water	DMF/SLS Mixture	Water
100	0	0	22.4	18.4	6.2	4.6	6.4	4.5
0	100	0	45.0	40.2	18.4	13.4	21.7	18.6
75	25	0	32.0	26.7	11.5	8.6	14.5	12.2
70	20	10	35.2	30.0	13.6	10.8	17.2	16.5
60	20	20	39.7	34.5	15.3	13.0	19.3	17.5
50	20	30	48.1	42.0	17.3	15.3	28.4	25.0

porosity, and pore density of the membranes were determined from the protein rejection experiments as reported in our earlier work [10], and the results were summarized in Table 2.5. The average pore radius of neat PES membrane was 22.4 Å. The pore radius increased slightly to 32.0 Å when 25% CA was blended in PES. Furthermore, by adding 10% of Al_2O_3 to the solid component (PES + CA + Al_2O_3) of the casting solution, the pore radius in the membrane increased to 35.2 Å, probably due to the weak attachment between polymers and Al_2O_3 particles. The number of pores per unit area (pore density) also increased from 14.7×10^9 to $17.2 \times 10^9/cm^2$ when 10% of Al_2O_3 was added to the polymer blend. However, the porosity remained in the range of 11.5 and 17.3×10^{-2} for the PES/CA blend regardless of Al_2O_3 content, indicating that morphologies of all the blend membranes are nearly equal. Therefore, it can be concluded that not increase in porosity but increase in pore radius and pore density lead to the increase in the water flux of the PES/CA/Al_2O_3 membrane at 15/20/30% composition. Ultimately, the pore density increased from 6.4×10^9 to $28.4 \times 10^9/cm^2$ for neat PES membrane with an addition of concentration of Al_2O_3 from 0 to 30%. These results show that the addition of Al_2O_3 particles may disrupt the polymer chain packing and increase the pores between the polymer chains and thus enhance permeability. Distribution of Al_2O_3 particles may disrupt the polymer chain packing, thus increasing the free volume and thereby enhancing the porosity [37].

2.4.5 Effect of Al_2O_3 Concentration on Morphology

The top surface and cross section of membranes with compositions 75/25/0, 70/20/10, 50/20/30% PES/CA/Al_2O_3 are shown in Figures 2.2a–c and 2.3a–c. The comparison of top views shown in Figure 2.3a,b reveals that addition of Al_2O_3 to the PES/CA blend leads to the formation of larger voids. From Figure 2.3b,c, the pores in the microporous skin further increased in size as the Al_2O_3

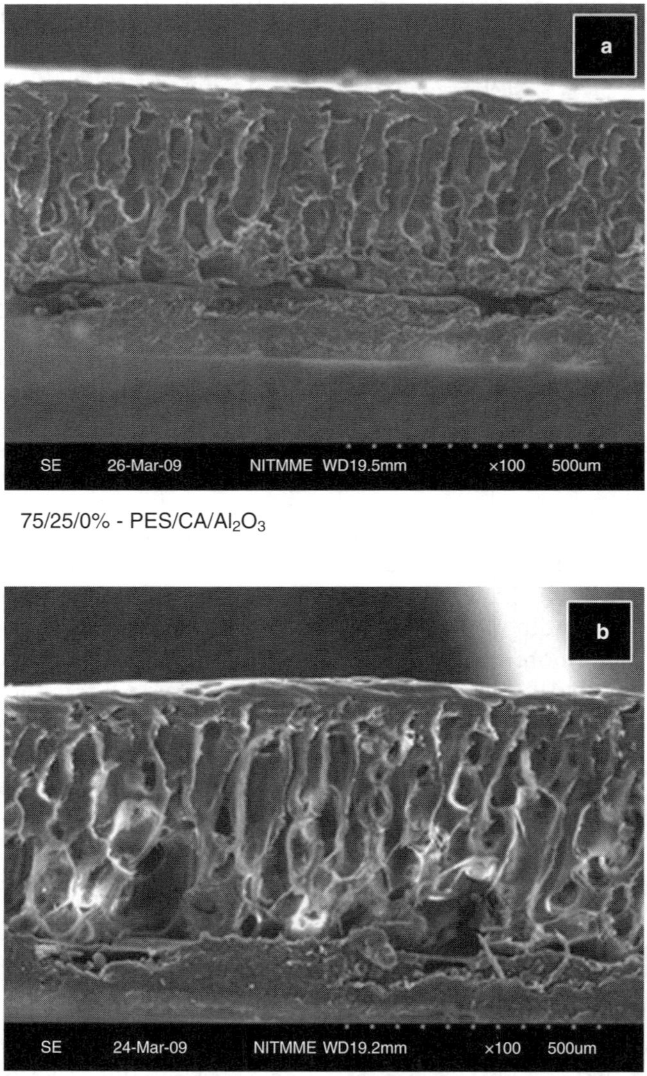

75/25/0% - PES/CA/Al_2O_3

70/20/10% - PES/CA/Al_2O_3

Figure 2.2. Cross-sectional view of PES/CA blend membranes with different Al_2O_3 concentrations.

content was increased from 20% to 30%. In the presence of the Al_2O_3 in the casting solution, the mechanism of phase separation in the system and the structure of the membrane surface are altered. This may be due to the detachment of polymer chains from the Al_2O_3 surface, causing the interfacial voids. The poor polymer/inorganic particle contact could result in interfacial voids, which presumably is the major cause for the improved performance in water

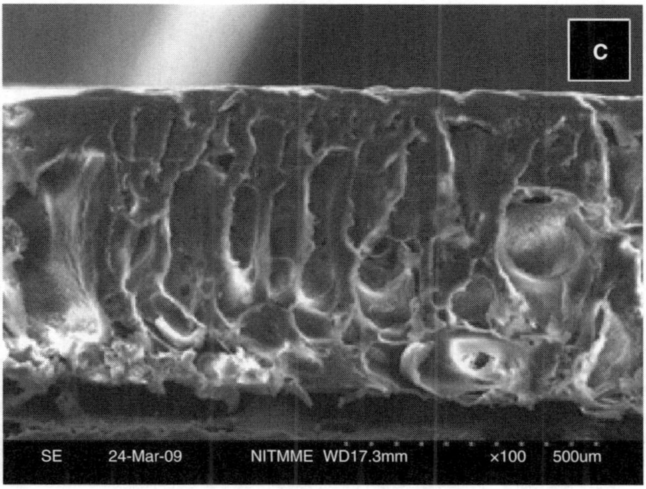

50/20/30% - PES/CA/Al_2O_3

Figure 2.2. (*Continued*)

permeation [38]. Other possible causes for the interfacial voids formation include repulsive force between polymer and inorganic particles [39]. The SEM results can confirm our previous findings such as increase in pure water flux, water content, pore size, and pore density with an increase in Al_2O_3 content in the PES/CA blend.

From the cross-section images, typical asymmetric structures are observed with finger-like pores formed in the sublayer. Moreover, the pores and voids in the cross section become smaller when the 75/25% PES/CA blend membrane without Al_2O_3 is compared to those with Al_2O_3. Similar observations have been reported by Zhao [40] using Pluronic F127 as modifier and pore-forming agent in PES UF membranes. According to top surface and cross-section observation, with the increase in Al_2O_3 content in the polymer matrix, a slightly altered membrane morphology was obtained. The possible reason is the formation of larger pores and looser structure in the inorganic and organic networks in the presence of Al_2O_3 particles.

2.5 CONCLUSIONS

In this work we investigated the effects of Al_2O_3 particles as modifying agent during the formation of the PES/CA blend membranes through a phase inversion method. In particular, the effect of Al_2O_3 on the performance of PES/CA blend membranes has been studied. The physical incompatibility was observed for the PES/CA blend when CA content was above 25%. The UF performance

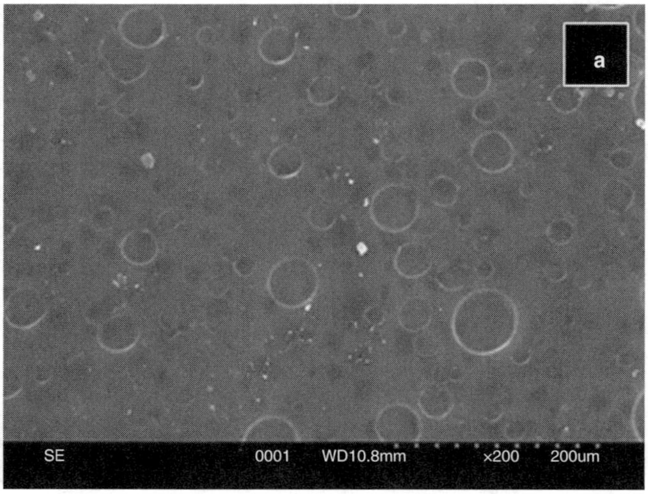

70/20/10% - PES/CA/Al_2O_3

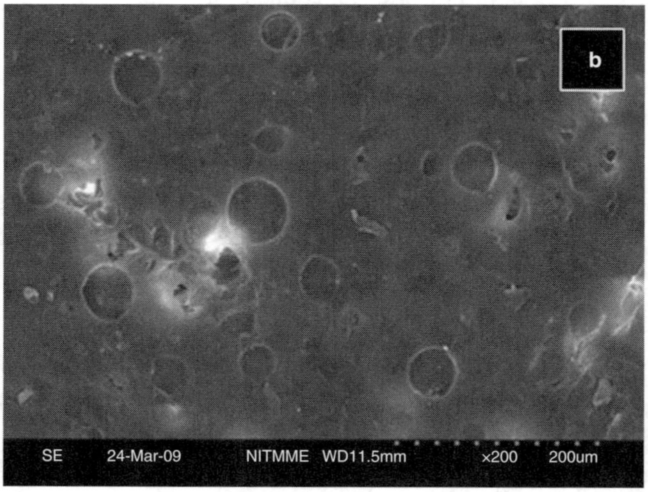

70/20/10%- PES/CA/Al_2O_3

Figure 2.3. Top surface view of PES/CA blend membranes with different Al_2O_3 concentrations.

of the membrane was improved when 30% Al_2O_3 was added to the 75/25% PES/CA blend; that is, the pure water flux increased from about 33.87 to 103.22 L/m^2/h at the operating pressure of 414 kPa with the increase of Al_2O_3 particles from 10% to 30% in the 75/25% of PES/CA blend membrane. Partial substitution of Al_2O_3 in the place of PES to form PES/CA/Al_2O_3 blends resulted in substantial and systematic increases in the membrane permeability,

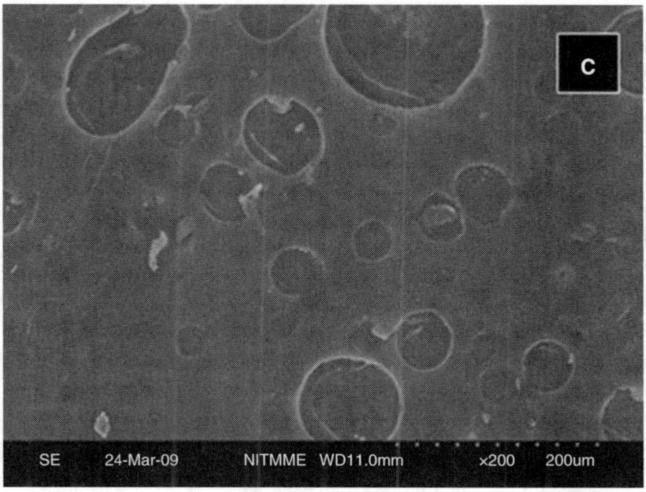

50/20/30% - PES/CA/Al_2O_3

Figure 2.3. (*Continued*)

water content, and pore size, while decreasing membrane resistance. The formation of macrovoids and asymmetric membrane structures observed by SEM agree well with the results of pore size distribution and water flux studies. The experimental results also showed that the addition of DMF/SLS mixture to the coagulation bath played a major role in altering the properties of the resultant membranes. This approach not only allows the control of the morphology and the pore organization of the membrane by the addition of Al_2O_3 particles but also provides a method to form novel polymer–inorganic membrane materials.

REFERENCES

[1] Baker R. W. *Membrane Technology and Applications*. John Wiley and Sons, England, 2004.

[2] Matsuura T. *Synthetic Membranes and Membrane Separation Processes*. CRC Press Inc., Boca Raton, FL, 1994.

[3] Loeb G. S. and S. Sourirajan. Sea water demineralization by means of osmotic membrane. *Adv. Chem. Ser.* 38 (1963) 117.

[4] Zeng M., Z. Fang, and C. Xu. Effect of compatibility on the structure of the microporous membrane prepared by selective dissolution of chitosan/synthetic polymer blend membrane. *J. Memb. Sci.* 230 (2004) 175.

[5] Li J., S. Wang, K. Nagai, T. Nakagawa, and A. W.-H. Mau. Effect of polyethyleneglycol (PEG) on gas permeabilities and permselectivities in its cellulose acetate (CA) blend membranes. *J. Memb. Sci.* 138 (1998) 143.

[6] Chen Z., M. Deng, Y. Chen, G. He, M. Wu, and J. Wang. Preparation and performance of cellulose acetate/polyethyleneimine blend microfiltration membranes and their applications. *J. Memb. Sci.* 235 (2004) 73.

[7] Sivakumar M., D. Mohan, R. Rangarajan, and Y. Tsujita. Studies on cellulose acetate–polysulfone ultrafiltration membranes: I. Effect of polymer composition. *Polym. Int.* 54 (2005) 956.

[8] Malaisamy R., R. Mahendran, D. Mohan, M. Rajendran, and V. Mohan. Cellulose acetate and sulfonated polysulfone blend ultrafiltration membranes. I. Preparation and characterization. *J. Appl. Polym. Sci.* 86 (2002) 1749.

[9] Mahendran R., R. Malaisamy, and D. Mohan. Preparation, characterization and effect of annealing on performance of cellulose acetate/sulfonated polysulfone and cellulose acetate/epoxy resin blend ultrafiltration membranes. *Eur. Polym. J.* 40 (2004) 623.

[10] Arthanareeswaran G., D. Mohan, and M. Raajenthiren. Preparation and performance of polysulfone-sulfonated poly(ether ether ketone) blend ultrafiltration membranes. Part I. *Appl. Surf. Sci.* 253 (2007) 8705.

[11] Mahendran R., R. Malaisamy, and D. Mohan. Cellulose acetate and polyethersulfone blend ultrafiltration membranes: Part I. Preparation and characterizations. *Polym. Adv. Technol.* 15 (2004) 149–157.

[12] Vijayalakshmi A., D. Lawrence Arockiasamy, A. Nagendran, and D. Mohan. Separation of proteins and toxic heavy metal ions from aqueous solution by CA/PC blend ultrafiltration membranes. *Sep. Purif. Tech.* 62 (2008) 32.

[13] Nagendran A., D. Vijayalakshmi, K. Lawrence Arockiasamy, H. Shobana, and D. Mohan. Toxic metal ion separation by cellulose acetate/sulfonated poly(ether imide) blend membranes: effect of polymer composition and additive. *J. Hazard. Mater.* 155 (2008) 477.

[14] Sivakumar M., D. Mohan, and R. Rangarajan. Studies on cellulose acetate-polysulfone ultrafiltration membranes: II. Effect of additive concentration. *J. Memb. Sci.* 268 (2006) 208.

[15] Pan F., Q. Cheng, H. Jia, and Z. Jiang. Facile approach to polymer–inorganic nanocomposite membrane through a biomineralization-inspired process. *J. Memb. Sci.* 357 (2010) 171.

[16] Yan L., Y. S. La, C. B. Xiang, and S. Xianda. Effect of nano-sized alumina particle addition of PVDF ultrafiltration membrane performance. *J. Memb. Sci.* 276 (2006) 162.

[17] Bottino A., G. Capannelli, V. Dasti, and P. Piaggio. Preparation and properties of novel organic–inorganic porous membranes. *Sep. Purif. Tech.* 22 (2001) 269.

[18] Bottino A., G. Capannelli, and A. Comite. Preparation and characterization of novel porous PVDF-ZrO2 composite membranes. *Desalination* 146 (2002) 35.

[19] Genne S., R. Kuypers, and R. Leysens. Effect of addition of ZnO_2 to polysulfone based ultrafiltrsation membranes. *J. Memb. Sci.* 113 (1996) 343.

[20] Oh S. J., N. Kim, and Y. T. Lee. Preparation and characterization of PVDF/TiO_2 organic–inorganic composite membranes for fouling resistance improvement. *J. Memb. Sci.* 345 (2009) 13.

[21] Finken H. Bentonite-stabilized CDA/CTA membranes: I. Improved long-term transport properties. *Desalination* 48 (1983) 207.

[22] Adoor S. G., M. Sairam, L. S. Manjeshwar, K. V. S. N. Raju, and T. M. Aminabhavi. Sodium montmorillonite clay loaded novel mixed matrix membranes of poly(vinyl alcohol) for pervaporation dehydration of aqueous mixtures of isopropanol and 1,4-dioxane. *J. Memb. Sci.* 285 (2006) 182.

[23] Qiu S., L. Wu, X. Pan, L. Zhang, H. Chen, and C. Gao. Preparation and properties of functionalized carbon nanotube/PSF blend ultrafiltration membranes. *J. Memb. Sci.* 342 (2009) 165.

[24] Jian P., H. Yahui, W. Yang, and L. Linlin. Preparation of polysulfone–Fe_3O_4 composite ultrafiltration membrane and its behavior in magnetic field. *J. Memb. Sci.* 284 (2006) 9.

[25] Maximous N., G. Nakhla, W. Wan, and K. Wong. Preparation, characterization and performance of Al_2O_3/PES membrane for wastewater filtration. *J. Memb. Sci.* 341 (2009) 67.

[26] Sarbolouki M. N. General diagram for estimating pore size of UF and RO membranes. *Sep. Sci. Technol.* 17 (1982) 381.

[27] Velicangil O. and J. A. Howell. Estimation of the properties of high flux ultrafiltration membranes. *J. Phys. Chem.* 84 (1980) 2991.

[28] Wu L., S. Junfen, and W. Qingrui. Poly(vinylidene fluoride)/polyethersulfone blend membranes: effects of solvent sort, polyethersulfone and polyvinylpyrrolidone concentration on their properties and morphology. *J. Memb. Sci.* 285 (2006) 290.

[29] Keurentjes J. T. F., G. I. Doornbusch, and K. van't Riet. The removal of fatty acids from edible oil. Removal of the dispersed phase of a water-in-oil dispersion by a hydrophilic membrane. *Sep. Sci. Technol.* 26 (1991) 409.

[30] Elizabeth A., K. Denis, and G. Vitaly. Impact of chemical cleaning on properties and functioning of polyethersulfone membranes. *J. Memb. Sci.* 305 (2007) 176.

[31] Peng Z., L. X. Kong, and S. D. Li. Non-isothermal crystallisation kinetics of self-assembled polyvinylalcohol/silica nano-composite. *Polymer* 46 (2005) 1949.

[32] Tan J. M. A., S.-H. Noh, G. Chowdhurry, and T. Matsuura. Influence of surface tensions of solvent/non-solvent mixtures in membrane casting solutions on the performance of poly(2,6-dimethyl-1,4-phenylene) oxide membranes for gas separation applications. *J. Memb. Sci.* 174 (2000) 225.

[33] Sikder J., C. Pereira, S. Palchoudhury, K. Vohra, D. Basumatary, and P. Pal. Synthesis and characterization of cellulose acetate-polysulfone blend microfiltration membrane for separation of microbial cells from lactic acid fermentation broth. *Desalination* 249 (2009) 802.

[34] Rahimpour A. and S. S. Madaeni. Polyethersulfone (PES)/CA phthalate(CAP) blend ultrafiltration membranes: preparation, morphology, performance and antifouling properties. *J. Memb. Sci.* 305 (2007) 299.

[35] Mu C., Y. Su, M. Sun, W. Chen, and Z. Jiang. Remarkable improvement of the performance of poly(vinylidene fluoride) microfiltration membranes by the additive of cellulose acetate. *J. Memb. Sci.* 350 (2010) 293–300.

[36] Mulder M., *Basic Principles of Membrane Technology*, 2nd ed. Kluwer Academic Publisher, Dordrecht, The Netherlands, 1996, pp. 214–225.

[37] Ye S. H., J. Watanabe, Y. Iwasaki, and K. Ishihara. Antifouling blood purification membrane composed of cellulose acetate and phospholipid polymer. *Biomaterials* 24 (2003) 4143.

[38] Aerts P., I. Genne, S. Kuypers, R. Leysen, I. F. J. Vankelecom, and P. A. Jacobs. Polysulfone–aerosil composite membranes: Part 2. The influence of the addition of aerosil on the skin characteristics and membrane properties. *J. Memb. Sci.* 178 (2000) 5–9.

[39] Jia M., K. V. Peinemann, and R. D. Behling. Preparation and characterization of thin-film zeolite–PDMS composite membranes. *J. Memb. Sci.* 73 (1992) 119.

[40] Zhao W., Y. L. Su, C. Li, Q. Shi, X. Ning, and Z. Y. Jiang. Fabrication of antifouling polyethersulfone ultrafiltration membranes using Pluronic F127 as both surface modifier and pore-forming agent. *J. Memb. Sci.* 318 (2008) 405.

3

CHARACTERIZATIONS OF NEWLY DEVELOPED BACTERIAL CELLULOSE–CHITOSAN MEMBRANE WITH PYRROLINE

Siti Nur Hidayah Mohamad, Norhayati Pa'e,
Abdul Halim Mohammad Yusof, and Ida Idayu Muhamad

3.1 INTRODUCTION

Membranes play an essential role not only in nature but also in today's modern industrial society. Biological membranes have an essential function in the metabolism of all living species. Through exchange processes, a single cell can be supplied with nutrients, and it can excrete waste products through its outer cell membrane. Moreover, these exchange processes enable communication between cells, by exchange of, for example, hormones, which makes the development of complex life forms possible.

Considering that the first membrane experiments dated from the 18th century using membranes (from Latin membrane, a thin sheet) of biological origin [1], industrial membrane separation with synthetic membranes has been strongly developed since the introduction of the asymmetric polymeric membrane [2] in the early 1960s. Current biotechnological separation applications (e.g., food and pharmaceutical industries) are fast developing applications of artificial membranes. Also there are new membrane applications, such as affinity membrane, biosensors, biohybrid organs, and artificial tissue structures.

Sustainable Membrane Technology for Energy, Water, and Environment, First Edition.
Edited by Ahmad Fauzi Ismail and Takeshi Matsuura.

In the chemical process industry, membranes can, in principle, carry out most of the separation processes and can complement or form an alternative for processes like distillation and extraction. Advantages of membrane filtration among others are low energy consumption, continuous operation, relatively simple upscaling, and versatility of the process.

The success of using membranes is closely related to the intrinsic properties of the membrane. Interfacial interactions between membrane surface, surrounding environment, and solutes govern membrane performance to a great extent. These interactions have considerable impact on transport characteristics such as selectivity, fouling propensity, and bio- and hemo-compatibility of the membrane. Especially in biotechnological and medical applications, where highly adsorptive solutes such as proteins are present, adsorption of proteins often results in fouling of the membrane, which leads to considerable losses in flux, selectivity, and performance [3].

Compared with the hydrophobic membranes, cellulose or derived cellulose membranes, hydrophilic in nature, have very low nonspecific binding. Microbial cellulose as synthesized by *Gluconacetobacter xylinum* consists of elementary fibrils with an average cross section of 16·58 Å [4]. Typically 46 such elementary fibrils make up a flat ribbon-like microfibril [4]. Small angle X-ray scattering revealed microfibril dimensions of 10·160 Å in cross section, whereas ribbons of up to 500 Å width observed in the emission scanning electron miscroscope (ESEM) are considered aggregates of several microfibrils [5]. Microfibril diameters of 24–86 nm [3] and 72–175 nm [6] are also reported in the literature. Presumably due to uninterrupted cellulose synthesis during cell division, the cellulose microfibrils in bacterial cellulose (BC) films are branched, and their length has been estimated to multiples of 100 μm [7]. With 60–80%, a large part of BC is arranged in crystallites [6]. The most substantial use of BC is currently in the foodstuff industry [7], but BC shows considerable potential for other technical applications. Medical usage of BC comprises its use as temporary substitute for human and animal skin [8] and artificial blood vessels for microsurgery [9]. Furthermore, BC membranes were utilized in combination with palladium to produce experimental fuel cells for the generation of electricity [10]. BC was also added to the production of paper, resulting in considerably improved mechanical properties [11]. Finally, synthesized in the presence of pectin and/or xyloglucan [3, 5, 12, 13], BC serves as model system for the study of the interaction of polymers during the formation of plant cell walls.

Chitosan molecules have both amino and hydroxyl groups that can be used to couple with ligands under mild conditions besides being a good flocculation agent. But their poor mechanical properties prevented them from being used widely. In order to develop a membrane with good mechanical and chemical properties, this study proposes to make a bacterial cellulose-chitosan pyrroline (BCCP) membrane, which combines the advantages of cellulose, chitosan, and unique pyrroline properties.

BC, an inexpensive naturally occurring material produced from fruits waste, has superior mechanical strength and high water holding capacity. Over the last decades, chitosan has been receiving high research and industrial interest for application as chelating and flocculation agent for wastes treatment [14, 15]. Chitosan has good flocculation capability and also antimicrobial properties [16] while pyrroline has ketones which can serve as adsorption sites for binding heavy metals and good coagulation capabilities besides its fresh odor; these properties make microbial cellulose like chitosan and pyrroline potential sustainable membranes for food processing, pharmaceutical industries, and wastewater treatment plants. Owing to the high hydrophilicity of chitosan and BC material, the membrane swells more significantly in a feed solution with higher water content; the membrane needs to be esterified with pyrroline in order to stabilize and increase hydrophobic surface character so that it will not swell easily.

3.2 METHODOLOGY

3.2.1 Preparation of the BC Membrane

BC membrane is obtained by incubating *Acetobacter xylinum* in pineapple waste juice supplemented with 4% sucrose (w/v) at pH 5.0. A stainless steel round shallow tray of 39 cm diameter is used to grow the cellulose-producing bacteria at 30°C at the surface of culture medium under static conditions. The culture volume is 500 mL and the effective area from membrane growth is 20 cm^2. Buffered Schamm and Hestrin's medium (BSH medium), which is employed as the preculture medium, is composed of 2.0% (w/v) glucose, 0.5% (w/v) yeast extract, 0.5% (w/v) peptone, 0.033% (w/v) $Na_2HPO_4 \cdot 2H_2O$, and 0.11% (w/v) citric acid·H_2O. *A. xylinum* is grown in 50 mL of BSH medium for 3 days to be used as preculture. The pellicles of BC formed on the surface of this medium surface are harvested aseptically.

The harvested pellicle is rinsed with distilled water and soaked in 1% NaOH for 24 hours for removing the incorporated cells and culture medium components, then thoroughly washed with distilled water to neutrality. The purified pellicle is placed on a glass plate and allowed to dry under ambient conditions. The formed membrane was removed from the plate by soaking in distilled water.

3.2.2 Preparation of the Bacterial Cellulose–Chitosan (BCC) Membrane

The BCC nanocomposite membranes are prepared by phase inversion method with polyethylene glycerol as the porogen: The chitosan solution of 5% (w/v) chitosan and 15% (w/v) polyethylene glycerol in a 1% (v/v) aqueous acetic acid solution is obtained. The chitosan solution is filtered by a fritted silica glass Buchner filter of pore size of 40–60 μm and treated in ultrasonic cleaner for 1 hour to remove undissolved substances and air bubbles.

Chitosan coating on BC membrane is prepared as follows: The BC membrane was flushed with the chitosan solution, before being subsequently soaked in the chitosan solution overnight. The chitosan solution of 6.5 mL is then poured onto BC membrane placed in a petri dish (100 mm, diameter) and allowed to evaporate for optimized 24 hours at ambient temperature. Subsequently, the membrane is immersed overnight in 1 M NaOH to extract the porogen and form a microporous membrane. The resulting composite membrane is washed with a large amount of water until the washing solution becomes neutral and then it is treated with 10% glycerol solution before drying in air to avoid shrinking.

The Pyrroline-BCC membrane is prepared by casting 2 mL pyrroline solution that has been extracted from *Pandanus amaryllifolius* using ethanol onto BCC membrane. The membrane is then allowed to dry in the ambient air. The resulting membrane is removed from the plate after it is dried.

3.2.3 Membrane Characterization

3.2.3.1 Microscopic Observation. Microscopic observation was carried out by a scanning electron microscope (LEO 1430VP, UK) which directly provides the visual information of the top surface as well as a cross-sectional morphology of the membranes. Computerized analysis of scanning electron microscope (SEM) image is a standard and widely used method for the investigation of perforated materials [17,18]. However, in the present study, the morphological parameters such as pore size and pore number could not be measured from the SEM photographs as almost all the pores are in the ultrafiltration range. Such practice would only give rough approximation of the pore size by overestimating the smallest pores on the surface and also by considering dead-end (blocked) pores along with the open pores [19]. Liquid displacement method was adopted to compare the morphology of different membranes.

3.2.3.2 Equilibrium Water Content (EWC). EWC is considered to be an important characterization parameter as it indirectly indicates the degree of hydrophilicity or hydrophobicity of a membrane [19]. Also, it is related to the porosity of a membrane. Membranes were weighed in an electronic balance in wet state after mopping the surface water with a clean tissue paper. The wet membranes were dried by putting in a vacuum oven for 24 hours at a temperature of 50–60°C and again they were weighed in dry state. Then the EWC at room temperature was calculated as in Equation 3.1:

$$\text{EWC}\,(\%) = \frac{W_\text{w} - W_\text{d}}{W_\text{w}} \times 100 \qquad (3.1)$$

where W_w is weight of wet membranes (kg) and W_d is weight of dry membranes (kg).

3.2.3.3 Porosity. Porosity of the membrane plays an important role on permeation and separation. The membrane porosity is determined as in Equation 3.2 [20]:

$$\text{Porosity} = \frac{W_w - W_d}{\rho_w \times V} \tag{3.2}$$

Where, ρ_w is density (kg/m^3) of pure water at room temperature and V is volume of the membrane in wet state (m^3).

3.2.3.4 Permeation Experiment. The permeation test was conducted by using a simple dead-end permeation cell. Circular membrane discs were cut and mounted in a stainless steel cylindrical membrane test cell by a porous support and tightened by a rubber O-ring. Effective permeation area of each membrane was about 13.2 cm^2. Prior to testing, the pure water flux (PWF) was measured to ensure that the membranes used were stable. Feed pressure was varied from 100 to 700 kPa while the permeate side was opened to the atmosphere. Experiments were carried out at ambient temperature (27°C).

3.2.3.5 PWF. Membrane hydraulic permeability has significance particularly for membranes used in pressure-driven separation processes. Membrane permeability was determined by allowing deionized water to pass through the compact membrane. Flux values of pure water at different transmembrane pressures (ranging 0–240 kPa) were measured under steady-state condition using Equation 3.3:

$$J_w = \frac{Q}{A\Delta t} \tag{3.3}$$

where J_w is PWF (L/m^2 h), Q is volume of water permeated (L), A is effective membrane area (m^2), and Δt is sampling time (h).

3.3 RESULTS AND DISCUSSION

3.3.1 Morphological Structure of BC Chitosan Membrane Coated with Pyrroline

In the humid state the biofilm formed on the surface of the culture medium of the bacterium *A. xylinum* is a homogenous transparent, moldable, and handle-resistant gel (Fig. 3.1a,b). After drying, the thickness of the membranes varied between 150 and 200 μm. Figure 3.2a,b give an overview of the dried BC. The SEM image shows the microfibrillar structure of cellulose and aggregates of semicrystalline extended-cellulose chains in an ultrafine network structure. This structure consists of continuous nanofibers about 10 nm thick and 50 nm wide [21].

Microscopic study through SEM analysis was carried out to have qualitative information regarding surface and cross-sectional morphology of the membranes prepared. SEM analysis is an important technique for microscopic observations of the membrane morphology. In order to observe the morphological structure of BCCP membrane, the SEM test has been done. The results from this SEM test are expected to be dense.

(a) (b)

Figure 3.1. The humid state biofilm formed on the surface of the culture medium of the bacterium *A. xylinum*.

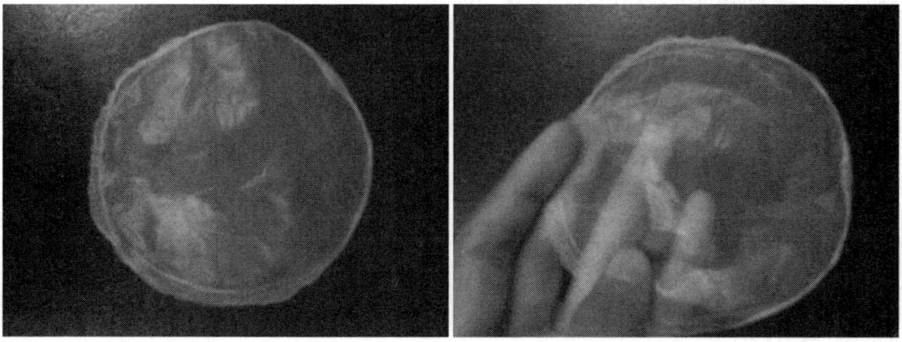

(a) (b)

Figure 3.2. Dried BCCP membrane.

A more detailed analysis by SEM revealed that the membranes after drying are composed of long fibers. Figure 3.3 shows the SEM micrograph of the dried membrane surface, highlighting details of the microstructure. The morphologies of these samples were remarkably different from each other: Whereas the original BC membrane is composed of many lamella-like layers having wide gaps in between, the pyrroline-treated membrane has a densely packed spongy structure. This difference could be noted during handling of the samples for permeability measurements: The original membrane was more highly swollen and softer than the latter. Thus, the pyrroline treatment seems to have a compacting effect to the BC membrane, without changing the basic structure of cellulose **I.** The film consists of a dense, flawless, and homogeneous cellulose matrix (Fig. 3.3a). The matrix is formed by an interlaced network of long cellulose nanofibers with high aspect ratio and mean diameters smaller than 100 nm (Fig. 3.3b). This high aspect ratio may affect the mechanical properties of the film such as elasticity and tensile strength. The reason for that fibrous morphology was subject of discussion in the last decade and is already

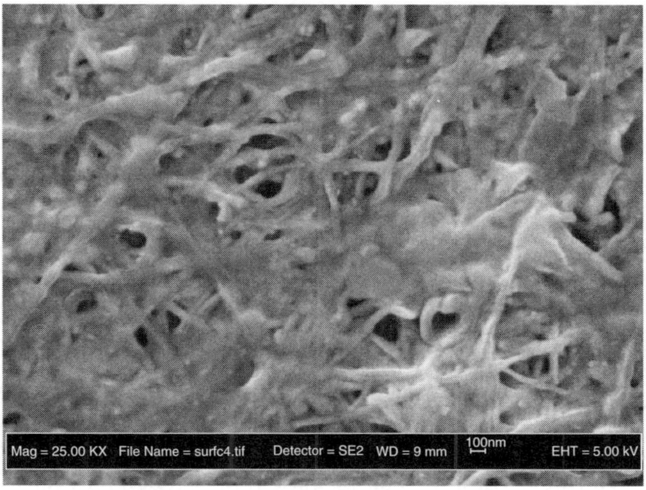

(a)

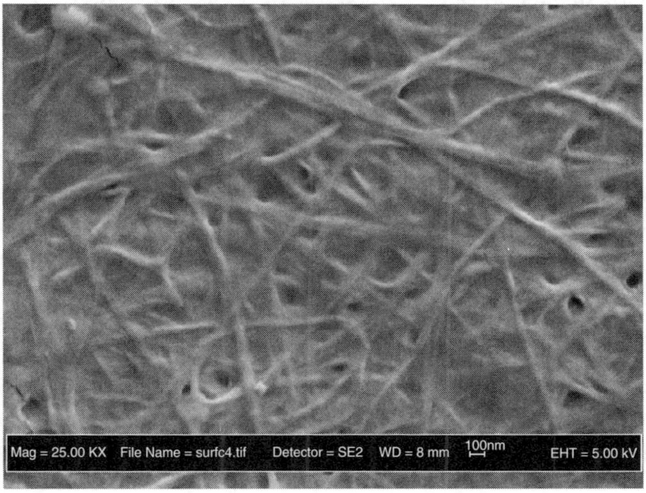

(b)

Figure 3.3. SEM images of prepared membranes: (a) BC membrane surface, (b) BCCP membrane surface.

established [22, 23]. Cellulose biosynthesis is characterized by unidirectional growth and crystallization, where glucose molecules are linearly bonded by $\beta(1 \rightarrow 4)$-glycosidic bonds. The union of glycosidic chains forms oriented microfibrils with intramolecular hydrogen bonds [24]. The preferential orientation of cellulose crystal growth during biosynthesis was already reported by Koyama et al. [25]. The cellulose is crystallized outward the organisms, particularly in *A. xylinum* that synthesizes cellulose chains by introducing glucose units to the reducing ends of the polymer. The growth mechanism during bacterial activity determines the morphology of the final cellulose.

Although the SEM test had been done to see the morphological structure of membranes, which could influence BC properties such as water holding capacity, thermal stability, and mechanical strength, the figure itself, however, cannot prove the performance of the membrane. Thus, the permeation tests still need to be run in a way to see the performance of the membranes.

3.3.2 Permeation Test

BCCP membrane shows a promising structure. It comprises a layer of pyrroline and chitosan and BC support that was prepared to increase the permeation flux of water with a minimal sacrifice of separation factor. The performance of this membrane can be shown by the value of permeation flux, J_w, obtained. The permeation experiment is done by using a simple dead-end permeation cell.

BCCP with surface area of 13.2×10^{-4} m^2 is used in this water permeation experiment. Theoretically, the permeation flux will increase with the increased feed pressure. The time required to collect 2 mL of permeate was measured. Table 3.1 shows all the data and result obtained from the permeation test. PWF, (J_w) and hydraulic permeability (P_m) have been calculated.

Membranes that have been prepared are tested to see the effect of coated pyrroline on their permeation behavior. The membranes are characterized in terms of PWF at different pressures. High values of P_m for higher pressure in BCCP are possibly due to relatively higher surface area for water to permeate of the former membrane although the porosity is lower. This result agrees with the findings of the EWC test that will be discussed in the following section.

Figure 3.4 shows the relationship of PWF (J_w) versus pressure. It seems that PWF increases almost linearly with an increase in transmembrane pressure for the membrane within the range of 0–700 kPa. This is due to the increase in effective driving force (transmembrane pressure) required for water permeation.

3.3.3 EWC (%) and Porosity

EWC is an important parameter for membrane characterization as it is closely related to PWF. The EWC of different membranes is presented in Table 3.2. It may be found from the table that the pyrroline coating decreases the EWC of the BCC membrane. Moreover, porosity decreases as expected.

3.4 CONCLUSION

In this work, the BC membrane was integrated with chitosan composite layer and modified with polyethylene glycol as the porogen and pyrroline as an

TABLE 3.1. Pure Water Flux and Hydraulic Permeability Data of BCCP Membrane

P (kPa)	Q (L)	Δt (h)	J_w (L/m^2 h)	P_m (L/m^2 h kPa)
		0		
100	0.002	0.272	5.570	0.057
	0.004	0.558	5.431	0.054
	0.006	0.824	5.516	0.055
	0.008	1.087	5.576	0.056
	0.010	1.408	5.381	0.054
200	0.002	0.101	15.002	0.075
	0.004	0.218	13.900	0.070
	0.006	0.325	13.986	0.070
	0.008	0.442	13.712	0.069
	0.010	0.559	13.552	0.068
300	0.002	0.059	25.681	0.086
	0.004	0.138	21.959	0.073
	0.006	0.207	21.959	0.073
	0.008	0.275	22.039	0.073
	0.010	0.351	21.583	0.072
400	0.002	0.041	36.955	0.092
	0.004	0.091	33.300	0.083
	0.006	0.151	30.102	0.075
	0.008	0.200	30.303	0.076
	0.010	0.251	30.182	0.075
500	0.002	0.034	44.563	0.089
	0.004	0.073	41.511	0.083
	0.006	0.117	38.850	0.078
	0.008	0.154	39.355	0.079
	0.010	0.200	37.879	0.076
600	0.002	0.027	56.117	0.094
	0.004	0.069	43.917	0.073
	0.006	0.103	44.131	0.074
	0.008	0.139	43.601	0.073
	0.010	0.175	43.290	0.072
700	0.002	0.024	63.131	0.090
	0.004	0.058	52.247	0.075
	0.006	0.093	48.876	0.070
	0.008	0.135	44.893	0.064
	0.010	0.171	44.303	0.063

additive to increase hydrophobic surface character of BCCP membrane. This newly developed membrane, BCCP, offers an alternative for existing polymeric membrane with its promising characteristics. The potential of BCCP membrane is corresponding to its chemical structure having free amine group of this material as well as morphological aspects. This membrane can stand higher pressure, that is, over 7 bar compared to the BCC membrane

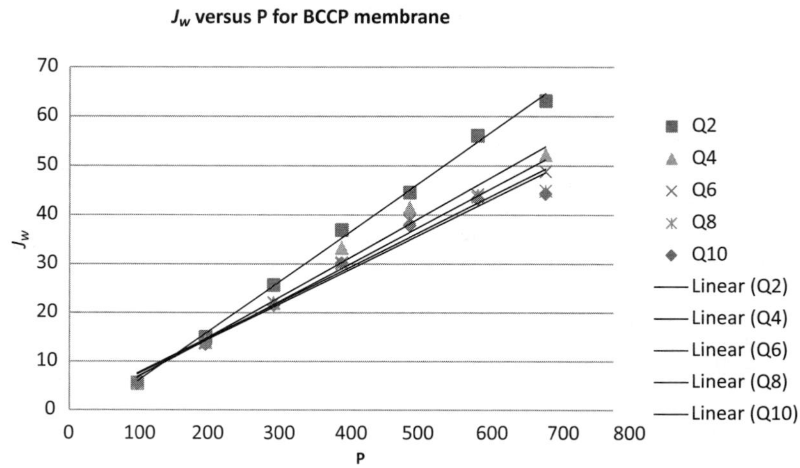

Figure 3.4. Pure water flux versus pressure for BCCP membrane.

TABLE 3.2. EWC and Porosity for BCC and BCCP Membrane

Samples	EWC (%)	Porosity
BCC membrane	71.36	0.29
BCCP membrane	61.69	0.11

without pyrroline. The highest permeation flux given by BCCP membrane is 63.13 L/m^2 h at 700 kPa. The EWC for the BCCP is 61.69%, that is, lower than the BCC membrane because the BCCP membrane is denser than the BCC membrane. In addition, the surface porosity of the BCCP membrane is 0.11, that is, lower than the BCC membrane. Besides using the low-cost, sustainable, and environmental friendly material, this membrane has a very good potential to be used in separation processes in the future.

REFERENCES

[1] Bledzki A. K. and J. Gassan. Composites reinforced with cellulose based fibres. *Prog. Polym. Sci.* 24 (1999) 221–274.

[2] Eichhorn S. J., C. A. Baillie, N. Zafeiropoulos, L. Y. Mwaikambo, M. P. Ansell, and A. Dufresne. Current international research into cellulosic fibres and composites. *J. Mater. Sci.* 36 (2001) 2107–2131.

[3] Chanliaud E., K. M. Burrows, G. Jeronimidis, and M. J. Gidley. Mechanical properties of primary plant cell wall analogues. *Planta* 215 (2002) 989–996.

[4] Brown R. M. Jr, J. H. M. Willison, and C. L. Richardson. Cellulose biosynthesis in *Acetobacter xylinum*: 1. Visualization of the site of synthesis and direct measurement of the in vivo process. *Proc. Natl. Acad. Sci. U.S.A.* 73 (1976) 4565–4569.

[5] Astley O. M., E. Chanliaud, A. M. Donald, and M. J. Gidley. Structure of *Acetobacter* cellulose composites in the hydrated state. *Int. J. Biol. Macromol.* 29 (2001) 193–202.

[6] Bohn A., H. P. Fink, J. Ganster, and M. Pinnow. X-ray texture investigations of bacterial cellulose. *Macromol. Chem. Phys.* 201 (2000) 1913–1921.

[7] Iguchi M., S. Yamanaka, and A. Budinho. Bacterial cellulose—a masterpiece of nature's arts. *J. Mater. Sci.* 35 (2000) 261–270.

[8] Jonas R. and L. F. Farah. Production and application of microbial cellulose. *Polym. Degrad. Stability* 59 (1998) 101–106.

[9] Klemm D., D. Schumann, U. Udhardt, and S. Marsch. Bacterial synthesized cellulose—artificial blood vessels for microsurgery. *Prog. Polym. Sci.* 26 (2001) 1561–1603.

[10] Evans B. H., H. M. O'Neill, V. P. Malyvanh, I. Lee, and J. Woodward. Palladium-bacterial cellulose membranes for fuel cells. *Biosens. Bioelectron.* 18 (2003) 917–923.

[11] Yamanaka S., K. Watanabe, N. Kitamura, M. Iguchi, Y. Mitsuhashi, and Y. Nishi. The structure and mechanical properties of sheets prepared from bacterial cellulose. *J. Mater. Sci.* 24 (1989) 3141–3145.

[12] Kacurakova M., A. C. Smith, M. J. Gidley, and R. H. Wilson. Molecular interactions in bacterial cellulose composites studied by 1D FT-IR and dynamic 2D FT-IR spectroscopy. *Carbohydr. Res.* 337 (2002) 1145–1153.

[13] Tokoh C., K. Takabe, J. Sugiyama, and M. Fukita. Cellulose synthesized by *Acetobacter xylinum* in the presence of plant cell wall polysaccharides. *Cellulose* 9 (2002) 65–74.

[14] An H. K., B. Y. Park, and D. S. Kim. Crab shell for the removal of heavy metals from aqueous solutions. *Water Res.* 35 (2001) 3551–3556.

[15] Gamage A. and F. Shahidi. Use of chitosan for the removal of metal ion contaminants and proteins from water. *Food Chem.* 104 (2007) 989–996.

[16] Renault F., B. Sancey, P. M. Badot, and G. Crini. Chitosan for coagulation/flocculation processes—an eco-friendly approach. *Eur. Polym. J.* 45 (2009) 1337–1348.

[17] Palacio L., P. Pradanos, J. I. Calvo, and A. Hernandez. Porosity measurements by a gas penetration method and other techniques applied to membrane characterization. *Thin Solid Films* 348 (1999) 22–29.

[18] Kim K. J., A. G. Fane, C. J. D. Fell, T. Suzuki, and M. R. Dickson. Quantitative microscopic study of surface characterization of ultrafiltration membranes. *J. Memb. Sci.* 54 (1990) 89–102.

[19] Mulder M. *Basic Principles of Membrane Technology*. Kluwer Academic Publishers, Dordrecht, The Netherlands, 1991.

[20] Chakrabarty B., A. K. Ghoshal, and M. K. Purkait. Effect of molecular weight of PEG on membrane morphology and transport properties. *J. Memb. Sci.* 309 (2008) 209–221.

[21] Chianella I., S. A. Piletsky, I. E. Tothill, B. Chen, and A. P. F. Turner. MIP-based solid phase extraction cartridges combined with MIP-based sensors for the detection of microcystin-LR. *Biosens. Bioelectron.* 18 (2003) 119–127.

[22] Barud H. S., C. Barrios, T. Regiani, M. Verelst, J. Dexpert-Ghys, R. F. C. Marques, Y. Messaddeq, and S. J. L. Ribeiro. *Mater. Sci. Eng. C.* 28 (2008) 515–518.

[23] O'Sullivan A. C. The structure slowly unravels. *Cellulose* 4 (1997) 173–207.

[24] Helbert W., J. Sugiyama, S. Kimura, and T. Itoh. High-resolution electron microscopy on ultrathin sections of cellulose microfibrils generated by glomerulocytes in Polyzoa vesiculiphora. *Protoplasma* 203 (1998) 84–90.

[25] Koyama M., W. Helbert, T. Imai, J. Sugiyama, and B. Henrissat. Parallel-up structure evidences the molecular directionality during biosynthesis of bacterial cellulose. *Proc. Natl. Acad. Sci. Biophys.* 94 (1997) 9091–9095.

4

EFFECT OF POSTTREATMENT TO ENHANCE THE PERFORMANCE OF NANOFILTRATION ASYMMETRIC MEMBRANE IN ATRAZINE-HERBICIDE REMOVING PROCESS

NORA'AINI ALI, NURBAITI ABDUL HANID, ASMADI ALI, AHMAD JUSOH, AND MARINAH ARIFFIN

4.1 INTRODUCTION

Pesticides application raises a great environmental concern. Over 98% of insecticides and 95% herbicides reach nontarget species, air, water, sediment, and food due to pesticides drift [1]. The amount of pesticides that migrate from the intended application area is influenced by particular chemical properties: its propensity for binding to soil, its vapor pressure, its water solubility, and its resistance to being broken down over time [2]. There are four major routes through which pesticides reach the water: it may drift outside of the intended area when it is sprayed; it may percolate, or leach, through the soil; it may be carried to the water as runoff; or it may be spilled, for example, accidentally or through neglect [1]. They may also be carried to the water by eroding soil [2]. Factors that affect a pesticide's ability to contaminate water include its water solubility, the distance from an application site to a body of water, weather, soil type, presence of a growing crop, and the method used to apply the chemical [3]. Atrazine is a leading agricultural chemical and is so

Sustainable Membrane Technology for Energy, Water, and Environment, First Edition.
Edited by Ahmad Fauzi Ismail and Takeshi Matsuura.

extensively used that it has been identified as a significant pollutant in surface water, groundwater, in offshore areas, and in the atmosphere.

Recently, there were many studies that have been conducted based on performance in removing atrazine. In general, conventional treatment processes, specifically coagulation-flocculation, sedimentation, and conventional filtration, are ineffective in removing pesticides from drinking water. Disinfection and water softening, however, may lead to pesticide transformation and in some cases, may facilitate their removal or degradation [4]. It is well known that membrane technique has been broadly applied in many industrial sectors. Nanofiltration (NF) is a promising membrane technique with a growing number of applications for the treatment of drinking water and wastewater [5] as it offers good rejections for micropollutants. Apart from pesticides, natural and synthetic hormones can also be mentioned [6–10]. Additionally, removal of micropollutants by NF possesses another significant advantage since the softening and removal of natural organic matter (NOM) can take place simultaneously. By considering that the molecular weight of most pesticides is in the range of 200–400 Da, NF membrane can be effectively applied to remove both hardness and pesticides [11–15].

The exploitation of membrane properties could be made by the alteration of fabrication parameters such as polymer concentration, shear rate, force convection residence time, and posttreatment [16]. In order to achieve a better performance of pesticide removal, a modification of the membrane properties has been attempted. The aim of this study is to investigate the effect of posttreatment to the membranes' morphology to remove atrazine from wastewater samples and also to investigate the effect of solute concentration and solution pH on the rejection of the atrazine.

4.2 MATERIALS AND METHOD

4.2.1 Membrane Preparation

A casting dope formulation of 21% polyethersulfone (PES) (Udel—P1700), 75% *N*-methyl-2 pyrrolidone (NMP) (Merck Malaysia, Petaling Jaya, Selangor, Malaysia), and 4% water was cast by a dry/wet phase inversion process using a semiautomated electrically controlled flat sheet membrane casting machine. The membrane then was immersed in water for 24 hours. After that, the membrane was soaked for 3 hours in different types of posttreatment which were methanol (PES-M), ethanol (PES-E), and 2-propanol (PES-P) solution to remove the solvent residues. The membrane was then kept in the ambient air for drying.

4.2.2 Performance Measurement

The fabricated membrane was characterized by the measurement of pure water permeation rate and flux. The membrane was then subjected to permeation experiments with the solution of 0.01 M atrazine. All permeation experiments were conducted using a dead-end permeation cell (Sterlitech HP 4750) at the operating pressure of 1–5 bars. The pure water flux was calculated using Equation 4.1 [17]:

$$J_w = \frac{V}{t \times A \times \Delta p}, \tag{4.1}$$

where V is the volume (L) of permeate collected; J_w is the water flux (L/m^2 h); ΔT is the sampling time (h); and A is the membrane area (m^2). The solute rejection was calculated by Equation 4.2 [18]:

$$\mathrm{R}\,(\%) = [1 - (C_p)/(C_f)] \times 100\%, \tag{4.2}$$

where C_f and C_p are concentrations of solute in the feed and permeate, respectively.

4.2.3 Herbicide

Atrazine was used as solute in this experiment since atrazine was detected with a great frequency in the surface water. Atrazine with 97.2% purity was purchased from Sigma-Aldrich. Atrazine analytical standard was purchased from Sigma-Aldrich. The standard stock solution (5 mg/L) was prepared by dissolving atrazine in the methanol (high performance liquid chromatography [HPLC]-grade from Merck) solution. The feed solution was prepared by the dilution of standard stock solution in ultra-pure water in the range of 0.2–1 mg/L. The molecular structure and properties of atrazine are presented in Table 4.1.

TABLE 4.1. Properties of Herbicide Used in This Work

Herbicide	Atrazine
Chemical structure	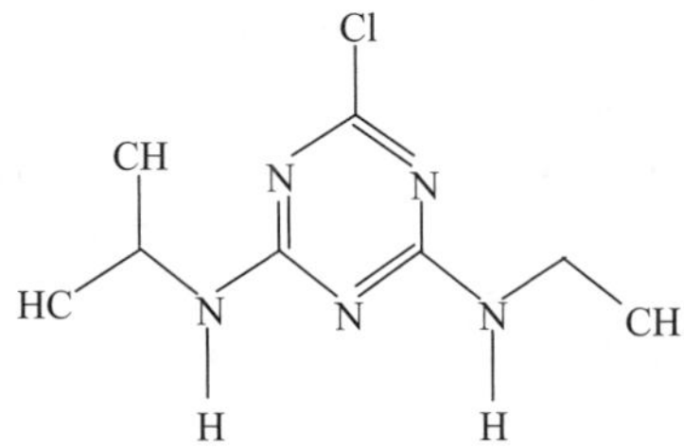
Molecular formula	$C_8H_{14}ClN_5$
Chemical class	Cl-triazine
Molecular weight (Da)	215.69
Molecular size(nm)	0.788
Log K_{ow}	2.68
Aqueous solubility (mg/L)	33

Source: Reprinted with permission from Van der Bruggen, B., J. Schaep, W. Maes, D. Wilms, and C. Vandecasteele, Nanofiltration as a treatment method for the removal of pesticides from groundwater. *Desalination* 117 (1998) 139–147 [22].

4.2.4 Analytical Method

The atrazine concentration was measured by HPLC. The HPLC column used was Platinum C18 column (5 μm, 250 × 4.6 mm long, Grace Davison Discovery Science). The mobile phase used was acetonitrile (Merck). The UV-detector was operated at a wavelength of 226 nm. The atrazine peak appeared at around 7 minutes. The rejection value was obtained using Equation 4.2.

The membrane morphology, such as skin layer, pore structure, and surface porosity was observed by the scanning electron microscopic model JSM 6360 LA. The samples were fractured in liquid nitrogen without deformation on the cross-sectional surface. The samples were cut into small dimension and were mounted on the stubs. The membrane samples were coated with thin gold layer using JFC 1600 coater to obtain a more clear view of the scanning image and to save the samples from damage due to the electron beam.

4.3 RESULTS AND DISCUSSION

4.3.1 Effect of Posttreatment

Figure 4.1a,b show the pure water flux and atrazine rejection, respectively, by the membranes subjected to the posttreatment with different solvents. The results in Figure 4.1a indicate that the flux for PES-M and PES-E was much alike. However, for PES-P, the flux became very low. It shows that a high carbon content in the propanol has a significant effect in reducing the flux rate.

Figure 4.1b shows that PES-M exhibited the lowest solute rejection. This phenomenon is due to the formation of more open pores during the posttreatment in methanol. The solute rejection increased considerabaly for PES-E.

PES-P demonstrated the highest percentage of atrazine rejection, which was in the range of 40–70%. This finding showed that propanol affected the membrane pore size and led to the formation of smaller pore, which contributed to the increment of atrazine rejection and low flux. During the membrane making process, the pore size distribution of the membrane became more uniform through the interaction between the casting solution and the propanol coagulation medium.

This occurrence is due to the fact that as the molar mass of the alcohol increases from C1 to C3 as in Table 4.2, it tends to promote a delayed mechanism of liquid–liquid phase separation; therefore, the voids of the asymmetric membrane can be suppressed [18]. Moreover, as the voids were suppressed, the permeation rate was decreased. This finding corresponds well with the results indicated in Figure 4.1b.

From the results obtained, it could be concluded that even though PES-P exhibited the highest atrazine rejection, the flux was the lowest. Meanwhile, PES-M demonstrated the highest flux but the atrazine rejection was the lowest. Hence, the membrane immersed in ethanol, PES-E, exhibited the optimum combination of flux and rejection performance.

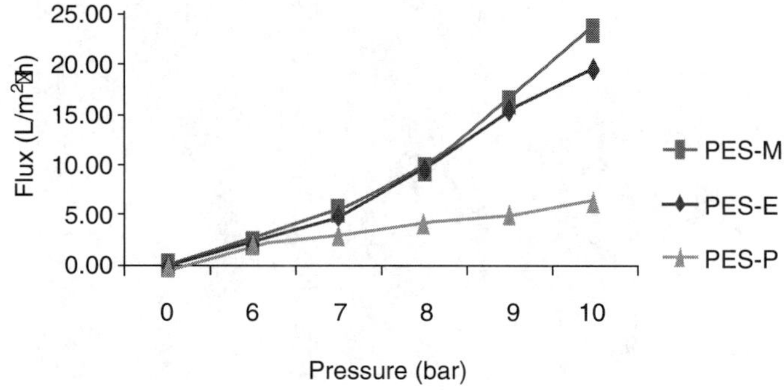

(a)

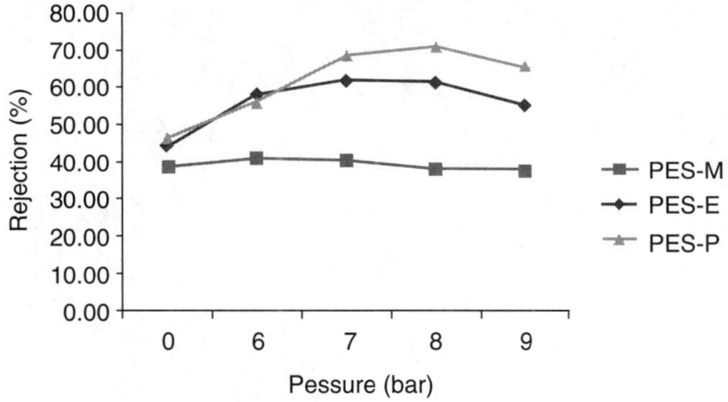

(b)

Figure 4.1. (a) Pure water flux and (b) retention of atrazine for different types of posttreatment.

TABLE 4.2. Molecular Formula and Molar Mass of Posttreatment Medium

Posttreatment Medium	Molecular Formula	Molar Mass (g/mol)
Methanol	CH_3OH (C_1)	32.04
Ethanol	C_2H_6OH (C_2)	46.07
Propanol	C_3H_7OH (C_3)	60.09

4.3.2 Membrane Morphology

The structure of membrane is often described in terms of the presence or absence of connecting voids. Membranes are described as being porous or nonporous. Porous membranes are defined as membranes that have visible and measurable pores traversing the membrane from one face to the other while the visible pores can be visualized using SEM. Figure 4.2a–c show

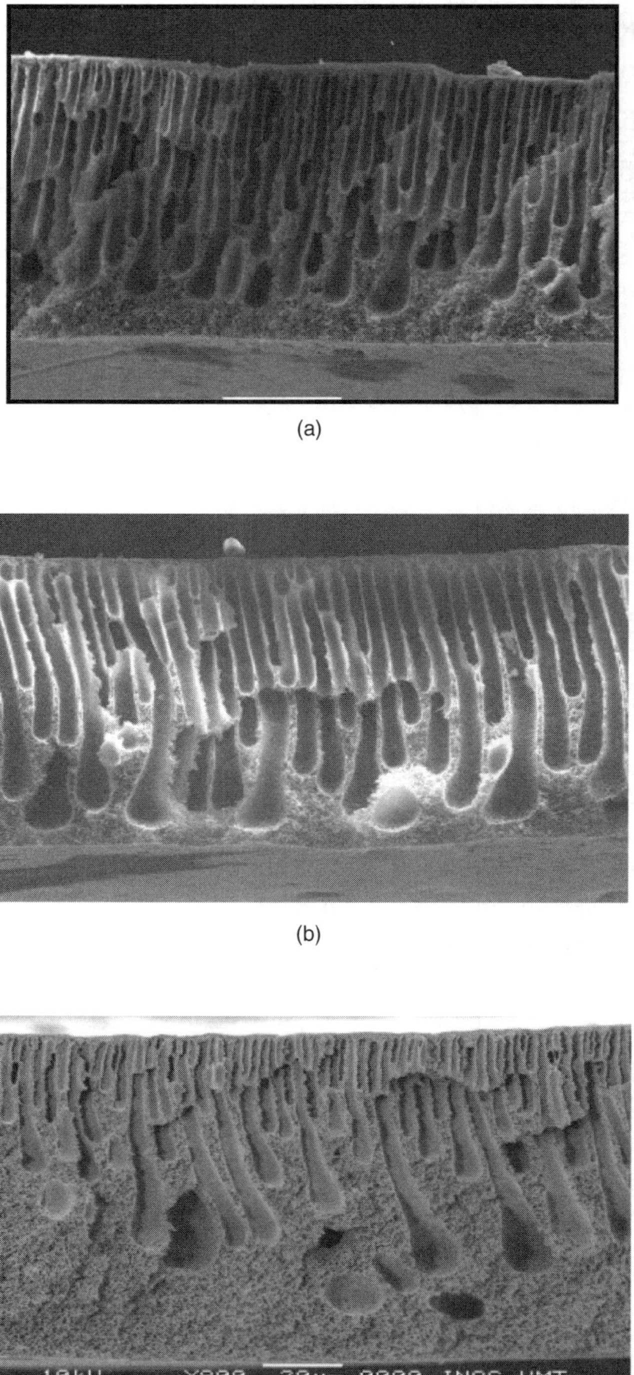

(a)

(b)

(c)

Figure 4.2. SEM micrograph of the cross section of the posttreated PES membranes: (a) PES-M, (b) PES-E, and (c) PES-P.

scanning electron microscopy (SEM) pictures of the cross sections of the asymmetric flat sheet membranes posttreated in methanol, ethanol, and propanol, respectively.

There is a significant amount of finger-like voids penetrating through the cross section of the membranes, and channels can be observed that are projected from the top to the bottom layer of the membrane for all cases of posttreatment. The cross sections of the membrane PES-M and PES-E look similar (Fig. 4.2a,b). Both figures consist of finger-like voids, which lead from the top to the bottom of the membrane. These structures have significantly affected in the pure water flux evaluation. The water flux rate was higher. The indistinct image of teardrop-like structure can be spotted in Figure 4.2c, which also contains a few sponge structure regions. This supported the findings that membrane immersed in propanol produced a tighter pore compared to methanol and ethanol. Therefore, the water flux was much lower.

4.3.3 Effect of concentration

Figure 4.3 shows atrazine rejection in a range of atrazine concentrations from 0.2 to 1 mg/L. As can be seen, the removal efficiency did not vary greatly in the studied concentration range. PES-E showed the highest rejection among all the studied membranes, and the rejection percentage was in the range from 70% to 85%. PES-M had slightly higher rejections compared to PES-P, with atrazine rejection up to 40–60%.

Due to the structure of atrazine that has branched functional groups, the effect of steric hindrance on the solute transport through the membrane pore is a governing factor. The solute rejection increases as the membrane pore size decreases. The size of the molecules might be small, but given the fact that they have a size comparable to the pore size of the membrane, the relative difference proves to be significant. Other aspects, such as the hydrophobicity of the dissolved component, may also have influence on the rejection [19].

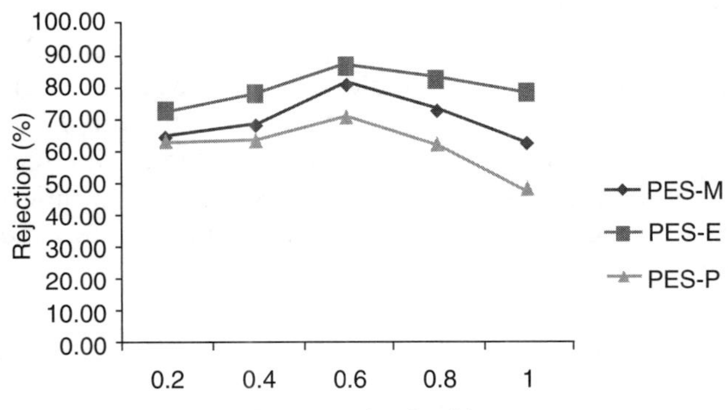

Figure 4.3. Retention of pesticide at different concentrations.

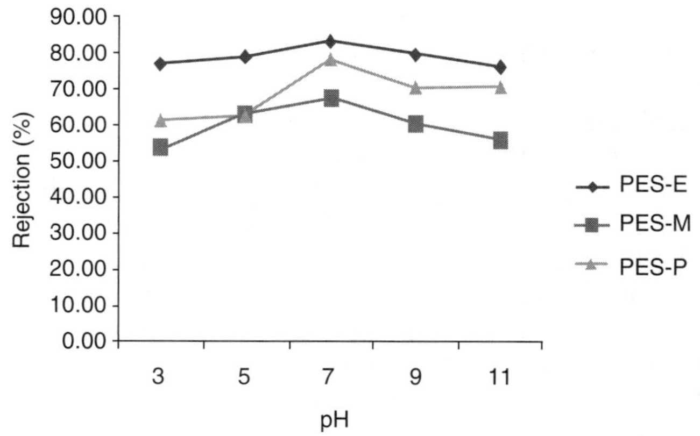

Figure 4.4. Retention of pesticide at different pH.

The concentration effect was not significant on the rejection of atrazine as there was only a slight increment of atrazine rejection with the increase of solute concentration. This finding is in agreement with the work done by Causserand et al. [20] and Zhang et al. [21]. This indicates that atrazine retention performance around 80% was obtained by Plakas et al. [19] when the concentration of atrazine was between 0.150 and 0.300 mg/L. This suggests that while effect of concentration did not pose much impact, it was still a governing factor for transport of solute through membrane.

4.3.4 Effect of pH

Figure 4.4 shows the effect of pH on atrazine removal. Both feed and permeate had the pH value in the range of 3–11. As the pH value of the feed increased, the rejection increased gradually up to pH = 7, then decreased from pH 9 to 11. PES-E had the highest rejection ranging from 70% to 80%, compared to PES-P and PES-M. The decrease in solute rejection at high pH may have been caused by ion adsorption: at higher pH, OH^- ions were absorbed, resulting in an increase of the membrane charge.

Polar components such as atrazine are poorly rejected when the membrane charge increases because they are dipoles which can be a preferential oriented toward the membrane in the sense that the side of the dipole with a charge opposite to the membrane charge is closely attracted to the membrane [21]. Only in the approximately neutral solution is the concentration of H^+ and OH^- minimal, which lessens their influence on the binding interactions between the target molecule and charged sites.

4.4 CONCLUSION

This study proves that the performance and stability of NF membranes are influenced by the posttreatment of the membranes in the coagulation bath medium. When the membrane was immersed in ethanol as the posttreatment solution, the membrane pore size was greatly affected. The results indicated that the posttreatment was an important parameter that affected the NF membrane performance. The results clearly showed that the rejection of atrazine increased with the increase of carbon atoms in the solvent, from methanol to propanol. In conclusion, the posttreatment has resulted in a high performance membrane that can be effectively used for atrazine removal. Moreover, concentration and pH of the solute showed only a little effect on atrazine rejection.

ACKNOWLEDGMENT

The authors would like to thank the Ministry of Science Technology and Innovation for funding this research (vot 59112) and Universiti Malaysia Terengganu for the their facilities and cooperation.

REFERENCES

[1] Miller G. T. *Sustaining the Earth*, 6th ed. Thompson Learning, Inc., Pacific Grove, CA, 2004.

[2] Papendick R. I., L. F. Elliott, and R. B. Dahlgren. Environmental consequences of modern production agriculture: how can alternative agriculture address these issues and concerns? *American Journal of Alternative Agriculture* 1 (1986) 3–10.

[3] Pedersen T. L. Pesticide residues in drinking water. 1997. Retrieved on September 15, 2007.

[4] Plakas K. V. and A. J. Karabelas. Membrane retention of herbicides from single and multi-solute media: the effect of ionic environment. *J. Memb. Sci.* 320 (2008) 325–334.

[5] Van der Bruggen B. and C. Vandercasteele. Modeling of retention of uncharged molecules with nanofiltration. *Water Res.* 36 (2002) 1360–1368.

[6] Ericsson B. and G. Tragardh. Treatment of surface water rich in humus-membrane filtration vs. conventional treatment. *Desalination* 108 (1997) 117–128.

[7] Dewitte J. P. Surface water potabilisation by means of a novel nanofiltration element. *Desalination* 108 (1997) 153–157.

[8] Ozaki H. and H. F. Li. Rejection of organic compounds by ultra-low pressure reverse osmosis membranes. *Water Res.* 36 (2002) 123–130.

[9] Schafer A. I., A. G. Fane, and T. D. Waite. Cost factors and chemical pretreatment effects in the membranes filtration of waters containing natural organic matter. *Water Res.* 35 (2001) 1509–1517.

[10] Nghiem L. D., A. I. Schafer, and T. D. Waite. Membrane filtration in water recycling: removal of natural hormones. *Water Supply* 3 (2003) 155–160.

[11] Reinhard M., N. L. Goodman, P. L. Mc Carty, and D. G. Argo. Removing trace organics by reverse osmosis using cellulose acetate and polyamide membranes. *J. AWWA* 78 (1986) 163.

[12] Baier J. H., B. W. Lyskins, C. A. Fronk, and S. J. Kramer. Using reverse osmosis to remove agricultural chemicals from groundwater. *J. AWWA* 79 (1987) 55.

[13] Hofman J. A. M., T. H. M. Noji, and J. C. Schippers. Removal of pesticides and other organic micropollutants with membrane filtration. *Water Supply* 11 (1993) 101.

[14] Berg P., G. Hagmeyer, and R. Gimbel. Removal of pesticides and other micropollutants by nanofiltration. *Desalination* 113 (1997) 205.

[15] Hofman J. A. M. H., E. F. Beerendonk, H. C. Folmet, and J. C. Kruithof. Removal of pesticides and other micropollutants with cellulose-acetate, polyamide and ultra-low pressure reverse osmosis membranes. *Desalination* 113 (1997) 209.

[16] Vasarhelyi K., J. A. Ronner, M. H. V. Mulder, and C. A. Smolders. Development of dry-wet reversible reverse osmosis membranes with high performance from cellulose acetate and cellulose triacetate blends. *Desalination* 61 (1987) 211–235.

[17] Ahmad A. L., M. Sarif, and S. Ismail. Development of an integrally skinned ultrafiltration membrane for wastewater treatment: effect of different formulations of Psf/NMP/PVP on flux and rejection. *Desalination* 179 (2005) 257–263.

[18] Fan S. C., Y. C. Wang, C. H. Li, K. R. Lee, D. L. Liaw, H. P. Huang, and J. Y. Lai. Effect of coagulation media on membrane formation and vapor permeation performance of novel aromatic polyimide membrane. *J. Memb. Sci.* 204 (2002) 67–79.

[19] Plakas K. V., A. J. Karabelas, T. Wintgens, and T. Melin. A study of selected herbicides retention by nanofiltration membranes—the role of organic fouling. *J. Memb. Sci.* 284 (2006) 291–300.

[20] Causserand C., P. Aimar, J. P. Cravedi, and E. Singlande. Dichloroaniline retention by nanofiltration membranes. *Water Res.* 39 (2005) 1594–1600.

[21] Zhang Y., B. Van der Bruggen, G. X. Chen, L. Braeken, and C. Vandecasteele. Removal of pesticides by nanofiltration: effect of the water matrix, separation and purification. *Technology* 38 (2004) 163–172.

[22] Van der Bruggen, B., J. Schaep, W. Maes, D. Wilms, and C. Vandecasteele. Nanofiltration as a treatment method for the removal of pesticides from groundwater. *Desalination* 117 (1998) 139–147.

5

POLYACRYLONITRILE NANOFIBER ASSEMBLED BY ELECTROSPINNING: EFFECT OF DOPE CONCENTRATIONS ON THE STRUCTURAL AND PORE CHARACTERIZATIONS

AGUNG MATARAM, AHMAD FAUZI ISMAIL, AND TAKESHI MATSUURA

5.1 INTRODUCTION

Microfiltration is an important operation for biopharmaceutical processes, wastewater pretreatment, water and air purification, and food and beverage applications. The majority of commercial microfiltering media are inherently inhomogeneous (nonuniform in mass and thickness) at all locations. This affects the operational performance of the filtering media. Often the developmental objectives during fabrication of efficient microfiltration membranes are permeability, filtration performance, and attaining uniformity in structure [1–3]. Recently, there have been a few attempts to prepare microfilter from nanofibers [4–6]. Nanofiber mats offer unique properties such as high specific surface area (ranging from 1 to 35 m^2/g depending on the diameter of the fibers), good interconnectivity of the pores, and the potential to incorporate active chemistry or functionality on a nanoscale [7, 8]. Unlike conventional fiber spinning techniques (wet spinning, dry spinning, melt spinning, gel

Sustainable Membrane Technology for Energy, Water, and Environment, First Edition.
Edited by Ahmad Fauzi Ismail and Takeshi Matsuura.

spinning), which are capable of producing polymer fibers with diameters down to the micrometer range, electrostatic spinning or "electrospinning" is a process capable of producing polymer fibers in the nanometer diameter range [9–12].

Electrospinning is an efficient fabrication process that can be utilized to assemble fibrous polymer mats composed of fiber diameters ranging from several microns down to fibers with diameters lower than 100 nm. This electrostatic processing method uses a high-voltage electric field to form solid fibers from a polymeric fluid stream (solution or melt) delivered through a millimeter-scale nozzle. Nanofibers are the ultrafine solid fibers notable for their very small diameters (lower than 100 nm), their large surface area per unit mass, and small pore size. Due to the inherent properties of the electrospinning process, which can control the deposition of polymer fibers onto a target substrate, nanofibers with complex and seamless three-dimensional shapes could be formed. Construction of nanoscale composite fibers by electrospinning from a mixture of rigid rod polymers and flexible polymers is also feasible. The electrospun nanofibers can even be aligned to construct unique functional nanostructures such as nanotubes and nanowires. Furthermore, depending on the specific polymer being used, a wide range of fabric properties such as strength, weight and porosity, and surface functionality can be achieved. The challenges confronted during fabrication of nanofiber mat by electrospinning are attaining (1) homogeneity in the size (diameter) distribution of fibers in the mat, (2) uniformity in the deposition and orientation of fibers in the mat (thickness and structural indexes), and (3) durability of the fiber layers in the nanofibrous mat. Understanding the distribution, deposition, and orientation of nanofibers would be extremely useful for preparation of uniform nanofibrous microfilters.

Hence, this investigation is aimed at studying the orientation of nanofibers in the electrospun mat and correlating the diameter of the nanofibers with the pore size of the mat.

5.2 EXPERIMENTAL

5.2.1 Materials and Dope Preparation

Polyacrylonitrile (PAN) powder of 150,000 molecular weight (53.06 g/mol), *N,N*-dimethylformamide (DMF) (73.10 g/mol) and acrylamide (AM) (71.08 g/mol) were obtained from Aldrich Chemical and were used without further purification. Dope solutions were prepared by dispersing predetermined amount of silica nanoparticles (60.08 g/mol) (1 wt % to PAN) into 14, 16, and 18 wt % PAN solution in DMF. The mixture was mechanically stirred for at least 24 hours at 60°C in order to obtain homogeneous silica-dispersed PAN solutions [13]. The viscosity of the solutions was determined by using Brookfield Dial Viscometer.

5.2.2 Electrospinning

The experimental setup used for the preparation of nanofiber mat is as follows. A 100-mL reservoir was used to hold the electrospinning solution. The PAN solution prepared in Section 5.2.1 was pumped at a constant rate of 2 mL/h with the help of a metering pump through a stainless steel needle of inner diameter 0.8 mm. A drum of 15 cm diameter and 35 cm circumference, connected to a variable speed motor, was used to collect the nanofibers. A high DC voltage was applied to the needle with the help of high voltage regulated DC power supply (Model ES 30P-5W, Gamma High Voltage Research, Ormond Beach, FL). The applied voltage was set at 21 kV and distance between the tip of the needle and the surface of the drum at 10 cm. The collecting drum was ground so as to generate the desirable electric field strength between the tip of the spinneret and the collector surface. The nanofibrous mat was carefully removed from the collector, and the residual solvent associated with nanofiber mat was removed by keeping the mat in an oven for at least 2 days at 40°C. The dried electrospun mats were stored in desiccators.

5.2.3 Fiber Characterizations

Scanning electron microscopy (SEM) was used to investigate the structure of the PAN nanofibers. The pore size of the PAN nanofiber membrane was determined using the bubble point method. It is based on the measurement of pressure necessary to blow air through a liquid-filled membrane. The schematic of the equipment used is shown in Figure 5.1.

The wet membrane, after being used in the permeation experiment, was placed in the cell supporting 5 cm^3 of distilled water and connected to a bubble

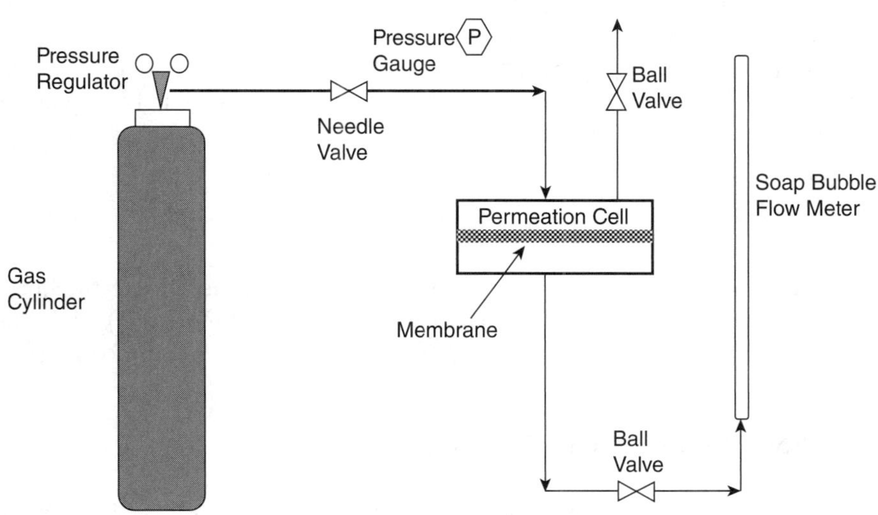

Figure 5.1. Equipment for bubble point measurement.

flowmeter. Pressure was applied to the membrane base. The pressure at which the air flow starts to occur is called the bubble point, and the relationship between the pore size, R, and the bubble point is given by the Young–Laplace equation [14]:

$$R = \frac{2\gamma}{\Delta P}\cos\theta,$$

where ΔP is the cross-sectional pressure difference at the bubble point, γ is the surface tension of the wetting agent, and θ is the wetting angle.

The static contact angle membrane was measured using a contact angle analysis system (Video Contact Angle [VCA] Optima Surface Analysis, AST Products Inc., Billerica, MA). The measurement was carried out by dropping 5 μL of water on membrane with $2 \times 2\ \text{cm}^2$ dimension and observing within 10 seconds. To minimize experimental error, the measurement was repeated 10 times to obtain the average and standard deviation.

In order to determine membrane surface porosity, membranes were immersed in distilled water for 4 hours at 25°C. Membrane in wet state was weighed in an electronic balance after carefully wiping the surface with a clean tissue. This wet membrane was dried in an oven at 50–60°C for 24 hours. Then, the membrane was weighed again in dry state. The membrane surface porosity, ε, was calculated using the following equation [15–17]:

$$\varepsilon = \frac{w_w - w_d}{\rho_w V},$$

where w_w is weight of the wet membrane (g), w_d is weight of the dry membrane (g), ρ_w is the density of pure water at room temperature (g/cm^3), and V is the volume of membrane in wet state (cm^3). (Measurement of V was made using a micrometer and is subject to large errors.)

Mean pore radius, r_m (μm), on the other hand, was determined using the filtration velocity method. According to the Guerout–Elford–Ferry equation, r_m could be experimentally determined by [17, 18]

$$r_m = \sqrt{\frac{(2.9 - 1.75\varepsilon) \times 8\eta l Q}{\varepsilon A \Delta P}},$$

where η is the water viscosity (8.9×10^{-4} Pa·s), l is the membrane thickness (m), Q is the volume of the permeate water per unit time (m^3/s), A is the membrane effective area (m^2), and ΔP is the operational pressure (0.1 MPa).

5.3 RESULTS AND DISCUSSION

In electrospinning, the PAN concentration was increased from 14 to 18 wt % while keeping the solution flow rate (2 mL/h), rotational speed of collector

TABLE 5.1 Concentration of Dope, Contact Angle on the Fiber Diameter and Pore Size

No	Dope concentration (wt.%)	Contact angle (°)	Fiber diameter (μm)	Pore size (μm)
1	14	58 (1.41)[a]	0.25 (0.38)[a]	0.074
2	16	52 (0.93)[a]	0.37 (0.93)[a]	0.129
3	18	47 (0.78)[a]	0.52 (0.38)[a]	0.280

[a] Inside the bracket is standard deviation.

(1.8 m/s), screen distance (0.1 m), and electric field (21 kV) constant. The viscosities of 14, 16, and 18 wt % PAN solutions were, respectively, 18,000, 29,000, and 32,000 centipoises at room temperature. As the PAN concentration in the solution increased, the viscosity also increased, resulting in the increment of the fiber diameter as tabulated in Table 5.1. At a low polymer concentration of 14 wt %, fiber diameter of 0.25 μm was obtained whereas polymer concentrations of 16 and 18 wt % resulted in fiber diameter of 0.37 and 0.52 μm, respectively. The reason for the increment of fiber diameter is probably due to the swelling effect during the phase transformation, which increases with an increase in polymer concentration. The solvent outward diffusion predominated at higher polymer concentration as exposed to the screen distance of 0.1 m and contributed to bigger fiber diameter. Meanwhile at low polymer concentration the air inward diffusion predominated, which contributed to smaller fiber diameter [19].

The average fiber diameters were calculated from 3 to 5 measurements using at least three SEM fractographs of 20,000× magnification (see Fig. 5.2). A shifting trend in the fiber distribution toward a larger fiber size is noted, and this also is the reason for the slight increase of the fiber diameter as the PAN concentration increases. The effect of the dope concentration on fiber diameter is expected to be prominent in the case of highly viscous feed solutions. The morphological changes in the membranes due to variation of the PAN concentration are shown in Figure 5.2. Reduced number of beaded fibers and small-sized beads are observed when the dope concentration is 14 wt % (Table 5.1). This finding is consistent with the reports in the literature, which confirmed that a low dope concentration favors the whipping instability and suppresses the axisymmetric instabilities, thereby suppressing the formation of beaded fibers [20].

The drawing rate of the nanofibers can be enhanced, under a given applied voltage, by decreasing the dope concentration. It was hypothesized that an enhanced drawing rate would increase the number of fiber crossings and that the high extent of fiber crossing would reduce the pore size and improve the interconnectivity of pores. The pore sizes were measured to examine this hypothesis by the bubble point method and results are shown in Table 5.1. The pore sizes were also measured by the other technique, filtering velocity

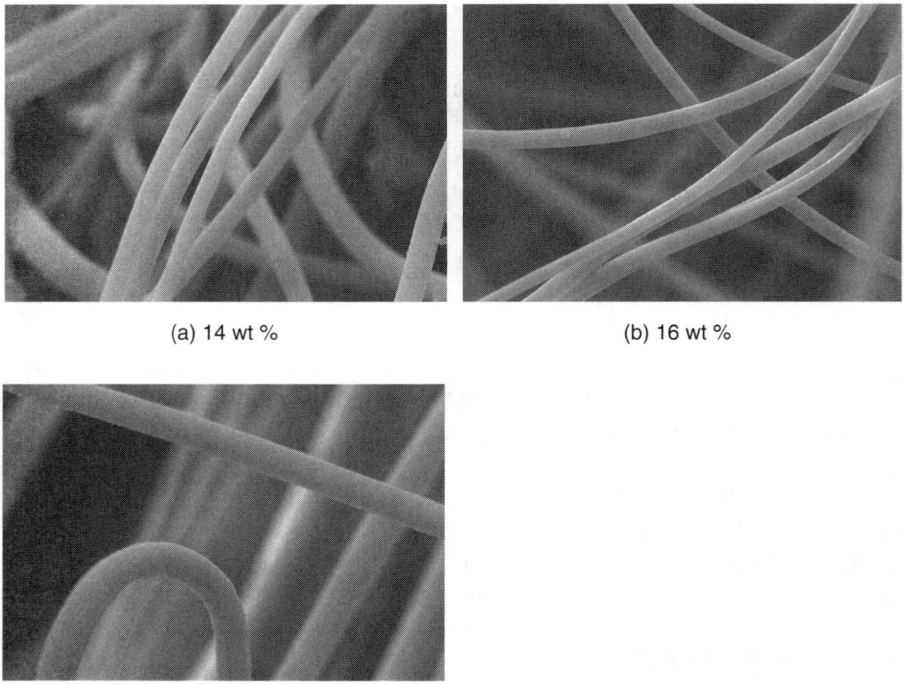

(a) 14 wt % (b) 16 wt % (c) 18 wt %

Figure 5.2. SEM micrographs of nanofiber membranes (magnification 20,000×) fabricated from dopes of different PAN concentrations: (a) 14 wt %, (b) 16 wt %, and (c) 18 wt %.

TABLE 5.2 Other Technique to Validate the Pore Size

No	Dope concentration (wt.%)	Porosity	Pore size (μm)
1	14	0.1792	0.087
2	16	0.1811	0.133
3	18	1.001	0.298

technique, and results are shown in Table 5.2. The pore sizes measured by these two different methods are coincidentally very similar. This seems somewhat surprising since the pore size measured by the former method corresponds to the maximum value while those measured by the latter method corresponds to the mean value. Probably, the obtained results indicate the difference in the pore sizes at the nanofiber surface and at the bulk. From both methods, however, it was observed that the pore size increased as the dope concentration increased. The increase in dope concentration enhances repulsion between the fibers and changes the trend in fiber arrangement as evidenced from the increase in the pore size values. Furthermore, a considerable increase in accumulation of charge on electrospun PAN nanofibers has been noted when the dope concentration was increased [21, 22].

5.4 CONCLUSIONS

The fiber crossing and pore size can be optimized to attain improved structure (pore size distribution, pore interconnectivity, and porosity) of the electrospun filtering media. Reduced number of beaded fibers and smaller sized beads were observed when the dope concentration was at 14 wt %. This finding is consistent with the reports in the literature which confirmed that a low polymer concentration favors the whipping instability and suppresses the axisymmetric instabilities, thereby suppressing the formation of beaded fibers. The pore size increased as the PAN concentration increased. The increase in polymer concentration enhances repulsion between the fibers and changes the trend in fiber arrangement as evidenced from the increase in the pore size values. The results elucidated that increasing the PAN polymer concentration enhanced the nanofiber diameter and pore diameter of nanofiber mats.

ACKNOWLEDGMENT

The authors would like to acknowledge the Ministry of Science, Technology and Innovation of Malaysia (MOSTI) for funding this research under project no. 03-01-06-SF0258 79136 E-science and 03-02-06-0060-PR0072/08-03-74539 Carbon Fiber Top Down.

REFERENCES

[1] Ma Z. and S. J. Ramakrishna. *J. Membr. Sci.* 319 (2008) 23–28.

[2] Sang Y., Q. Gu, T. Sun, F. Li, and C. Liang. *J. Hazard. Mater.* 153 (2008) 860–866.

[3] Bjorge D., N. Daels, S. D. Vrieze, P. Dejans, T. V. Camp, W. Audenaert, J. Hogie, P. Westbroek, K. D. Clerck, and S. W. H. Van Hulle. *Desalination* 243 (2009) 942–948.

[4] Bazargan A. M., M. Keyanpour-rad, F. A. Hesari, and M. E. Ganji. *Desalination* 265 (2010) 148–152.

[5] Gopal R., S. Kaur, Z. Ma, C. Chan, S. Ramakrishna, and T. Matsuura. *J. Membr. Sci.* 281 (2006) 581–586.

[6] Homaeigohar S. S., K. Buhr, and K. Ebert. *J. Membr. Sci.* 365 (2010) 68.

[7] Ramakrishana S., K. Fujihara, W. E. Teo, T. C. Lim, and Z. Ma. *An Introduction to Electrospinning and Nanofibers*. World Scientific Publishing Co., Singapore, 2005.

[8] Reneker D. H. and H. Fong. Mechanical Behavior of Nonwoven Electrospun Fabrics and Yarns, Polymer Nanofibers ACS Symposium Series 918, Oxford University Press, 2006.

[9] Chronakis I. S. *J. Mater. Process. Technol.* 167 (2004) 283–293.

[10] Theron S. H., E. Zussman, and A. L. Yarin. *Polymers* 45 (2004) 2017–2030.

[11] Pant H. R., M. P. Bajgai, C. Yi, R. Nirmala, K. T. Nam, W. I. Baek, and H. Y. Kim. *Colloids Surf. A* 379 (2010) 87–94.

[12] Homayoni H., S. H. H. Ravandi, and M. Valizadeh. *Polymers* 77 (2009) 656–661.

[13] Mataram A., A. F. Ismail, D. S. A. Mahmod, and T. Matsuura. *Mater. Lett.* 64 (2010) 1875–1878.

[14] Mulder M. *Basic Principles of Membrane Technology*, 2nd ed. Kluwer Academic Publishers, London, 2003, pp. 165–168.

[15] Chen Z., M. Deng, Y. Chen, G. He, M. Wu, and J. Wang. *J. Membr. Sci.* 235 (2004) 73–86.

[16] Chakrabarty B., A. K. Ghoshal, and M. K. Purkait. *J. Membr. Sci.* 309 (2008) 209–221.

[17] Basri H., A. F. Ismail, and M. Azis. *Desalination* 273 (2011) 72–80.

[18] Li J. F., Z. L. Xu, H. Yang, C. D. Feng, and J. H. Shi. *J. Appl. Polym. Sci.* 107 (2008) 4100–4108.

[19] Barhate R. S., C. K. Loong, and S. Ramakrishna. *J. Membr. Sci.* 283 (2006) 209–218.

[20] Zuo W., M. Zhu, M. Yang, H. Yu, Y. Chen, and Y. Zhang. *Polym. Eng. Sci.* 45 (2005) 704–709.

[21] Ignatova M., N. Manolova, and I. Rashkov. *Eur. Polym. J.* 43 (2007) 1112–1122.

[22] Dali L., W. Guolei, D. Biao, B. Xue, W. Yu, S. Hongwei, and X. Lin. *Solid State Sci.* 12 (2010) 1837–1842.

PART II

APPLICATIONS IN GAS AND VAPOR TREATMENT

6

POLYMER STRUCTURES AND CARBON DIOXIDE PERMEATION PROPERTIES IN POLYMER MEMBRANES

Shinji Kanehashi, Shuichi Sato, and Kazukiyo Nagai

6.1 INTRODUCTION

To date, a large number of polymers have been synthesized and investigated as membrane materials for gas separation. Recent polymer science research has focused on membrane materials that have precise and highly ordered structures, as well as those that are crystalline or have a well-ordered orientation. Polymers are generally amorphous or semicrystalline. The latter is also called partially crystalline polymer and consists of a complex combination of crystalline and amorphous regions. In the crystalline region, polymer segments are packed parallel to each other and form a phase of finite size. Small molecules only pass through the amorphous region because the polymer chain mobility is considerably restricted in the crystalline phase, leading to difficult diffusion and dissolution [1]. Therefore, gas permeation behavior in dense polymer membranes strongly affects the molecular structures not only in terms of the chemical structure, but also microstructure of crystalline, liquid crystalline, amorphous, and molecular ordering [2].

Generally, it is known that the gas transport properties (i.e., permeability, solubility, and diffusivity) in crystalline and liquid crystalline polymers decrease with increasing crystallinity. Since the gas permeability of crystalline

Sustainable Membrane Technology for Energy, Water, and Environment, First Edition.
Edited by Ahmad Fauzi Ismail and Takeshi Matsuura.

and liquid crystalline polymer membranes is usually lower than that of amorphous polymers, the crystalline polymer membranes show higher gas barrier. For example, the carbon dioxide (CO_2) permeability of amorphous poly(dimethylsiloxane) (PDMS) is 3.2×10^{-7} (cm^3(STP)cm/(cm^2 s cmHg)) [3], crystalline poly(acrylonitrile) (PAN) is 1.8×10^{-13} (cm^3(STP)cm/(cm^2 s cmHg)) [3], and liquid crystalline poly(*p*-hydroxybenzoic acid-*co*-6-hydroxy-2-naphthioic acid) (HBA/HNA) and poly(*p*-phenylene terephthalamide) (PPTA) is lower than that of poly(ethylene terephthalate) (PET) or polyvinylchloride(PVC) but is similar to that of polyacrylonitrile [4, 5].

Recently, we have reported that gas transport properties of poly(lactic acid) (PLA) were independent of its crystallinity [6–8]. This result suggested that gas transport properties are strongly affected by the size and distribution of the crystalline phase rather than by the polymer crystallinity [6]. Furthermore, we have also reported that unlike common crystalline polymer membranes, the permeability of crystalline PLA membranes is higher than that of amorphous PLA membranes. One possible explanation based on this result is that there might be a continuous space for gas diffusion around the interface between the crystalline and amorphous regions [8].

Recently, CO_2 capture using membrane has attracted more attention because it has the potential to reduce CO_2 that causes the global warming. Basically, CO_2 ($T_b = 195$ K) is a more condensable gas in air, including nitrogen (77 K) and oxygen (90 K) [9]. Therefore, the sorption property in dense polymer membranes usually favors CO_2 compared to other gases such as hydrogen, oxygen, nitrogen, and methane. However, the higher CO_2 solubility in the dense polymer membranes could cause critical problems for gas separation applications.

It is known that high-pressure CO_2 can plasticize polymer membranes in a similar manner to propane, butane, and other vapors [1]. Membrane plasticization behavior induced by CO_2 in polymer membranes is still a catastrophic problem in polymeric membrane-based gas separation applications [10]. This undesired behavior has exerted negative impact on the gas separations that involve condensable components, such as natural gas purification (CO_2/CH_4) [11], the removal of global warming gases from power plant emissions (CO_2/N_2) [12], and hydrogen purification (CO_2/H_2) for use of fuel cells and oil refining [13]. The absorbed CO_2 swelled and plasticized the polymer membranes, hence increasing the polymer segmental mobility. This phenomenon has increased the gas permeability and a loss in gas selectivity and caused a loss of membrane performance. The plasticization affects the gas transport properties in several aspects; for example, the time dependence [14, 15], pressure dependence [15–17], temperature dependence [18], and thickness dependence [19] of CO_2 permeability above and below the plasticization pressure. Besides that, differential hysteresis for pressurization/depressurization cycles has also been observed [17]. Other aspects are the possibility to improve the degree of plasticization effects using the thermal technique [14, 15], crosslink approaches [16, 20]. This chapter investigates and discusses the polymer structures and the carbon dioxide permeation properties in dense polymer membranes.

6.2 BACKGROUND

Permeation of a gas molecule in a dense polymer membrane is usually governed by a solution–diffusion mechanism. The permeability, P, is the product of solubility, S, and diffusivity, D, when the downstream pressure is significantly lower than that of upstream pressure:

$$P = S \times D \tag{6.1}$$

Gas permeation in partially crystalline and liquid crystalline polymers is illustrated through a dual-phase model because these polymers have a combination of amorphous and crystalline regions [1].

Rubbery polymers are in a hypothetical thermodynamic equilibrium liquid state, and their gas solubility obeys Henry's law. On the other hand, glassy polymers are typically assumed to be in a nonequilibrium state containing two components: a hypothetical liquid state and a solid state. Over the past four decades, sorption of gas molecules in glassy polymers has been described and analyzed by the dual-mode sorption model [21]. This model is based on two types of sorption sites that obey (1) Henry's law solution and (2) Langmuir-type sorption. The former is related to sorption of gases into rubbery polymers and low molecular weight liquids. The latter is related to Langmuir-type sorption in porous solids. The Langmuir sorption site in a glassy polymer corresponds to holes or "microvoids" which arise from the nonequilibrium nature of glassy polymers. The dual-mode sorption model is expressed by

$$C = C_D + C_H = k_D p + \frac{C'_H bp}{1 + bp}, \tag{6.2}$$

where C is the total gas concentration in a glassy polymer, C_D is the gas concentration based on Henry's law sorption, C_H is the gas concentration based on Langmuir sorption, k_D is the Henry's law coefficient, and b and C'_H are the Langmuir hole affinity parameter and the capacity parameter, respectively. The k_D parameter represents the penetrant dissolved in the polymer matrix at equilibrium and b characterizes the sorption affinity for a particular gas-polymer system. C'_H is often used to measure the amount of the nonequilibrium excess free volume in the glassy state. As the polymer glass transition temperature, T_g, increases, the nonequilibrium excess free volume increases at a given temperature, T.

The solubility of a gas molecule in glassy amorphous polymers can be expressed by the dual-mode sorption parameters:

$$S \equiv \frac{C}{p} = S_D + S_H = k_D + \frac{C'_H b}{1+bp}, \tag{6.3}$$

where S is the solubility of penetrant, S_D and S_H are solubility values based on Henry's law and Langmuir-type sorption, respectively. In general, gas solubility of glassy polymers is higher than that of rubbery polymers. As previously mentioned, glassy polymers have nonequilibrium excess free volume; therefore, higher gas solubility can be attributed directly to the existence of this additional free volume into which sorption can occur.

On the other hand, solubility in crystalline polymer membrane is described by the following equation:

$$S = \left(\frac{100 - X_c}{100}\right) S_a, \tag{6.4}$$

where X_c is the degree of crystallinity, and S_a is the solubility for the purely amorphous polymer. Based on this equation, small gases seemed to be difficult to dissolve into crystal region, that is, dissolved amorphous phase only.

It is inadequate to explain the diffusivity in crystalline and liquid crystalline polymers because the crystalline phase seems to include two factors [22]. Small molecules are assumed incapable of diffusing or permeating through the crystalline phase. Therefore, they must pass though an elongated path due to the presence of the crystalline phase. This factor is generally accounted for the tortuosity or geometric impedance factor, τ. The other factor is the chain-immobilization factor, β. The reciprocal of β, $(1/\beta)$, represents a fractional reduction in diffusivity due to the restriction of the polymer chain segmental mobility in the amorphous phase. The effect of the crystalline phase on diffusivity can be expressed by using τ and β:

$$D = \frac{D_a}{\tau\beta}. \tag{6.5}$$

6.3 CO_2 PERMEATION PROPERTIES

Figure 6.1 shows the CO_2 permeability in crystalline and liquid crystalline polymers as a function of crystallinity [2]. We employed the published CO_2 permeation property determined at 25–35°C with 1–10 atm of feed pressure for crystalline, those determined at 20–40°C with 1–10 atm for liquid crystalline, and those determined at 35°C with 10 atm for glassy amorphous polymers. Liquid crystalline polymers have higher gas barrier properties (i.e., lower gas permeability) compared with crystalline polymers. Gas permeability is known to decrease with increasing crystallinity in crystalline and liquid crystalline polymers. Generally, this is because gases cannot diffuse and permeate in the crystalline phase. However, this study shows that CO_2 permeability in the crystalline and liquid crystalline polymers are independent of their crystallinity at

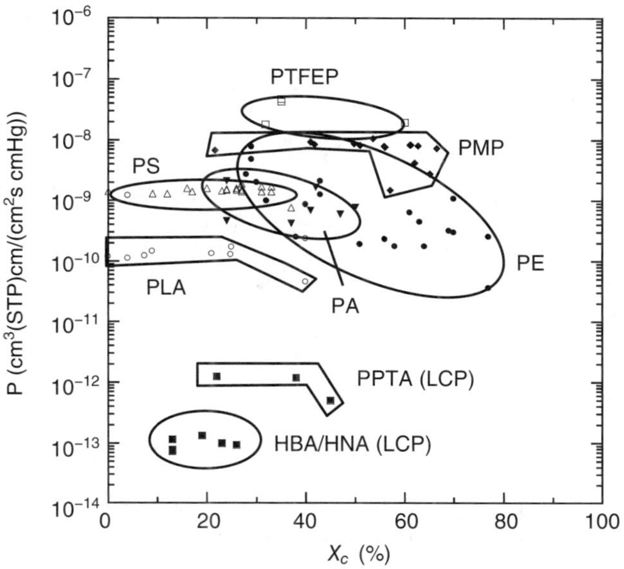

Figure 6.1. CO_2 permeability in crystalline and liquid crystalline polymers as a function of crystallinity, X_c. (Reprinted from Kanehashi S., et al. Analysis of permeability; solubility and diffusivity of carbon dioxide; oxygen; and nitrogen in crystalline and liquid crystalline polymers. *J. Membr. Sci.* 365 (2010) 40, with permission from Elsevier.)

lower crystallinity values and are almost the same or higher than in the amorphous polymers. For example, CO_2 permeability of crystalline and liquid crystalline polymers is almost constant from 20% to 50% crystallinity for poly(4-methyl-1-pentene) (PMP), 30% to 60% for poly[bis (trifluoroethoxy phosphazene)] (PTFEP), 10% to 30% for liquid crystalline HBA/HNA, and 20% to 40% for PPTA. In contrast, permeability of PMP, PLA, and PPTA decreases from 50% to 55%, 25% to 40%, and 40% to 50%, respectively. Reduction in the permeability from 20% to 50% and 30% to 80% is also observed in polyacrylate (PA) and polyethylene (PE), respectively. Interestingly, CO_2 permeability slightly increases from 0% to 25% for PLA, and from 0% to 35% crystallinity for PS. As shown in PMP and PLA, gas permeability begins to decrease beyond a certain crystallinity value. This result suggests that the phase transitions of crystalline structure in crystalline and liquid crystalline polymers occur at these crystallinity values. Hence, CO_2 permeability of crystalline and liquid crystalline polymers may decrease over phase transitions.

Figure 6.2 presents the relationship between CO_2 diffusivity and solubility in crystalline and liquid crystalline, and glassy amorphous polymers [2]. Gas permeability is a product of the solubility and diffusivity as shown in Equation 6.1. Regardless of the membrane structures such as crystalline, liquid crystalline, and amorphous, the CO_2 diffusivity in dense polymer membranes are distributed widely compared to solubility. For example, CO_2 diffusivity varies from 5.00×10^{-11} to 5.40×10^{-7} for glassy amorphous polymers, from 5.66×10^{-9}

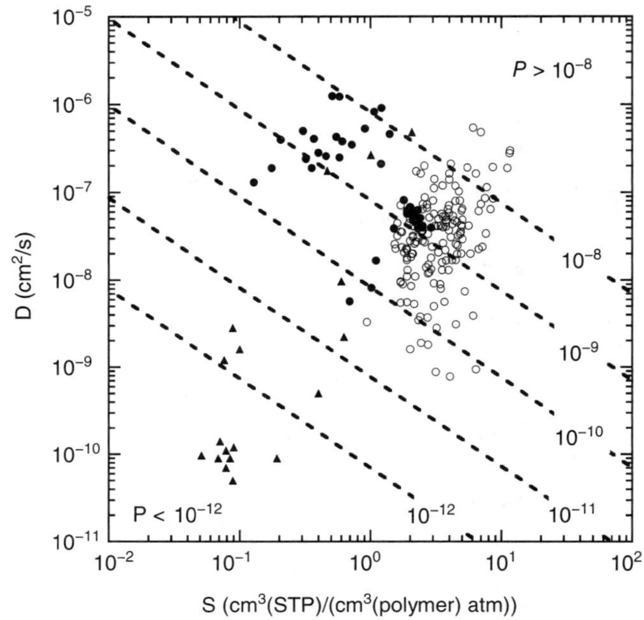

Figure 6.2. Relationship between CO_2 diffusivity and solubility in the amorphous phase compared to crystalline and liquid crystalline polymers. The represented symbols are crystalline polymers (●), liquid crystalline polymers (▲), and conventional amorphous glassy polymers (○), respectively. (Reprinted from Kanehashi S., et al. Analysis of permeability; solubility and diffusivity of carbon dioxide; oxygen; and nitrogen in crystalline and liquid crystalline polymers. *J. Membr. Sci.* 365 (2010) 40, with permission from Elsevier.)

to 1.22×10^{-6} for crystalline polymers, and from 5.00×10^{-11} to 4.81×10^{-7} cm^2/s for liquid crystalline polymers. CO_2 solubility varies from 8.90×10^{-2} to 6.08 for amorphous polymers, from 0.128 to 2.99 for crystalline polymers, and from 5.10×10^{-2} to 1.01 cm^3 (STP)/(cm^3 atm) for liquid crystalline polymers. Therefore, the permeability in dense polymer membranes is seemed to depend more on the CO_2 diffusivity relative to the solubility, regardless of their membrane structures.

As shown in Figure 6.2, at the same CO_2 permeability, the CO_2 solubility in crystalline and liquid crystalline polymers is lower; meanwhile, the diffusivity is higher than that of glassy amorphous polymers. This result indicates that the CO_2 transport behavior in crystalline and liquid crystalline polymers can be different from glassy amorphous polymers attributed to the presence of crystal structures assuming they follow similar solution/diffusion mechanisms. Therefore, CO_2 transport properties in dense polymer membranes could be strongly affected by the molecular ordering structures such as crystalline, liquid crystalline, and orientation.

As mentioned before, since CO_2 solubility in dense polymer membranes is larger than the other inert gases, CO_2 induced membrane plasticization under

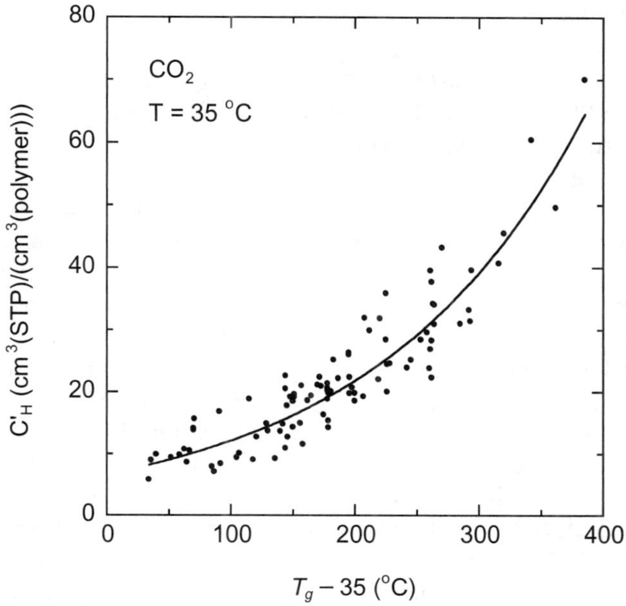

Figure 6.3. The Langmuir capacity parameter C'_H at 35°C as a function of T_g–35°C. (Reprinted from Kanehashi S. and K. Nagai. Analysis of dual-mode model parameters for gas sorption in glassy polymers. *J. Membr. Sci.* 253 (2005) 117, with permission from Elsevier.)

the high pressure of CO_2. A CO_2 sorption isotherm in a glassy polymer is generally concave to the pressure axis. This behavior is the characteristic for dual-mode sorption model composed of Henry's law dissolution in an equilibrium region and Langmuir-type sorption in a nonequilibrium region [21,23]. This nonequilibrium region is directly related to the excess free volume or unrelaxed free volume in a glassy polymer. The excess free volume reflects the nonequilibrium structure of a glassy amorphous polymer, which results from the quenching of the polymer from rubbery to glassy state.

Figure 6.3 shows the experimental C'_H values determined from the equilibrium sorption isotherms as a function of (T_g—35°C) [23]. As mentioned previously, C'_H presents the maximum amount of penetrant sorbed into microvoids [24]. The C'_H value decreases with increasing experimental temperature up to T_g. At T_g, the C'_H value becomes zero because the polymer transforms from a glassy to a rubbery state [25]. As T_g increases, the overall C'_H values tend to increase as demonstrated by Figure 6.3. Based on this result, increase in T_g indicates the rigidity or decrease in the mobility of polymer segments with increasing the excess free volume in the dense polymer membranes. Hence, variety membrane structure produced large different CO_2 sorption property attributed to different rigidity or mobility of polymer segments. Although Toi et al. reported for CO_2 [26], C'_H increases linearly with increasing polymer T_g, C'_H in Figure 6.3 tends to increase exponentially with increasing T_g. However, it is difficult to explain the reason for this behavior.

6.4 CO_2-INDUCED PLASTICIZATION

CO_2-induced plasticization of glassy polymers can have effects on their gas sorption properties. In general, plasticization results from sorption of condensable gases into glassy polymer membranes. The sorbed gas molecules reduce the segment–segment interactions and enhance chain segmental mobility (i.e., the intersegmental mobility). Upon plasticization of a glassy polymer, the selectivity generally decreases. In the past, this problem has limited the use of glassy gas separation membranes for mixtures containing condensable components. Various researchers have investigated plasticization and antiplasticization phenomena in glassy polymer membranes [10]. In the last decade, it was reported that polyimides composed of 4,4-(hexafluoroisopropylidene) diphthalic anhydride (6FDA) show higher CO_2 permeabilitiy and selectivitiy than other polyimides or conventional glassy polymers. However, 6FDA-based polyimides were especially liable to plasticize when exposed to high pressure CO_2 [15, 16]. Interestingly, CO_2-induced plasticization for glassy polyimide membranes tends to begin at about CO_2 pressure of 10 atm (relative pressure: 0.125) at 35°C [15, 16]. We focused on the CO_2 concentration in dense polyimide membranes at this point as the minimum pressure for plasticization pressure.

The total number of CO_2 molecules, N (mol(CO_2)/(mol(polymer repeat unit)), dissolved in a polymer repeat unit at 10 atm is given by:

$$N = \frac{C_{10}}{22{,}414V}, \tag{6.6}$$

where C_{10} is the CO_2 concentration in polymers (cm^3(STP)/cm^3(polymer)) at 10 atm calculated from the dual-mode parameters, and V is the polymer specific volume (cm^3/mol) estimated from the molecular weight of a polymer repeat unit and geometric membrane density. Figure 6.4 shows the N value in various glassy polymers as functions of T_g– 35°C [23]. As the T_g– 35°C increases, the N value increases exponentially. The results suggest that N value increases with decreasing polymer segmental motions or increasing rigidity of polymer segments. Therefore, rigid and low mobility of polymer segments seems to require larger CO_2 molecules to be plasticized at the plasticization pressure of 10 atm.

Among the 6FDA-based polyimides, tetramethyl-substitued phenylendiamine, such as 2,3,5,6-tetramethyl-1,4-phenylene-diamine (TeMPD), are extremely permeable glassy polymers compared to other 6FDA-based polyimides [27]. Using this 6FDA-TeMPD polyimide, we investigated the fundamentals of CO_2-induced plasticization with focus on the different molecular ordering structures. Variety molecular ordering structure of 6FDA-TeMPD dense membranes was prepared by different casting solvent and dry condition. Generally, polymer chains make a variety of conformation in different casting solvents [15]. This provides different types of polymer membrane morphology

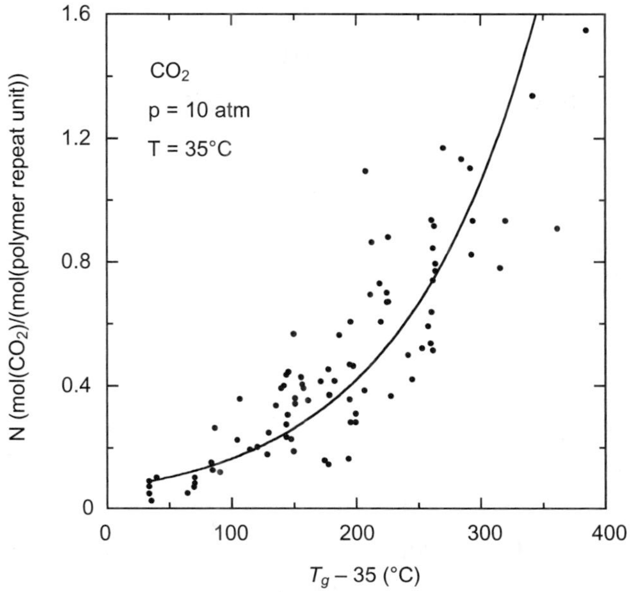

Figure 6.4. The number of carbon dioxide molecules, N sorbed in a polymer repeat unit at 35°C and 10 atm as a function of T_g–35°C. (Reprinted from Kanehashi S. and K. Nagai. Analysis of dual-mode model parameters for gas sorption in glassy polymers. *J. Membr. Sci.* 253 (2005) 117, with permission from Elsevier.)

such as chain packing behavior. Thermal treatment induces densification of glassy polymer membranes [28]. The cooling step, especially, can considerably affect the relaxation of polymer chains; that is, different cooling protocols produce different amounts of nonequilibrium excess free volume because of different relaxation environments [29]. Consequently, the gas permeability also strongly affects the membrane preparation protocol.

The pressure dependence of CO_2 permeability in 6FDA-TeMPD membranes at various measurement times and 35°C is shown in Figure 6.5 [15]. In general, the gas permeability in glassy polymers exhibits pressure dependence for condensable gases such as carbon dioxide. As the gas pressure increases, CO_2 permeability in glassy polymers initially decreases and then gradually increases. This behavior is characteristic for the dual-mode transport theory with plasticization behavior. The pressure at the minimum gas permeability before the polymer begins to plasticize was termed as the critical pressure for plasticization [10]. Beyond this critical pressure, the gas permeability begins to increase gradually. It is believed that the molecular structure of the plasticized polymer is changed as compared to that of the unplasticized polymer. The CO_2 permeability of both the as-cast membranes and the heat-treated membranes at 250°C is almost constant at the feed pressure of 1 atm over 720 minutes.

This result indicates that the relaxation of polymer segments does not occur at a low pressure (1 atm). However, when the feed pressure is over 10 atm,

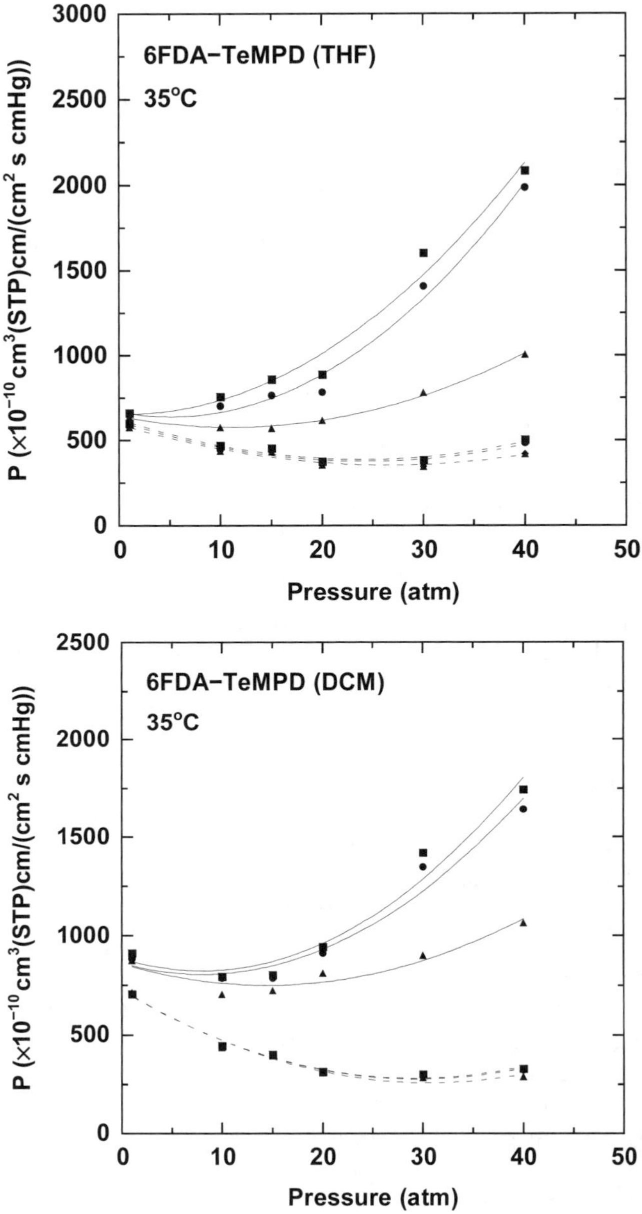

Figure 6.5. Pressure dependence of CO_2 permeability at a given time at 35°C of 6FDA-TeMPD membranes prepared from tetrahydrofuran (THF) or dichloromethane (DCM) solutions: (a) 5 minutes later (▲), (b) 360 minutes later (●), and (c) 720 minutes later (■). The solid lines are for as-cast membranes, and the dotted lines are for membranes heat-treated at 250°C. (Reprinted from Kanehashi S., et al. Effects of carbon dioxide-induced plasticization on the gas transport properties of glassy polyimide membranes. *J. Membr. Sci.* 298 (2007) 147, with permission from Elsevier.)

the permeability in the as-cast membranes increases gradually and it does not become constant during the experimental time. This behavior suggests the typical CO_2-induced plasticization previously documented for some other glassy polymers [17]. The pressure of 10 atm at 35°C could be defined as the *critical* plasticization pressure of the 6FDA-TeMPD as-cast membranes. On the other hand, regardless of the casting solvent, the heat-treated membranes showed resistance to plasticization until nearly 30 atm. The critical pressure of CO_2 plasticization for heat-treated membranes could be defined as 30 atm. The thermal treatment caused a tighter packing of polymer segments in the membrane. As a result, the critical plasticization pressure shifted from nearly 10–30 atm with the thermal treatment of membranes at 250°C. Interestingly, regardless of the membrane preparation protocol, there is a critical CO_2 diffusivity of $73 \pm 5 \times 10^{-8}$ cm^2/s at the plasticization pressure in 6FDA-TeMPD membranes.

6.5 CONCLUSION

As described throughout this chapter, the dense polymer membrane structures including crystalline, liquid crystalline, amorphous, and molecular ordering structures are closely related to the CO_2 permeation property and CO_2-induced plasticization. CO_2 separation technology has attracted great attention for reduction of global warming, natural gas purification, and hydrogen recovery. Generally, polymer membranes for gas separation desire high gas permeability, selectivity, physical stability, processability, and low cost in industry. To enhance the CO_2 separation performance of dense polymer membranes, further investigations in the synthesis and chemical modifications of novel polymers as well as computer molecular dynamics and orbital simulation are fundamentally important.

REFERENCES

[1] Paul D. R. and Y. P. Yampol'skii. Polymeric Gas Separation Membranes. CRC, Boca Raton, FL, 1994.

[2] Kanehashi S., A. Kusakabe, S. Sato, and K. Nagai. Analysis of permeability; solubility and diffusivity of carbon dioxide; oxygen; and nitrogen in crystalline and liquid crystalline polymers. *J. Membr. Sci.* 365 (2010) 40.

[3] Allen S. M., M. Fujii, V. Stannett, H. B. Hopfenberg, and J. L. Williams. The barrier properties of polyacrylonitrile. *J. Membr. Sci.* 2 (1977) 153.

[4] Weinkauf D. H. and D. R. Paul. Gas transport properties of liquid crystalline poly(ethylene terephthalate-co-p-oxybenzoate). *J. Polym. Sci. Part [B]* 29 (1991) 329.

[5] Weinkauf D. H., H. D. Kim, and D. R. Paul. Gas transport properties of liquid crystalline poly (p-phenyleneterephthalamide). *Macromolecules* 25 (1992) 788.

[6] Komatsuka T., A. Kusakabe, and K. Nagai. Characterization and gas transport properties of poly(lactic acid) blend membranes. *Desalination* 234 (2008) 212.

[7] Komatsuka T. and K. Nagai. Temperature dependence on gas permeability and permselectivity of poly(lactic acid) blend membranes. *J. Polym.* 41 (2009) 455.

[8] Sawada H., Y. Takahashi, S. Miyata, S. Kanehashi, S. Sato, and K. Nagai. Gas transport properties and crystalline structures of poly(lactic acid) membranes. *Trans. Mat. Res. Soc. Jpn.* 35 (2001) 241.

[9] Poling B. E., J. M. Prausnitz, and J. P. O'Connell. The Properties of Gases and Liquids. McGraw-Hill, New York, 2000.

[10] Ismail A. F. and W. Lorna. Penetrant-induced plasticization phenomenon in glassy polymers for gas separation membrane. *Sep. Purif. Tech.* 27 (2002) 173.

[11] Staudt-Bickel C. and W. J. Koros. Improvement of CO_2/CH_4 separation characteristics of polyimides by chemical crosslinking. *J. Memb. Sci.* 155 (1999) 145.

[12] Duan S., T. Kouketsu, S. Kazama, and K. Yamada. Development of PAMAM dendrimer composite membranes for CO2 separation. *J. Memb. Sci.* 283 (2006) 2.

[13] Lin H., E. Van Wagner, B. D. Freeman, L. G. Toy, and R. P. Gupta. Plasticization-enhanced hydrogen purification using polymeric membranes. *Science* 311 (2006) 639.

[14] Duthie X., S. Kentish, S. J. Pas, A. J. Hill, C. Powell, K. Nagai, G. Stevens, and G. Qiao. Thermal treatment of dense polyimide membranes. *J. Polym. Sci. Part [B]* 46 (2008) 1879.

[15] Kanehashi S., T. Nakagawa, K. Nagai, X. Duthie, S. Kentish, and G. Stevens. Effects of carbon dioxide-induced plasticization on the gas transport properties of glassy polyimide membranes. *J. Memb. Sci.* 298 (2007) 147.

[16] Nagai K., N. Booker, A. Mau, J. Hodgkin, S. Kentish, G. Stevens, and A. Geertsema. Gas permeation properties of polyimide/epoxy composite materials. *Polym. Mater. Sci. Eng.* 85 (2001) 91.

[17] Puleo A. C., D. R. Paul, and S. S. Kelley. The effect of degree of acetylation on gas sorption and transport behavior in cellulose acetate. *J. Memb. Sci.* 47 (1989) 301.

[18] Duthie X., S. Kentish, C. Powell, K. Nagai, G. Qiao, and G. Stevens. Operating temperature effects on the plasticization of polyimide gas separation membranes. *J. Memb. Sci.* 294 (2007) 40.

[19] Zhou C., T. S. Chung, R. Wang, Y. Liu, and S. H. Goh. The accelerated CO_2 plasticization of ultra-thin polyimide films and the effect of surface chemical cross-linking on plasticization and physical aging. *J. Memb. Sci.* 225 (2003) 125.

[20] Duthie X. J., S. E. Kentish, C. E. Powell, G. G. Qiao, K. Nagai, and G. W. Stevens. Plasticization suppression in grafted polyimide-epoxy network membranes. *Ind. Eng. Chem. Res.* 46 (2007) 8183.

[21] Barrer R. M., J. A. Barrie, and J. Slater. Sorption and diffusion in ethyl cellulose: Part III. Comparison between ethyl cellulose and rubber. *J. Polym. Sci.* 27 (1958) 177.

[22] Michaels A. S. and H. J. Bixler. Flow of gases through polyethylene. *J. Polym. Sci.* 50 (1961) 413.

[23] Kanehashi S. and K. Nagai. Analysis of dual-mode model parameters for gas sorption in glassy polymers. *J. Memb. Sci.* 253 (2005) 117.

[24] Paul D. R. Gas sorption and transport in glassy polymers. *Ber. Bunsenges. Phys. Chem.* 83 (1979) 294.

[25] Koros W. J. and D. R. Paul. Observations concerning the temperature dependence of the langmuir sorption capacity of glassy polymers. *J. Polym. Sci.: Polym. Phys. Ed.* 19 (1981) 1655.

[26] Toi K., G. Morel, and D. R. Paul. Gas sorption and transport in poly(phenylene oxide) and comparisons with other glassy polymers. *J. Appl. Polym. Sci.* 27 (1982) 2997.

[27] Miyata S., S. Sato, K. Nagai, T. Nakagawa, and K. Kudo. Relationship between gas transport properties and fractional free volume determined from dielectric constant in polyimide films containing the hexafluoroisopropylidene group. *J. Appl. Polym. Sci.* 107 (2008) 3933.

[28] Kawakami H., M. Mikawa, and S. Nagaoka. Gas transport properties in thermally cured aromatic polyimide membranes. *J. Memb. Sci.* 118 (1996) 223.

[29] Hachisuka H., H. Takizawa, Y. Tsujita, A. Takizawa, and T. Kinoshita. Gas transport properties in polycarbonate films with various unrelaxed volumes. *Polymer* 32 (1991) 2382.

7

GAS PERMEABILITY AND ELECTRICAL PROPERTIES OF 6FDA-BASED POLYIMIDE MEMBRANES

SHUICHI SATO, SOU MIYATA, SHINJI KANEHASHI, AND KAZUKIYO NAGAI

7.1 INTRODUCTION

The optical and electrical properties of polymer materials are very important properties in designing optic and electric devices. Amorphous transparent polymer components are indispensable for such applications of optical fiber [1], compact discs [2], and display materials [3]. As a factor contributing to transparency, the reflection/absorption/dispersion properties of photon are necessary in designing optics polymer materials, particularly in the development of antireflection films that possess surfaces that are able to prevent optical reflection from the outside. The material characteristic of reflectance is important in next-generation displays. The effect of reflectance on the refractive index of materials may be evaluated by Fresnel's equation [4]. A low refractive index is necessary for next-generation antireflection films.

On the other hand, the dielectric constant of polymer materials is also a very important property in designing electric materials, and it is used for dielectric evaluation as well as refractive index. A low dielectric constant is necessary for faster signal propagation in microelectronic devices without cross talk for new multilevel high-density and high-speed electronic circuits [5]. The dielectric constant is a relative value of the vacuum state or air. The

Sustainable Membrane Technology for Energy, Water, and Environment, First Edition.
Edited by Ahmad Fauzi Ismail and Takeshi Matsuura.

dielectric constant, ε, of a general polymer material is larger than three [6]. If the ε value is less than three, the polymer is expected to be applied as an interlayer insulation film in large-scale integration (LSI) or in integrated circuits (ICs).

The permeability of small molecules is also a very important property for next-generation antireflection and interlayer insulation films in order to prevent material degradation (e.g., oxidation), caused by air exposure during use. The barrier property of water vapor and carbon dioxide is necessary for packing materials [7–9]. In designing antireflection and interlayer insulation films, considering the relationship between its electrical properties (i.e., refractive index and dielectric constant) and gas transport properties (i.e., permeability, diffusivity, and solubility) is crucial. However, there is no physical chemistry-based linkage between mass transport and the electrical properties.

Recently, considerable attention has been devoted to the fluorine-containing aromatic polyimides, especially trifluoromethyl (-CF_3) containing polyimides [7, 10–13]. Above all, the 2,2′-bis(3,4-dicarboxyphenyl) hexafluoropropanedianhydride (6FDA) is commercially available as a tetracarboxylic dianhydride component, which has led to its widespread use [14]. Besides that, 6FDA polyimides also provide other merits such as good thermal stability and low moisture absorption. Therefore, 6FDA-based polyimides are extensively investigated in the electronics industry [5, 15–17]. Furthermore, fractional free volume (*FFV*), which is standard of the space between polymer segments, increases upon the introduction of a substituent that has high volume among atoms, such as fluorine atom. In particular, 6FDA-based polyimides are known to be one of the most gas-permeable materials. 6FDA-based polyimides have higher heat resistance properties among all polyimides. It is therefore widely investigated as an electronics and gas separation material [7, 18–20]. Additionally, the introduction of the space between polymer segments into polyimides is an effective way of lowering their dielectric constant [7, 11–13].

Polymers with higher *FFV* are known to have greater permeabilities and diffusivities. Based on the free-volume theory, the logarithm of the permeability and diffusion coefficients of small molecules in a polymer is linearly correlated with the reciprocal of its *FFV* [21]. However, for the polymers with the same *FFVs*, their gas permeabilities and diffusivities are not identical. This is one of the reasons *FFV* is calculated using group contribution method. For instance, the carbon dioxide diffusion coefficient in various types of polymers with an *FFV* value of 0.18 varies from 10^{-6} to 10^{-8} cm^2/s [19]. This chapter summarizes the gas permeability and electrical properties in dense 6FDA-based polyimide membranes.

7.2 PHYSICAL AND ELECTRICAL PROPERTIES

In general, polyimides containing 6FDA have excellent solubility in polar solvents, gas permeability, and optical transparency together with a low

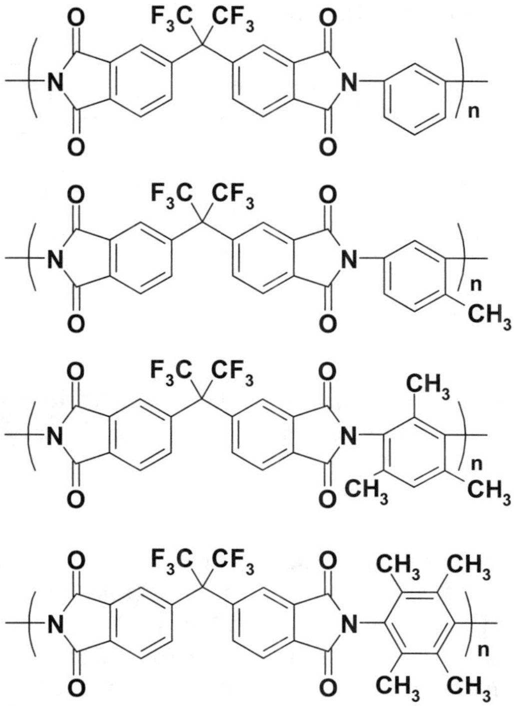

Figure 7.1. Chemical structures of fluorine-containing polyimides. 6FDA-based polyimides: 2,2′-bis (3,4-dicarboxyphenyl) hexafluoropropanedianhydride (6FDA), 1,3-phenylene diamine (mPD), 4-methyl-1,3-phenylene diamine (MPD), 2,4,6-trimethyl-1,3-phenylene diamine (TMPD), and 2,3,5,6-tetramethyl-1,4-phenylene diamine (TeMPD).

dielectric constant [22]. This has been attributed to the presence of the hexafluoroisopropylidene group (-$C(CF_3)_2$-), which imparts flexibility to the backbone and hinders intermolecular packing (and increase free volume). The chemical structures of 6FDA-based polyimides are shown in Figure 7.1. These are 6FDA-1,3-phenylene diamine (mPD), 6FDA-4-methyl-1,3-phenylene diamine (MPD), 6FDA-2,4,6-trimethyl-1,3-phenylene diamine (TMPD), and 6FDA-2,3,5,6-tetramethyl-1,4-phenylene diamine (TeMPD).

Table 7.1 summarizes the physical, optical, and electrical properties of the 6FDA-based polyimides. The membrane density of these polyimides varied from 1.34 to 1.47 g/cm^3. This density value strongly depends on the number of methyl groups in the diamine moiety. As the number of methyl groups increases, the density decreases.

The *FFV* can be determined by

$$FFV = \frac{V - V_0}{V}, \tag{7.1}$$

TABLE 7.1. Physical, Optical, and Electrical Properties of the 6FDA-Based Polyimides

Polyimide	Density[a] (g/cm^3)	*FFV*[a]	T_g^a (°C)	n_D^b	ε^a
6FDA-mPD	1.47	0.162	298	1.597	3.08
6FDA-MPD	1.42	0.171	335	1.572	2.82
6FDA-TMPD	1.36	0.177	380	1.544	2.59
6FDA-TeMPD	1.34	0.177	427	1.534	2.46

[a] Data from Reference 7.
[b] Data from Reference 8.

where V is the polymer molar volume and V_0 is the occupied volume at 0 K. The occupied volume can be estimated from

$$V_0 = 1.3V_W, \tag{7.2}$$

where V_w is the van der Waals volume, which was calculated from the group contribution method of van Krevelen [4]. The free-volume theory for gas transport is usually applied to rubbery polymers. However, it is also applied to various families of glassy polymers, such as polycarbonates, polysulfones, and polyimides [21]. The *FFV* value of these polymers varied from 0.162 to 0.177. As the number of the methyl group in the diamine moiety increased from one to three, the *FFV* also increased. The *FFV* values of 6FDA-TMPD and 6FDA-TeMPD were the same for both types of polymers.

The T_g value of 6FDA-mPD was 298°C. As the number of the methyl group in the diamine moiety increases, the T_g value also increases. The addition of four methyl groups has resulted in an increase of 129°C in T_g. Thus, it is seemed that the methyl groups restricted the mobility of polymer segments. These 6FDA-based polyimides provide good thermal stability.

The refractive index, n_D, and the dielectric constant, ε, values of the 6FDA-based polyimides are summarized in Table 7.1. As the number of the methyl side-chain groups in the diamine moiety increased, the refractive index decreased from 1.597 to 1.534 for n_D and varied from 2.46 to 3.08 for ε. Meanwhile as the number of the methyl group in the diamine moiety increased, the n_D and ε value decreased.

7.3 GAS AND VAPOR TRANSPORT PROPERTIES

The gas permeability coefficient, P (cm^3(STP)cm/(cm^2 s cmHg)), of a membrane was determined via constant volume-variable pressure method [7, 8]. The gases used in this study were hydrogen (H_2), oxygen (O_2), nitrogen (N_2), carbon dioxide (CO_2), and methane (CH_4). The gas permeation properties in the 6FDA-based polyimides were determined in a glassy state. The T_g value was much higher than the temperature employed in all the permeation measurements. As previously explained, based on the solution-diffusion mechanism, gas permeability is the product of gas solubility and gas diffusivity. Gas solubility is correlated with gas condensability. Based on the systematic data in this table, the ranking of the gas solubility coefficients in the 6FDA-based polyimide membranes was $CO_2 > CH_4 > O_2 > N_2$. This is the same order as the critical temperature, which is the measurement of gas condensability. The gas with a higher critical temperature dissolved in these polymers more than the gas with a lower critical temperature.

Gas diffusivity depends on gas size. The kinetic diameter indicates the diameter of the smallest cross section of a gas molecule, while the critical volume shows the average size of a gas molecule considering its three-dimensional structure. The ranking of the gas diffusion coefficient was $O_2 > N_2 > CO_2 > CH_4$. This is the same order as the critical volume. This is probably due to the gas diffusion among polymer segments that depends on both cross-section size of a gas molecule and the distance of the diffusion pass.

As a result, the ranking of the gas permeability coefficients was $H_2 > CO_2 > O_2 > N_2 > CH_4$ for 6FDA-mPD and 6FDA-MPD membranes and $CO_2 > H_2 > O_2 > N_2 > CH_4$ for 6FDA-TMPD and 6FDA-TeMPD membranes. This order is based on the combination of the solubility and diffusivity factors.

In Table 7.2, water vapor permeation properties in polymer membranes were summarized at infinite dilution [9]. The water vapor permeability and solubility of polyimide membranes were much higher than the other gases. Higher water vapor solubility depends on large cohesiveness of water vapor which shows higher critical temperature. However, water vapor diffusivity does not exhibit differences from the other gases; meanwhile, the water vapor permeability depends more on the solubility.

7.4 RELATIONSHIP BETWEEN GAS TRANSPORT AND ELECTRICAL PROPERTIES

Matsumoto et al. reported a linear relationship between carbon dioxide permeability and dielectric constant in several polymers [23]. Figure 7.2 presents the gas permeability coefficients of the 6FDA-based polyimides as a function of the refractive index and dielectric constant [7, 8]. For the 6FDA-based polyimides, as the refractive index and dielectric constant increased, the gas permeability coefficients of all gases decreased.

Unfortunately, there is no physical chemistry-based linkage between mass transport and electrical properties such as refractive index and dielectric constant. Hence, we described *FFV* as functions of the refractive index based on Lorentz–Lorenz Equation and the dielectric constant based on Clausius–Mossotti Equation. And we investigated the relationships between gas transport properties and the electrical properties-based polymer *FFV*. Equation 7.3 is derived from the Lorentz–Lorenz Equation, which is used to calculate the

TABLE 7.2. Gas and Water Vapor Permeation Data of the Nonporous Polyimide Membranes

Parameter	Polyimide	H_2^a	O_2^a	N_2^a	CO_2^a	CH_4^a	H_2O^b
$P \times 10^{10} \left(\frac{cm^3(STP)cm}{cm^2 \cdot s \cdot cmHg}\right)$	6FDA-mPD	43	3.1	0.47	12	0.18	650
	6FDA-MPD	63	6.7	1.2	43	0.92	1300
	6FDA-TMPD	239	46	11	309	9.8	12,000
	6FDA-TeMPD	529	130	31	615	28	14,000
$D \times 10^{8} \left(\frac{cm^2}{s}\right)$	6FDA-mPD	–	2.3	0.41	0.3	0.046	1.5
	6FDA-MPD	–	5	1	0.99	0.22	0.77
	6FDA-TMPD	–	22	6.5	5.2	1.8	7.8
	6FDA-TeMPD	–	52	15	9.9	4.2	16
$S \times 10^{2} \left(\frac{cm^3(STP)}{cm^3 \cdot cmHg}\right)$	6FDA-mPD	–	1.3	1.1	40	3.9	430
	6FDA-MPD	–	1.3	1.2	43	4.2	1700
	6FDA-TMPD	–	2.1	1.7	59	5.4	1500
	6FDA-TeMPD	–	2.5	2.1	62	6.7	880

[a] Data from Reference 7 at 30°C.

[b] Data at infinite dilution from Reference 9 at 35°C.

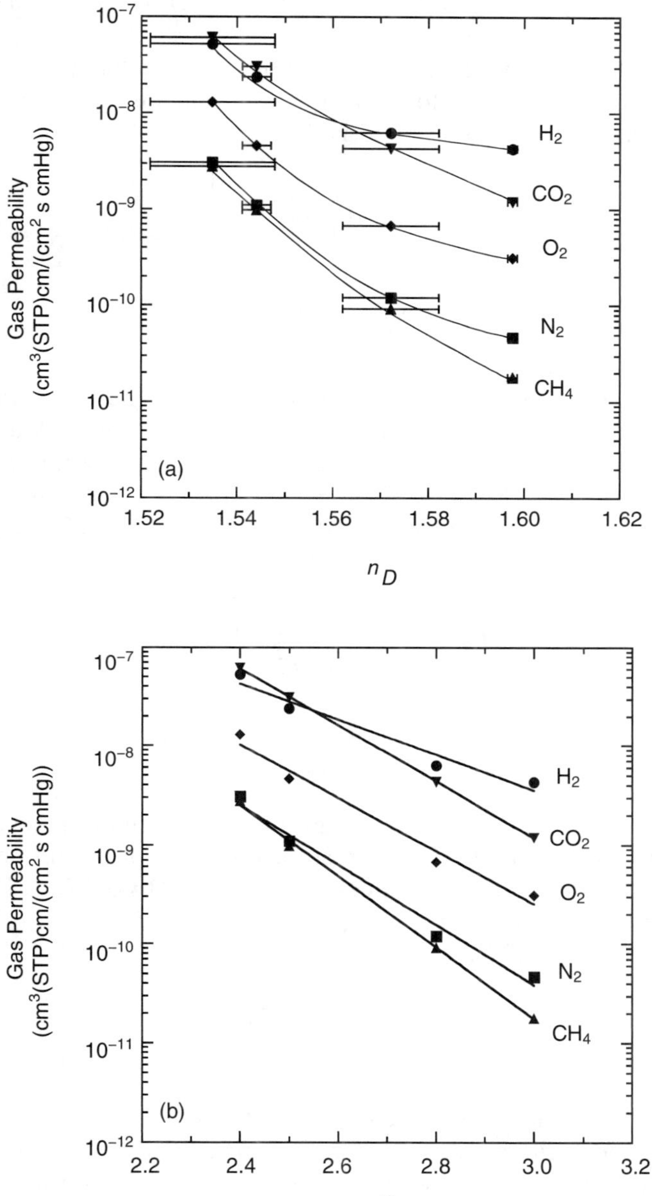

Figure 7.2. Permeability of various gases at 30°C in the 6FDA-based polyimides as functions of (a) refractive index and (b) the dielectric constant. Gases: hydrogen (●), nitrogen (■), oxygen (◆), methane (▲), and carbon dioxide (▼). (Reprinted from Miyata S., et al. Relationship between gas transport properties and fractional free volume determined from dielectric constant in polyimide films containing the hexafluoroisopropylidene group. *J. Appl. Polym. Sci.* 107 (2008) 3933, and Sato S., et al. Relationship between gas transport properties and refractive index in high free volume fluorine-containing polyimide membranes. *J. Appl. Polym. Sci.* 121 (2011) 2794, with permission from Wiley.)

refractive index of polymer material [4]. On the other hand, Equation 7.4 is derived from the Clausius–Mossotti Equation, which is used to calculate the dielectric constant of polymer material [4].

$$\frac{n_D^2-1}{n_D^2+2}=\frac{R_{LL}}{V} \tag{7.3}$$

$$\frac{(\varepsilon-1)}{(\varepsilon+2)}=\frac{P_{LL}}{V}, \tag{7.4}$$

where R_{LL} is the molar refraction (cm^3/mol), P_{LL} is the molar polarization (cm^3/mol), and V is the polymer molar volume (cm^3/mol). When polymer samples are isotropic dense amorphous homopolymer membranes without any additives, the V in Equation 7.1 is equal to the V in Equations 7.3 and 7.4. Hence, Equation 7.1 can be rewritten in Equations 7.5 and 7.6 as a function of the refractive index, n_D, and dielectric constant, ε.

$$FFV=1-1.3\frac{V_W}{R_{LL}}\frac{n_D^2-1}{n_D^2+2}=1-\varphi. \tag{7.5}$$

$$FFV=1-1.3\frac{V_W}{P_{LL}}\frac{\varepsilon-1}{\varepsilon+2}=1-\alpha. \tag{7.6}$$

As the number of the methyl side-chain groups in the diamine moiety increased, the gas permeability coefficients increased. The order of the gas permeability coefficients was 6FDA-TeMPD > 6FDA-TMPD > 6FDA-MPD > 6FDA-mPD.

The free-volume theory provides gas permeability, diffusivity, and solubility as a function of FFV [24, 25]:

$$P=A_P\,exp\left(\frac{-B_P}{FFV}\right), \tag{7.7}$$

where A_P and B_P, are adjustable constants. Generally, polymers with larger FFV have greater diffusivities as shown in Figure 7.3a [7, 8]. However, for the polymers with the same FFVs, their gas diffusivities are not exactly the same. For instance, the carbon dioxide diffusion coefficient in various types of polymers with an FFV value of 0.18 varies from 10^{-6} to 10^{-8} cm^2/s [19]. This indicates the limited usage of the current free-volume theory in 6FDA-TMPD and 6FDA-TeMPD with the same FFV value 0.177. Equation 7.7 can be described with Equations 7.5 and 7.6 as a function of the $1-\varphi$ and $1-\alpha$. Hence, Equation 7.7 can be rewritten as Equations 7.8 and 7.9.

$$P=A'_P\,exp\left(\frac{-B'_P}{1-\varphi}\right) \tag{7.8}$$

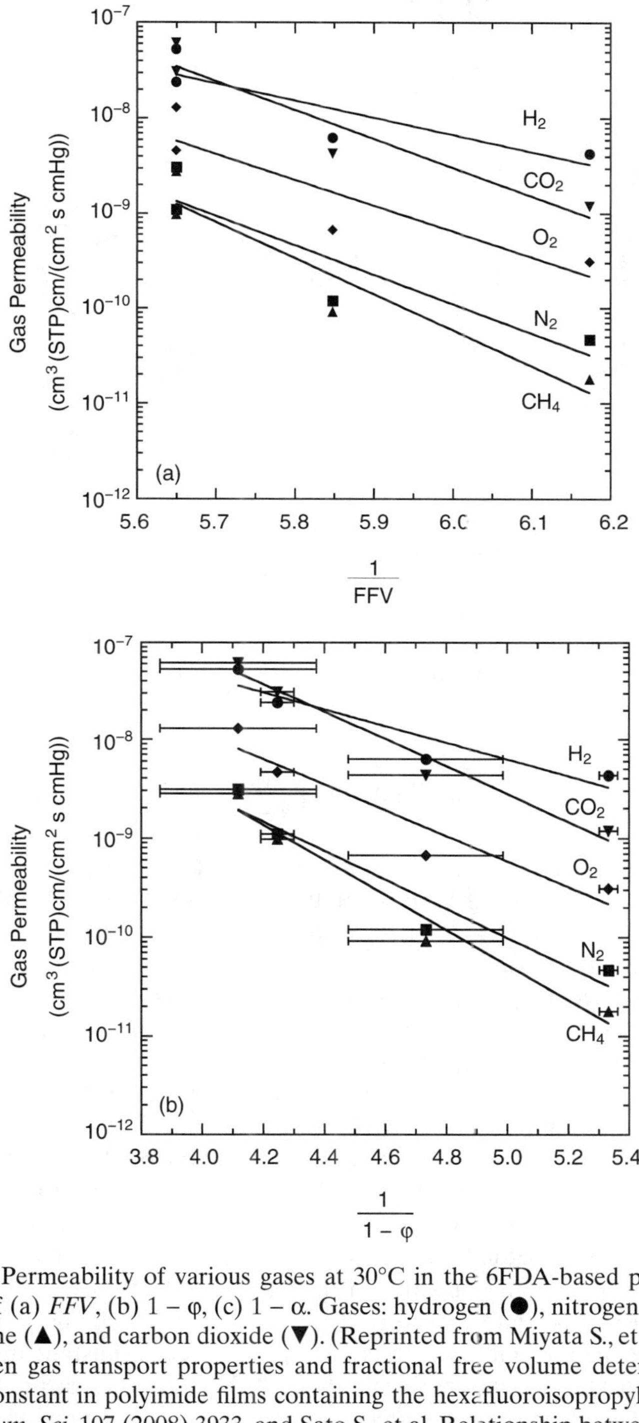

Figure 7.3. Permeability of various gases at 30°C in the 6FDA-based polyimides as functions of (a) *FFV*, (b) 1 – φ, (c) 1 – α. Gases: hydrogen (●), nitrogen (■), oxygen (◆), methane (▲), and carbon dioxide (▼). (Reprinted from Miyata S., et al. Relationship between gas transport properties and fractional free volume determined from dielectric constant in polyimide films containing the hexafluoroisopropylidene group. *J. Appl. Polym. Sci.* 107 (2008) 3933, and Sato S., et al. Relationship between gas transport properties and refractive index in high free volume fluorine-containing polyimide membranes. *J. Appl. Polym. Sci.* 121 (2011) 2794, with permission from Wiley.)

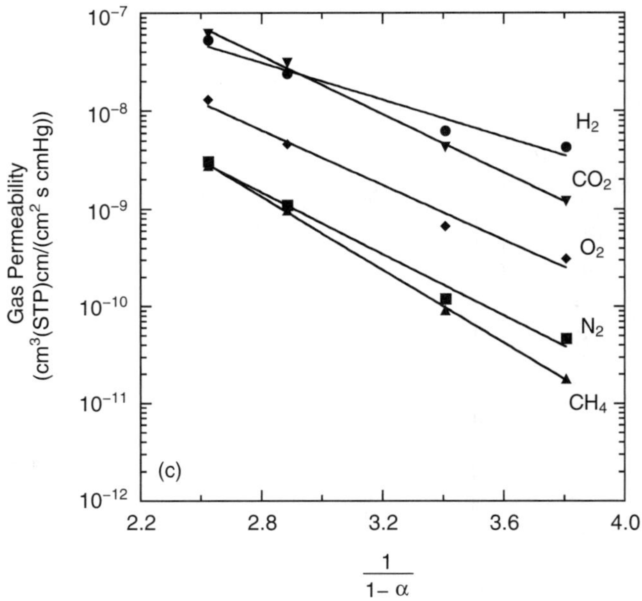

Figure 7.3. (*Continued*)

$$P = \gamma_P \exp\left(\frac{-\beta_P}{1-\alpha}\right), \tag{7.9}$$

where A'_P,B'_P, γ_P, and β_P are adjustable constants.

Figure 7.3 shows the gas permeability in the 6FDA-based polyimides as functions of the reciprocal of $1 - \varphi$ (Fig. 7.3b) and $1 - a$ (Fig. 7.3c) [7, 8]. As the $1 - \varphi$ and $1 - \alpha$ value increased, the gas permeability increased. Linear correlations were also observed between $1 - \varphi$, $1 - \alpha$, and the gas permeability; thus, the gas transport properties can be described as functions of the refractive index and dielectric constant.

Table 7.3 summarizes the group contribution-based *FFV* in Equation 7.1 and the refractive index-based *FFV*, $1 - \varphi$ and $1 - \alpha$ in Equations 7.5 and 7.6 in the 6FDA-based polyimides. The $1 - \varphi$ value of the 6FDA-based polyimides was 1.2–1.4 times larger than their *FFV* values. Additionally, the $1 - \alpha$ values of the 6FDA-based polyimides were 1.6–2.2 times larger than their *FFV* values. As predicted from Equation 7.1, the *FFV* was mainly dependent on free-volume space in a membrane. In contrast, the $1 - \varphi$ and $1 - \alpha$ (i.e., *FFV* determined from Eqs. 7.5 and 7.6) were dependent on free-volume space, optical, and electrical factors, such as refractive index, dielectric constant, molar refraction, and molar polarization, which affected the electronic structure and the interactions between the gas molecules and the polymer segments. This factor would provide more precise adjustment for gas transport in the free volume of the polymer membrane.

TABLE 7.3. Fractional Free Volume (*FFV*) Values Determined from the Group Contribution Method and the Refractive Index and Dielectric Constant of the 6FDA-Based Polyimides

	FFV[a]	$1-\varphi^b$	$1-\alpha^a$
Polyimide	From Equation 7.1	From Equation 7.5	From Equation 7.6
6FDA-mPD	0.162	0.188	0.263
6FDA-MPD	0.171	0.211	0.293
6FDA-TMPD	0.177	0.236	0.347
6FDA-TeMPD	0.177	0.243	0.381

[a] Data from Reference 7.
[b] Data from Reference 8.

7.5 CONCLUSION

As refractive index and dielectric constant decreased, gas permeability increased. Based on the Lorentz–Lorenz and Clausius–Mossoti equations, the relationship of the permeability and the *FFV* were determined from optical and electrical constants, such as refractive index, dielectric constant, molar refraction, and molar polarization. From this model, gas permeability coefficients of the 6FDA-based polyimide membranes were found to increase as their refractive index- and dielectric constant-based *FFV* increased.

REFERENCES

[1] Kaino T., M. Fujiki, and K. Jinguji. Preparation of plastic optical fibers. *Rev. Electr. Commun. Lab.* 32 (1984) 478.

[2] Birkett D. The chemistry of optical discs. *J. Chem. Educ.* 79 (2002) 1081.

[3] Nishimura S. and H. Mazaki. Development of liquid crystalline polymer film "Nisseki LC film" for viewing angle compensation of various LCD modes. *Mol. Cryst. Liq. Cryst.* 458 (2006) 35.

[4] van Krevelen D. W. Properties of Polymers, 4th ed. Elsevier B. V., Oxford, 2009.

[5] Vora R. H., P. S. G. Krishnan, S. H. Goh, and T. S. Chung. Synthesis and properties of designed low-k fluoro-copolyetherimides. Part 1. *Adv. Funct. Mater* 11 (2001) 361.

[6] Goosey M. Plastics for Electronics. Kluwer Academic, Dordrecht, The Netherlands, 1999.

[7] Miyata S., S. Sato, K. Nagai, T. Nakagawa, and K. Kudo. Relationship between gas transport properties and fractional free volume determined from dielectric constant in polyimide films containing the hexafluoroisopropylidene group. *J. Appl. Polym. Sci.* 107 (2008) 3933.

[8] Sato S., T. Ose, S. Miyata, S. Kanehashi, H. Ito, S. Matsumoto, Y. Iwai, H. Matsumoto, and K. Nagai. Relationship between gas transport properties and refractive

index in high free volume fluorine-containing polyimide membranes. *J. Appl. Polym. Sci.* 121 (2011) 2794.

[9] Sato S., M. Suzuki, S. Kanehashi, and K. Nagai. Permeability, diffusivity, and solubility of benzene vapor and water vapor in high free volume silicon- or fluorine-containing polymer membranes. *J. Memb. Sci.* 360 (2010) 352.

[10] Chung I. S. and S. Y. Kim. Soluble polyimides from unsymmetrical diamine with trifluoromethyl pendent group. *Macromolecules* 33 (2000) 3190.

[11] Miyata S. and K. Nagai. Block copolymerization of methyl methacrylate from fluorine-contained polyimide macroinitiator by atom-transfer radical polymerization. *Chem. Lett.* 36 (2007) 1114.

[12] Miyata S., H. Shirokura, and K. Nagai. Preparation of porous structure from ABA-type triblock copolymers derived from 6FDA-TeMPD polyimide macroinitiators. *Jpn. J. Polym. Sci. Tech.* 66 (2009) 170.

[13] Miyata S., K. Yoshida, H. Shirokura, M. Kashio, and K. Nagai. Solid and thermal properties of ABA-type triblock copolymers designed using difunctional fluorine-containing polyimide macroinitiators with methyl methacrylate. *Polym. Int.* 58 (2009) 1148.

[14] Chung C. L., C. P. Yang, and S. H. Hsiao. Organosoluble and colorless fluorinated poly(ether imide)s from 1,2-bis(3,4-dicarboxyphenoxy) benzene dianhydride and trifluoromethyl-substituted aromatic bis(ether amine)s. *J. Polym. Sci. [A1]* 44 (2006) 3092.

[15] Ichino T., S. Sasaki, T. Matsuura, and S. Nishi. Synthesis and properties of new polyimides containing fluorinated alkoxy side chains. *J. Polym. Sci. [A1]* 28 (1990) 323.

[16] Rogers F. E. Polyamide-acids and polyimides from hexafluoropropylidine bridged diamine, U.S. Patent 3356648, 1964.

[17] Hougham G., P. E. Cassidy, K. Johns, and T. Davidson. Fluoropolymers 1: Synthesis and Fluoropolymers 2: Properties. Kluwer Academic/Plenum Publishers, New York, 1999.

[18] Tanaka K., M. Okano, H. Toshino, H. Kita, and K. Okamoto. Effect of methyl substituents on permeability and permselectivity of gases in polyimides prepared from methyl-substituted phenylenediamines. *J. Polym. Sci., B: Polym. Phys.* 30 (1992) 907.

[19] Kanehashi S. and K. Nagai. Analysis of dual-mode model parameters for gas sorption in glassy polymers. *J. Memb. Sci.* 253 (2005) 117.

[20] Liu Y., R. Wang, and T. S. Chung. Chemical cross-linking modification of polyimide membranes for gas separation. *J. Memb. Sci.* 189 (2001) 231.

[21] Paul D. R. and Y. P. Yampolskii. Polymeric Gas Separation Membranes. CRC Press, Boca Raton, FL, 1994.

[22] Cassidy P. E., T. M. Aminabhavi, and J. M. Farley. Polymers derived from hexafluoroacetone. *J. Macromol. Sci., Rev. Macromol. Chem. Phys.* 29 (1989) 365.

[23] Matsumoto K., P. Xu, and T. Nishikimi. Gas permeation of aromatic polyimides: I. Relationship between gas permeabilities and dielectric constants. *J. Memb. Sci.* 81 (1993) 15.

[24] Lee W. M. Selection of barrier materials from molecular structure. *Polym. Eng. Sci.* 20 (1980) 65.

[25] Fujita H. Diffusion in polymer-diluent systems. *Fortschr. Hochpolymer. Forsch.* 3 (1961) 1.

8

POLYMERIC NANOCOMPOSITE MEMBRANES FOR GAS SEPARATION

Ghader Khanbabaei, Jamal Aalaei, and Ali Rahmatpour

8.1 INTRODUCTION

Membrane-based gas separation is an economical and environmentally friendly technology. However, when conventional membrane materials are used, the major obstacle that hinders the further development is the trade-off between permeability and selectivity in which as the selectivity increases, permeability decreases or vice versa. Development of hybrid or nanocomposite membranes is one of the recent contributions to overcome this trade-off. Organic/inorganic hybrid materials have the potential to combine the desired properties of inorganic and organic systems, thus combining the mechanical and thermal properties of inorganic phase with the flexibility and ductility of organic polymers. Organic–inorganic nanocomposite membranes have been the subject of interest in recent years [1–24]. These nanocomposite membranes can be obtained by adding fillers (e.g., zeolites, silica, alumina, molecular sieving carbons, others) to a polymer matrix. Solution blending is a simple way to fabricate polymer–inorganic nanocomposite membranes. During the preparation, the polymer is primarily dissolved in a solvent to form a solution, followed by the addition of inorganic nanoparticles under constant stirring. The nanocomposite membrane is formed upon the removal of solvent.

Sustainable Membrane Technology for Energy, Water, and Environment, First Edition.
Edited by Ahmad Fauzi Ismail and Takeshi Matsuura.

The permeability (*P*) of a gas through a membrane is proportional to the solubility (*S*) and diffusivity (*D*) of the gas in the membrane ($P = D \times S$). Therefore, addition of inorganic nanofillers to a polymer membrane may affect the gas separation properties in two ways: First, the interaction between polymer-chain segments and nanofillers can disrupt the polymer-chain packing and increase the voids (free volumes) between the polymer chains, thus promoting gas diffusion into the membrane; second, the hydroxyl and other functional groups on the surface of the inorganic phase may interact with polar gases such as CO_2 and SO_2, improving the penetrants' solubility in the nanocomposite membranes [24].

In this study, silica nanoparticles were added to a rubbery poly (dimethylsiloxane) (PDMS) and a glassy poly(1-trimethylsilyl-1-propyne) (PTMSP) membrane in various weight percents and permeations of N_2, CH_4, and CO_2 gases though the membranes were determined at different upstream pressures. The effect of the inorganic nanoparticles with different loading on gas permeability of polymeric membranes was evaluated.

8.2 MATERIALS AND METHODS

Commercial-grade two-component package of room temperature vulcanizing (RTV) PDMS was purchased from Wacker (Germany). The package contains part A which is PDMS and part B which is cross-linker and also contains platinum catalyst. PTMSP was purchased from Gelest (Morrisville, PA). Fumed silica (FS) nanoparticles (Aerosil) were supplied from Degussa (Germany) and solvents were obtained from Merck (Germany).

The membranes from pure PDMS were prepared by mixing the two components of RTV silicon package and casting the mixture on a Teflon-coated dish. Nanocomposite membranes of PDMS with silica were prepared in a similar way, but to overcome increased viscosity, normal hexane solvent was used. Silica nanoparticles were dispersed in solvent by mechanical mixing and sonication, prior to the addition of this mixture into the resin. Pure and nanocomposite membranes of PTMSP were prepared through casting on a glass substrate at room temperature. The membranes were dried for 24 hours and were then removed from the plate to allow further drying in a vacuum oven at 60°C for 3 hours. The average thickness of the membranes obtained was in the range of 300–350 μmfor PD MSand150 –200 μm for PTMSP membranes.

To investigate the effect of dispersion of inorganic particles on the phase behavior of organic matrices, X-ray diffraction (XRD) patterns were recorded by monitoring the diffraction angle 2θ from 1° to 40° with Philips PW1840 using nickel—filtered Cu—K_α radiation ($\lambda = 0.154$ nm) under a voltage of 40 kV and a current of 25 mA. The scanning speed and the step size were 5°/min and 0.05°, respectively.

The permeation properties of PDMS, PTMSP, and their nanocomposite membranes for pure gases of nitrogen, methane, and carbon dioxide were determined using a constant pressure/variable volume apparatus. The upstream pressure was varied from 2 to 10 bars, but in this chapter the results obtained at upstream pressure of 5 bars were reported. The downstream pressure was the atmospheric pressure. Gas flow rates were measured with a soap-film bubble flowmeter. The temperature was maintained at 25(±1)°C. Prior to the testing, both upstream and downstream sides of permeation cell were purged with penetrant gas. The permeability coefficients of gases were determined in the order of nitrogen, methane, and carbon dioxide. When steady-state conditions were achieved, the following expression was used to evaluate permeability:

$$P = \frac{q\ell}{A(p_1 - p_2)}, \tag{8.1}$$

where P is permeability expressed in Barrer (1 Barrer = 10^{-10} cm^3(STP)cm/cm^2 s cmHg), q is flow rate of the permeate gas passing through the membrane (cm^3/s), ℓ is membrane thickness (cm), p_1 and p_2 are the absolute pressures of feed side and permeate side, respectively (cmHg), A is the effective membrane area (cm^2).

The ideal selectivity, $\alpha_{A/B}$ (the ratio of pure gas permeabilities), of membranes was calculated from pure gas permeation experiments:

$$\alpha_{A/B} = \frac{P_A}{P_B}. \tag{8.2}$$

8.3 RESULTS AND DISCUSSION

XRD patterns of PTMSP-silica nanocomposite membrane containing 50 wt % silica was compared with that of pure PTMSP as shown in Figures 8.1 and 8.2. In general, when a polymer contains large crystalline region, the peak observed in the XRD pattern of the polymer is usually sharp and the intensity is strong, whereas that of amorphous polymer is rather broad. Pure PTMSP showed an X-ray pattern with strong crystalline peak at $2\theta = 12$, resulting mainly from the crystalline region of PTMSP. The location of corresponding peak of nanocomposite did not shift significantly by the incorporation of silica, but the intensity was decreased significantly upon the addition of silica, indicating an increase in amorphous regions. The presence of silica nanoparticles decreased the crystalline regions and therefore the packing density of the membrane. Increase in amorphous regions could in turn increase the gas permeability as the gas permeation in semicrystalline dense membranes takes place through the amorphous regions. On the other hand, decrease in chain packing has resulted in the increase in free volume and consequently increased the gas permeability.

Permeability coefficients of nitrogen and methane in pure PTMSP and nanocomposites containing 25 and 50 wt % FS are shown in Figure 8.3, and

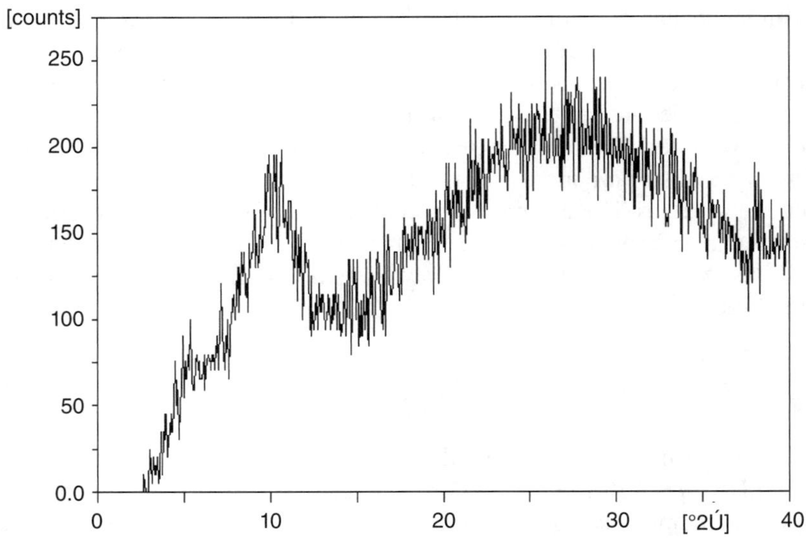

Figure 8.1. XRD pattern of pure PTMSP.

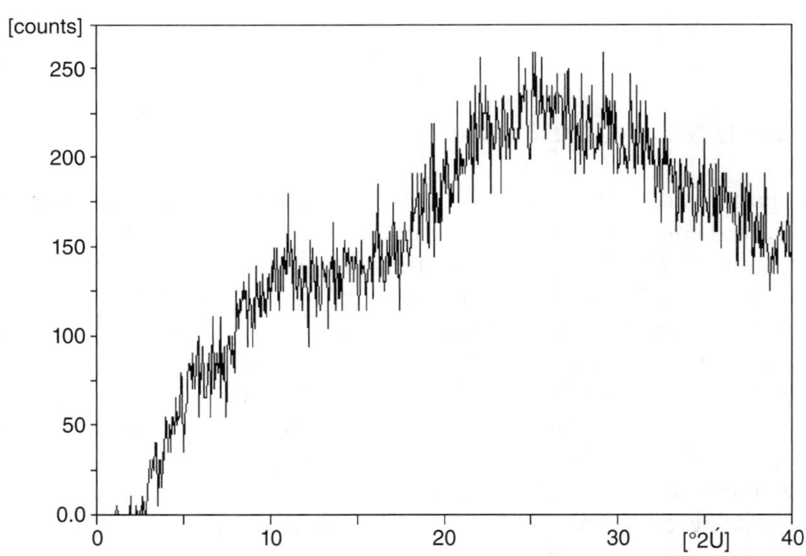

Figure 8.2. XRD pattern of PTMSP nanocomposite containing 50 wt % silica.

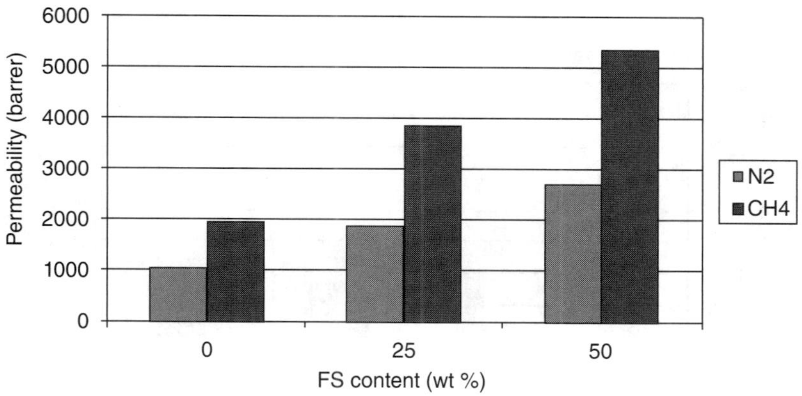

Figure8.3 . Permeabilityof N $_2$ and CH_4 in PTMSP and PTMSP-FS nanocomposite membranes.

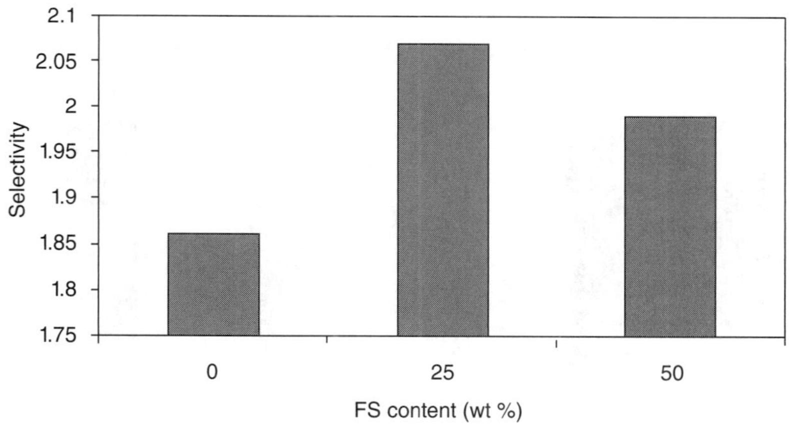

Figure 8.4. Permselectivity of CH_4 over N_2 in PTMSP and PTMSP-FS nanocomposite membranes.

the selectivity of membranes for methane over nitrogen is shown in Figure 8.4. The permeability of both gases increased by increasing silica content. Permeability of N_2 increased from 1040 Barrer in pure PDMS to 2807 Barrer in nanocomposite with 50 wt % silica, which is 2.6-folds of increment. For CH_4, the permeability increased from 1934 Barrer to 5331 Barrer (270% increase). As verified by XRD patterns, the main reason for the increase of gas permeability in nanocomposite membranes can be attributed by the decrease of chain packing density and corresponding increase in free volume and amorphous regions.

The results of permeability of nitrogen, methane, and carbon dioxide in PDMS and its nanocomposites containing 5, 10, and 15 wt % FS are presented

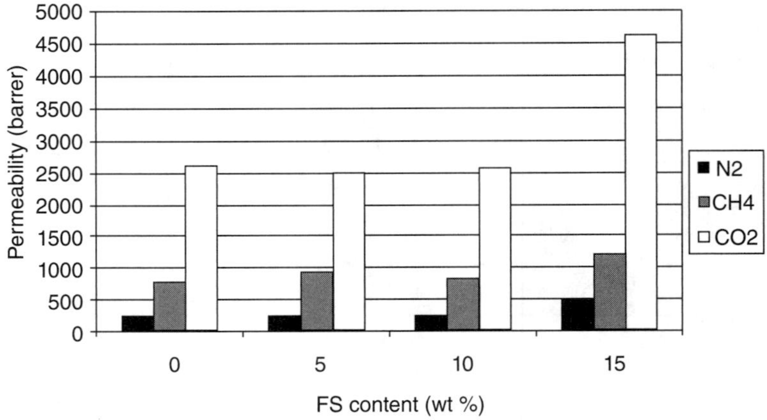

Figure 8.5. Permeability of N_2, CH_4, and CO_2 in PDMS and PDMS-FS nanocomposite membranes.

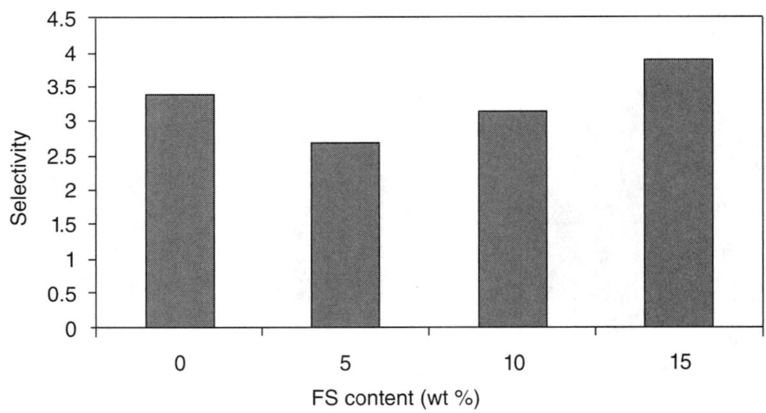

Figure 8.6. Permselectivity CO_2 over CH_4 in PDMS and PDMS-FS nanocomposite membranes.

in Figures 8.5 and 8.6. The addition of silica with loading more than 15 wt % was difficult to be carried out due to the high viscosity of PDMS. Furthermore, the resulted dry film exhibited serious defects which in turn resulted in poor mechanical properties.

In contrast to glassy PTMSP polymer, rubbery PDMS nanocomposite membranes demonstrated irregular trend in permeability upon the addition of silica. This phenomenon might be due to the low loading of the silica content in this resin. Increased permeability was recorded for nanocomposite membrane incorporated with 15 wt % silica particles. It can be concluded when PDMS was used that the increase in free volume of membrane through the addition of nanoparticles required more than 10 wt % of nanoparticles.

Nanocomposite membrane containing 15 wt % silica showed simultaneous increase in both permeability and selectivity for more condensable CO_2 gas compared to simple gases like CH_4 and N_2.

8.4 CONCLUSIONS

Addition of silica nanoparticles to the rubbery PDMS and glassy PTMSP membranes resulted in significant increase in permeability of all studied gases and simultaneous slight increase of selectivity for more condensable gases over simple gases in some of the prepared samples.

Permeability coefficients of nitrogen and methane in PTMSP nanocomposite membrane containing 50 wt % of silica were 2.5-fold greater than that of neat PTMSP membranes. By overcoming the usually encountered trade-off in those conventional membranes, nanocomposite membrane of PTMSP-silica exhibited 7% increment of CH_4/N_2 selectivity.

PDMS membrane containing 15 wt % of silica exhibited an increase of 80% in permeability of carbon dioxide. However, the trend of increase in permeability with increasing silica content was not in a regular manner.

REFERENCES

[1] Jia M., K. V. Peinemann, and R. D. Behling. Molecular sieving effect of the zeolite-filled silicone rubber membrane in gas separation. *J. Memb. Sci.* 57 (1991) 289–296.

[2] Moaddeb M. and W. J. Koros. Effects of colloidal silica incorporation on oxygen/nitrogen separation properties of ceramic-supported 6FDA-IPDA thin films. *J. Memb. Sci.* 111 (1996) 283–290.

[3] Nunes S. P., J. Schultz, and K. V. Peinemann. Silicone membranes with silica nanoparticles. *J. Mater. Sci. Lett.* 15 (1996) 1139–1141.

[4] Joly C., S. Goizet, J. C. Schrotter, J. Sanchez, and M. Escoubes. Sol–gel polyimide–silica composite membrane: gas transport properties. *J. Memb. Sci.* 130 (1997) 63–74.

[5] Moaddeb M. and W. J. Koros. Gas transport properties of thin polymeric membranes in the presence of silicon dioxide particles. *J. Memb. Sci.* 125 (1997) 143–163.

[6] Hu Q., E. Marand, S. Dhingra, D. Fritsch, J. Wen, and G. Wilkes. Poly(amide–imide)/TiO_2 nano-composite gas separation membranes: fabrication and characterization. *J. Memb. Sci.* 135 (1997) 65–79.

[7] Joly C., M. Smaihi, L. Porcar, and R. D. Noble. Polyimide–silica composite materials: how does silica influence their microstructure and gas permeation properties? *Chem. Mater.* 11 (1999) 2331–2338.

[8] Nunes S. P., K. V. Peinemann, K. Ohlrogge, A. Alpers, M. Keller, and A. T. N. Pires. Membranes of poly(ether imide) and nanodispersed silica. *J. Memb. Sci.* 157 (1999) 219–226.

[9] Smaihi M., J. C. Schrotter, C. Lesimple, I. Prevost, and C. Guizard. Gas separation properties of hybrid imide–siloxane copolymers with various silica contents. *J. Memb. Sci.* 16 (1999) 157–170.

[10] Cornelius C. J. and E. Marand. Hybrid silica–polyimide composite membranes: gas transport properties. *J. Memb. Sci.* 202 (2002) 97–118.

[11] He Z., I. Pinnau, and A. Morisato. Nanostructured poly(4-methyl-2-pentyne)/silica hybrid membranes for gas separation. *Desalination* 146 (2002) 11–15.

[12] Pinnau I. and Z. He. Filled superglassy membrane. U.S. Patent 6, 316,684, 2001.

[13] Merkel T. C., B. D. Freeman, R. J. Spontak, Z. He, I. Pinnau, P. Meakin, and A. J. Hill. Ultrapermeable reverse-selective nanocomposite membranes. *Science* 296 (2002) 519–522.

[14] Kong Y., H. Du, J. Yang, D. Shi, Y. Wang, Y. Zhang, and W. Xin. Study on polyimide/ TiO_2 nanocomposite membranes for gas separation. *Desalination* 146 (2002) 49–55.

[15] Park H. B., J. K. Kim, S. Y. Nam, and Y. M. Lee. Imide–siloxane block copolymer/ silica hybrid membranes: preparation, characterization and gas separation properties. *J. Memb. Sci.* 220 (2003) 59–73.

[16] Hibshman C., C. J. Cornelius, and E. Marand. The gas separation effects of annealing polyimide-organosilicate hybrid membranes. *J. Memb. Sci.* 211 (2003) 25–40.

[17] Iwata M., T. Adachi, M. Tomidokoro, M. Ohta, and T. Kobayashi. Hybrid sol–gel membranes of polyacrylonitrile–tetraethoxysilane composites for gas permselectivity. *J. Appl. Polym. Sci.* 88 (2003) 1752–1759.

[18] Merkel T. C., B. D. Freeman, R. J. Spontak, Z. He, I. Pinnau, P. Meakin, and A. J. Hill. Sorption, transport and structural evidence for enhanced free volume in poly(4-methyl-2-pentyne)/fumed silica nanocomposite membranes. *Chem. Mater.* 15 (2003) 109–123.

[19] Merkel T. C., Z. He, I. Pinnau, B. D. Freeman, P. Meakin, and A. J. Hill. Effect of nanoparticles on gas sorption and transport in poly(1-trimethylsilyl-1-propyne). *Macromolecules* 36 (2003) 6844–6855.

[20] Gomes D., S. P. Nunes, and K. V. Peinemann. Membranes for gas separation based on poly(1-trimetylsilyl-1-propyne)–silica nanocomposites. *J. Memb. Sci.* 246 (2005) 13–25.

[21] Winberg P., K. DeSitter, C. Dotremont, S. Mullens, I. F. J. Vankelecom, and F. H. J. Maurer. Free volume and interstitial mesopores in silica filled poly(1-trimethylsilyl-1-propyen) nanocomposites. *Macromolecules* 38 (2005) 3776–3782.

[22] Patel N. P., A. C. Miller, and R. J. Spontak. Highly CO_2-permeable and selective polymer nanocomposite membranes. *Adv. Mater.* 15 (2003) 729–733.

[23] Suzuki T. and Y. Yamada. Physical and gas transport properties of novel hyperbranched polyimide-silica hybrid membranes. *Polym. Bull.* 53 (2005) 139–146.

[24] Cong H., M. Radosz, B. F. Towler, and Y. Shen. Polymer–inorganic nanocomposite membranes for gas separation. *Sep. Purif. Technol.* 55 (2007) 281–291.

9

PREPARATION OF PEROVSKITE TITANIA CERAMIC MEMBRANE BY SOL-GEL METHOD

ABDUL LATIF AHMAD, SANI N. A. ABDULLAH,
AND SHARIF HUSSEIN SHARIF ZEIN

9.1 INTRODUCTION

Membranes play an important role in separation technology because membrane-based separation uses less energy and is more economically competitive than other conventional separation technologies, such as distillation, absorption, adsorption, solvent extraction, and cryogenics [1]. Among the various types of membranes, ceramic membranes are widely used due to their unique features, such as high temperature stability, high pressure resistance, good chemical stability, high mechanical resistance, long life, and good antifouling [2]. Presently, there are many researchers developing ceramic membranes to be used in oxygen separation. Applications that require oxygen separation include the oxidative coupling of methane (OCM) [3], partial oxidation of methane (POM) processes [4], and fuel cell systems [5].

Alumina, titania, zirconia, and silica are the common porous ceramic membrane materials. Among them, titania has gained considerable attention due to its high chemical resistance and several unique characteristics, such as its semiconductivity, catalytic and photocatalytic behaviors, and high water flux. The potential applications of titania membrane are numerous and include

Sustainable Membrane Technology for Energy, Water, and Environment, First Edition.
Edited by Ahmad Fauzi Ismail and Takeshi Matsuura.

ultrafiltration processes, catalytic/photocatalytic membrane reactors, gas separations/reactions, and gas sensing [6–8].

Although titania membranes have many advantages, they possess one drawback with respect to their anatase–rutile phase transformation. This phase transformation results in large volume and energy changes, leading to large variations in particle size, which directly affects the membrane pore size and porosity [9]. The transformation from the anatase to the rutile phase is not favored because the anatase phase has a higher specific surface area compared with the rutile phase. Thus, the anatase phase is a more useful membrane material in gas-separation applications [10]. In addition, titania membranes with porous structures have a high permeability but a low selectivity. Thus, to overcome the problems and improve the selectivity of the titania membrane with anatase phase, perovskite-type oxide $SrCO_{0.8}Fe_{0.2}O_3$ doping is utilized.

Perovskite is a typically mixed ionic, electronic conducting membrane material that has garnered a significant amount of attention and has been applied in chemical and petroleum industries. Presently, the use of perovskite as an oxygen separation membrane has received much attention due to its higher oxygen ion and electron conductivity. Perovskite can also operate steadily for longer operation times at high temperatures [11]. Usually, a perovskite membrane is used alone and is prepared by a conventional solid state method to form a perovskite powder. This powder is compressed into a disk and sintered before being applied in gas-separation processes [12].

Among several methods to create a titania membrane, the sol-gel process is considered the most practical due to its excellent compositional control, creation of a homogenous microstructure, and feasibility of producing thin films on complex shapes when dip-coating is used. Moreover, the sol-gel process is a low temperature operation and produces membranes that have a very high specific area; also, additional elements can be added to further modify its properties [8, 9]. A wet impregnation process was used to affix perovskite material within the titania membrane pores. Wet impregnation is a well-known method that develops heterogeneous catalysts and has been utilized to deposit nano-sized particles into the electrode structures of solid oxide fuel cells [13]. Thus, we hypothesized that the process would create a highly uniform perovskite dispersion within the titania matrix.

To the best of our knowledge, there are no papers on the preparation of a perovskite titania membrane. Hence, in this preliminary study, the feasibility of preparation perovskite titania membrane by the sol-gel method in combination with a wet impregnation process was studied. The effect of the calcination temperature on perovskite and the titania membrane was examined. The characteristics of the membrane support and titania sol are described in detail. The membranes were characterized by X-ray diffractometry (XRD), field emission scanning electron microscopy (FESEM), and thermogravimetric analysis (TGA).

9.2 EXPERIMENTAL

9.2.1 Preparation of Titania Support

A porous titania disc, which was used as a support for the perovskite titania membrane, was created from titanium (IV) oxide powder. The disc was 20 mm in diameter and 2 mm in thickness and was made by loading approximately 1.75 g of titania powder into a stainless steel mold. The mold was pressed using hydraulic press equipment under 20 MPa pressure. The disc was dried at 70°C for 4 hours before undergoing a calcination process at 965°C for 4 hours. In the calcination process, low heating and a cooling rate of 2°C/min was used to prevent cracks.

9.2.2 Preparation of Titania Sol

The sol-gel method was used to prepare the titania sol. The following starting materials were used in the sol preparation: titanium tetra-isopropoxide (TTIP), TTIP as the titania precursor, nitric acid as the catalyst for the peptization, distilled water as the dispersing media, and isopropanol as the solvent. Titania sol was prepared by hydrolysis of TTIP. A solution of TTIP in isopropanol (0.45 M) was added dropwise into a solution of water in isopropanol (4.5 M) under vigorous stirring. After the hydrolysis reaction was complete, the remaining white precipitate of the titanium hydroxide ($Ti(OH)_4$) was filtered and subsequently washed with water to remove the alcohol. Then, the filtrate was dispersed in distilled water (Ti^{4+}), and nitric acid was added to achieve a 0.5 M ratio of acid/alkoxide (H^+/Ti^{4+}). The ratio was varied from 0.5 to 2.5 M. Next, the solution was peptized for 2 hours at 70°C. The solution was peptized in a closed beaker to enhance the rate of peptization. To break the weak bonds of the agglomerated particles, the sol was treated ultrasonically for 30 minutes. The final product was a blue, semiopaque colloidal dispersion with a concentration of 0.325 M. Lower concentrations could be achieved simply by diluting the mixture with distilled water.

9.2.3 Preparation of Perovskite

Perovskite was prepared as described in a previous publication [14]. $Sr(NO_3)_2$, $Co(NO_3)_2.6H_2O$, and $FeCl_3.6H_2O$ were weighed in equimolar amounts and dissolved in distilled water (0.7 M). The solution was stirred for approximately 1 hour to ensure complete dissolution and mixing. The perovskite phase was only obtained after the membrane was calcined. The type of perovskite obtained in this study was $SrCo_{0.8}Fe_{0.2}O_3$.

9.2.4 Preparation of Perovskite Titania Membrane

The titania membrane was created from the titania sol; then, perovskite materials were affixed within the membrane pores using a wet impregnation process.

The dipping solution for the titania membrane contained 30 mL of 0.05 M titania sol mixed with 10 mL of polyvinyl alcohol (PVA) solution at a concentration of 0.1 g per 100 mL and 20 mL of hydroxypropyl cellulose (HPC) solution at a concentration 0.35 g/100 mL. An adequate amount of dipping solution was coated on a porous titania support disc by dip coating to form the supported titania membranes. Subsequently, the membrane was dried and calcined at 400°C for 2 hours.

During the perovskite impregnation process, the titania membrane was immersed in a perovskite solution for 10 minutes. Subsequently, the membrane was dried and then calcined in air at 400°C for 2 hours.

9.2.5 Membrane Characterization

The TGA was performed using a thermogravimetric analyzer (Perkin Elmer TGA 7) to determine the burnout temperature of the organic additives and to examine the calcination temperature. The phase transformation of the membrane during the calcination process was observed using an XRD (Philips PW1729 X-ray generator with a Philips PW1820 diffractometer) with Cu K_α radiation. The membrane's surface and thickness were observed by FESEM (Leo Supra 50VP).

9.3 RESULTS AND DISCUSSION

9.3.1 Titania Support

FESEM images of the titania support are shown in Figure 9.1a,b. The figures show that the support was composed of a uniform and homogenous microstructure with roughly spherical-shaped grains, which are responsible for the better mechanical strength. Sizes of the particles ranged from 210 to 230 nm. The images also reveal the presence of large voids in the support; due to these

Figure 9.1. FESEM image of the titania support with a high porosity microstructure: (a) surface and (b) cross section.

TABLE 9.1. Effect of Different H^+/Ti^{4+} Molar Ratios on Sol Characteristic Properties

Molar Ratio H^+/Ti^{4+}	pH	Sol Appearance after Peptization Process	Sol Stability
0.5	1.3	Semi-opaque blue	More than 6 months
1.0	1.0	Lighter semi-opaque blue	Less than 3 days
1.5	0.8	Almost clear	Less than 1 day
2.0	0.6	Clear	3.5 hours
2.5	0.4	Clear	3 hours

voids, it is impossible for the titania support to give any separation performance. Titania was chosen as the support material because it has a high chemical stability in strong acidic and basic environments and has a high resistance to harsh environments [6]. Also, it could replace conventional powder support materials (presintered alumina), which are relatively weaker and have a poorer surface quality [15].

In this study, nitric acid was used as the peptizing agent. Sol characteristics and its stability was studied at different H^+/Ti^{4+} molar ratios in the range 0.5–2.5 M. The final pH value of the resulting sol was measured using a pH meter. This study was conducted without the presence of PVA and HPC.

Table 9.1 shows the titania sol characteristic data with different H^+/Ti^{4+} molar ratios. It can be seen that the acidity of titania sol increased (the final pH value decreased from 1.3 to 0.4) as the molar ratio of H^+/Ti^{4+} increased. The optimal molar ratio was 0.5 with a 2-hour peptization process at 70°C. These conditions were sufficient to give a blue (semiopaque) colloidal dispersion as the final product. During the peptization process, H^+ ions from the nitric acid get adsorbed by the surface of the sol particles, repelling each other to form a stable colloidal sol. This electrostatic repulsion also prevents particles from sticking together and forming aggregates [16]. Therefore, by simple observation by the naked eye, the resulting sol with a 0.5 molar ratio was found to be stable and deflocculated evenly for several months. Forming a good membrane depends greatly on the condition of the sol used. It is expected that membranes that have a smaller pore diameter and a narrower pore size distribution could be made from a sol with small particle sizes [17].

9.3.2 Structural Properties

Figure 9.2 shows the TGA curve of the dried titania membrane gel. There are two regions of weight loss: below 105°C and between 120 and 364°C. In the first region, the weight loss of 0.7% is caused by the evaporation of the physically absorbed water from the membrane pores. In the second region, the weight loss of 5.3% is caused by the combustion of HPC and PVA compounds and the crystallization from titanium hydroxide to oxide. There is no weight loss at temperatures above 364°C. This result agrees with the finding of Mohammadi et al. [18] and Kermanpur et al. [19]. Thus, from the TGA

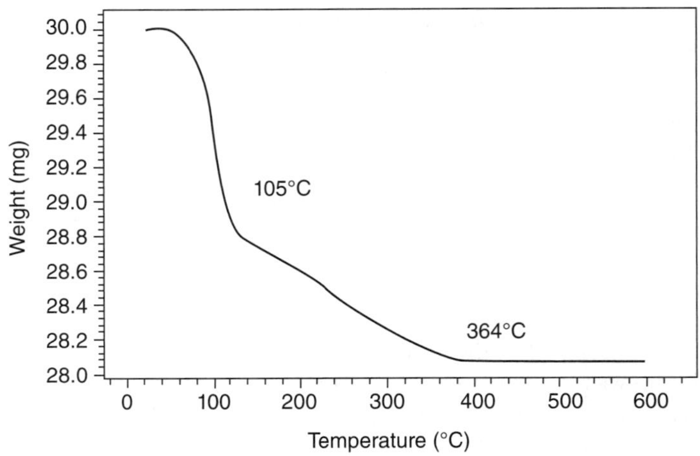

Figure 9.2. TGA curve of the dried titania membrane gel with 3°C/min heating rate in air.

observations, the membrane calcined at 400°C was considered to have HPC and PVA completely removed and contained only the anatase phase.

Figure 9.3 shows the XRD patterns of perovskite calcined at 300, 400, and 500°C. It can be seen that the perovskite calcined at 300°C consisted of $SrCO_3$ (strontium carbonate), Fe_2O_3 (iron oxide), and CoO (cobalt oxide). To prepare the $SrCo_{0.8}Fe_{0.2}O_3$ perovskite, the perovskite needed to be calcined until the perovskite phase was fully achieved. The observations at 300°C were due to an incomplete oxidation process of the perovskite. At 400 and 500°C, transformation of the perovskite had occurred based on the dominant XRD peak at $2\theta = 40.7°$. Based on the experimental results, the presence of a perovskite-type $SrCo_{0.8}Fe_{0.2}O_3$ structure readily occurred at 400°C. This finding is consistent with results obtained by Mohammadi and Fray [20], who prepared a low-temperature perovskite-type cadmium titanate membrane. Therefore, to inhibit the antase–rutile phase transformation, the calcination temperature of perovskite at 400°C was chosen as the calcination temperature of the titania.

9.3.3 Microstructure

FESEM images of a single coating of a titania layer on a titania support are shown in Figure 9.4a. It can be seen that the titania layer is formed on the surface of the titania support with some uncoated regions. Cracks were also found on the titania layer's surface. This occurred due to the drying and calcination process that removed water that was inside the titania layer. The release of water vapor and the shrinkage of the layer during calcination tends to create cracks. Figure 9.4b shows that the surface of the membrane layers contains irregularities on the titania support surface. This is due to the quality of the surface on top of which the membrane formed. The rough and uneven

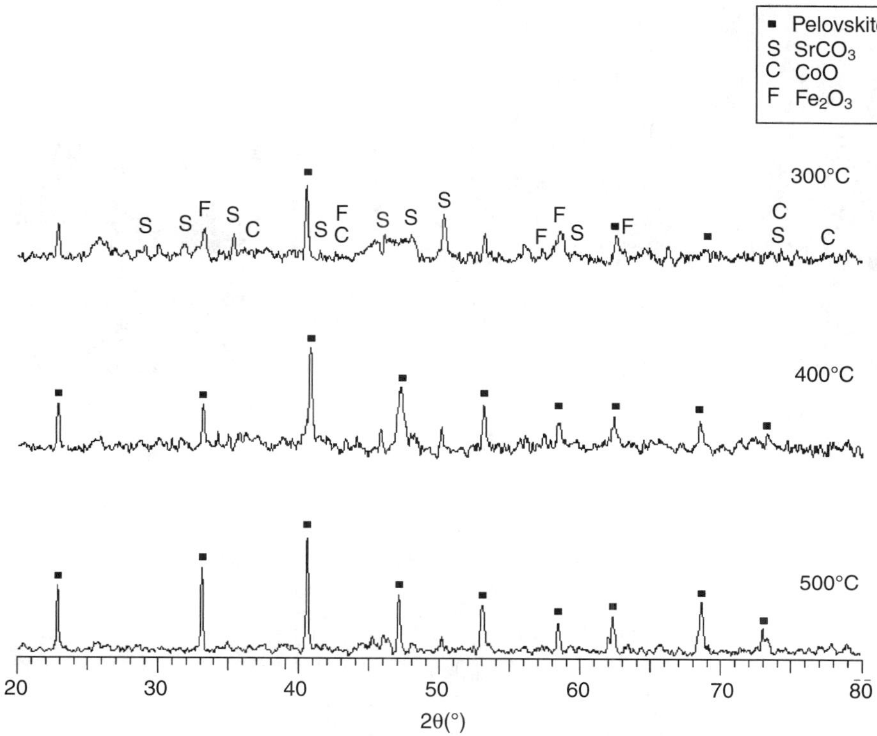

Figure 9.3. XRD patterns of $SrCo_{0.8}Fe_{0.2}O_3$perovskite oxide calcined at 300, 400, and 500°C.

Figure 9.4. FESEM images of titania membrane (a) surface, and (b) cross-section area calcined at 400°C after first time dip-coating process.

Figure 9.5. FESEM images of titania membrane (a) surface, and (b) cross-section area calcined at 400°C after five dip-coating processes.

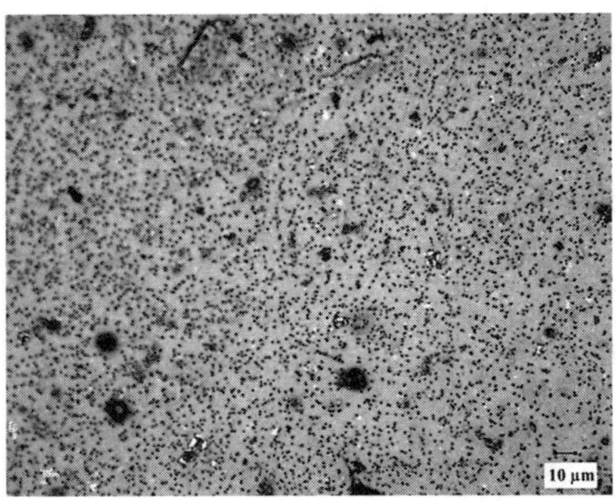

Figure 9.6. FESEM image of a perovskite titania membrane calcined at 400°C.

support surface transfers the roughness to the membrane layers. This is an extra stress-generation factor that might promote the nucleation of cracks and could result in a defective membrane [13].

Figure 9.5a,b show FESEM images of the surface and cross section of the titania layer after five dip-coating processes, respectively. It can be seen that the surface of the titania layer after five coatings became relatively denser and created a more complete coated layer than the single coating (Fig. 9.5a). Continuous contact exists in the interface, and the thickness of the layer was found to be approximately 0.75 μm, which is shown in Figure 9.5b. This multiple coating technique can reduce the pore size and cover up the surface of the porous titania support. After a crack-free titania layer was obtained, the perovskite wet impregnation process was carried out. Figure 9.6 shows an image of the perovskite titania membrane, which confirmed the presence of perovskite in the titania membrane. The surface area of the membrane showed a homogeneous dispersion of perovskite particles on the titania layers without cracking. This perovskite material contains several transition metal oxides (iron and cobalt oxides), which have been reported to be high-temperature adsorbents for gas separation, and theoretically have an infinitely high selectivity for O_2 over N_2 or other non-oxygen species [21]. Thus, perovskite in a titania membrane, which acts as a filler phase, will have the potential to be applied in gas-separation processes.

9.4 CONCLUSION

A titania membrane containing a dispersion of perovskite was successfully prepared using the sol-gel method and wet impregnation process. Based on our findings, the optimal peptization ratio of H^+/Ti^{4+} was found to be 0.5, which results in the formation of a stable and deflocculated sol over several months. The titania is fully transformed to the anatase phase, and the oxidation process of perovskite is complete when the calcination temperature for the titania and perovskite was at 400°C. A single coating of the titania membrane on the titania support resulted in cracks and an uneven surface due to the quality of the surface on which the membrane was formed. A crack-free and homogenous surface of the perovskite titania membrane was obtained after multiple dip coatings of the titania layer followed by the perovskite wet impregnation process. Therefore, it is believed that the perovskite titania membrane prepared using the present method has a great potential to be used in applications, such as in the separation of oxygen and oxygen enrichment membrane.

REFERENCES

[1] Pandey P. and R. S. Chauhan. Membranes for gas separation. *Prog. Polym. Sci.* 26 (2001) 853–892.

[2] Lin Y. S. and A. J. Burggraaf. Preparation and characterization of high temperature thermally stable alumina composite membrane. *J. Am. Ceram. Soc.* 74 (1991) 219–224.

[3] Tan X., Z. Pang, Z. Gu, and S. Liu. Catalytic perovskite hollow fibre membrane reactors for methane oxidative coupling. *J. Memb. Sci.* 302 (2007) 109–114.

[4] Bayraktar D., F. Clemens, S. Diethelm, T. Graule, J. V. Herle, and P. Holtappels. Production and properties of substituted $LaFeO_3$-perovskite tubular membranes for partial oxidation of methane to syngas. *J. Eur. Ceram. Soc.* 27 (2007) 2455–2461.

[5] Ding H., B. Lin, X. Liu, and G. Meng. High performance protonic ceramic membrane fuel cells (PCMFCs) with $Ba_{0.5}Sr_{0.5}Zn_{0.2}Fe_{0.8}O_{3-\delta}$ perovskite cathode. *Electrochem Comm.* 10 (2008) 1388–1391.

[6] Wang Y. H., X. Q. Liu, and G. Y. Meng. Preparation and properties of supported 100% titania ceramic membranes. *Mater. Res. Bull.* 43 (2008) 1480–1491.

[7] Ahmad A. L., M. R. Othman, and H. Mukhtar. H_2 separation from binary gas mixture using coated alumina–titania membrane by sol-gel technique at high-temperature region. *Int. J. Hydrogen Energy* 29 (2004) 817–828.

[8] Mohammadi M. R., D. J. Fray, and M. C. Cordero–Cabrera. Sensor performance of nanostructured TiO_2 thin films derived from particulate sol-gel route and polymeric fugitive agents. *Sens. Actuators B Chem.* 124 (2007) 74–83.

[9] Agoudjil N. and T. Benkacem. Synthesis of porous titanium dioxide membranes. *J. Desalinat.* 206 (2007) 531–537.

[10] Mohammadi, M. R., J. Fray, and A. Mohammadi. Sol-gel nanostructured titanium dioxide: controlling the crystal structure, crystallite size, phase transformation, packing and ordering. *Micro. Meso. Mat.* 112 (2008) 392–402.

[11] Tong J., W. Yang, R. Cai, B. Zhu, and L. Lin. Titanium-base perovskite-type mixed conducting ceramic membranes for oxygen permeation. *Mater. Lett.* 56 (2002) 958–962.

[12] Etchegoyen G., T. Chartier, and P. Del-Gallo. An architectural approach to the oxygen permeability of a $La_{0.6}Sr_{0.4}Fe_{0.9}Ga_{0.1}O_{3-\delta}$ perovskite membrane. *J. Eur. Ceram. Soc.* 26 (2006) 2807–2815.

[13] Zaspalis V. T., W. Van Praag, K. Keizer, J. R. H. Ross, and A. J. Burggraaf. Synthesis and characterization of primary alumina, titania and binary membranes. *J. Mater. Sci.* 27 (1992) 1023–1035.

[14] Ahmad A. L., N. F. Idrus, and M. R. Othman. Preparation of perovskite alumina ceramic membrane using sol-gel method. *J. Membr. Sci.* 262 (2005) 129–137.

[15] Luyten J., J. Cooymans, C. Smolders, S. Vercauteren, E. F. Vansant, and R. Leysen. Shaping of multilayer ceramic membranes by dip coating. *J. Eur. Cer. Soc.* 17 (1996) 273–279.

[16] Winardi S., R. R. Mukti, K. N. P. Kumar, J. Wang, W. Wunderlich, and T. Okubo. Critical nuclei size, initial particle size and packing effect on the phase stability of sol-peptization-gel-derived nanostructured titania. *Langmuir* 26(7) (2010) 4567–4571.

[17] Bae D. S., K. S. Han, and S. H. Choi. Preparation and thermal stability of doped TiO2 composite membranes by the sol-gel process. *Solid State Ionics* 109 (1998) 239–245.

[18] Mohammadi M. R., M. C. Cordero-Cabrera, D. J. Fray, and M. Ghorbani. Preparation of high surface area titania (TiO_2) films and powders using particulate sol-gel route aided polymeric fugitive agents. *Sens. Actuators B Chem.* 120 (2006) 86–95.

[19] Kermanpur A., E. Ghassemali, and S. Salemizadeh. Synthesis and characterization of microporous titania membranes by dip-coating of anodised alumina substrates using sol-gel method. *J. Alloys Comp.* 461 (2008) 331–335.

[20] Mohammadi M. R. and D. J. Fray. Low-temperature perovskite type cadmium titanate thin films derived from a simple particulate sol-gel process. *Acta Mater.* 57 (2009) 1049–1059.

[21] Hao H., L. Zhao, J. Hu, X. Hu, and H. Hou. Oxygen adsorption/desorption behavior of $YBaCo_4O_{7+\delta}$ and its application to oxygen removal from nitrogen. *J. Rare Earths* 27(5)(2009) 815–818.

PART III

APPLICATIONS IN WATER TREATMENT

10

FOULING CHARACTERISTICS AND CLEANING STRATEGIES OF A PVDF TUBULAR ULTRAFILTRATION MEMBRANE IN NATURAL RUBBER SKIM LATEX CONCENTRATION PROCESS

DEVARAJ VEERASAMY AND ZAIROSSANI MOHD NOR

10.1 INTRODUCTION

10.1.1 Natural Rubber Latex (NRL) as an Industrial Raw Material

NRL exudes from the natural rubber (NR) tree when it is tapped and is known as field latex. The concentrated latex can be obtained by increasing the dry rubber content (DRC) of the field latex from ca. 30% to ca. 60%. It serves as the main raw material for the NRL product manufacturing factories that produce examination gloves, condoms, and latex thread. Currently, three methods of concentration are being carried out, which include centrifugation, creaming, and evaporation. Centrifugation is the preferred method and accounts for some 95% of the total latex concentrate produced in Malaysia [1]. Centrifugation produces natural rubber skim latex (NRSL) as a by-product. Skim latex processing involves the recovery of about 4–5% of dry rubber by cheap grade sulfuric acid coagulation. Once the skim rubber is recovered, the effluent consists of sulfuric acid-contaminated serum with a high biological

Sustainable Membrane Technology for Energy, Water, and Environment, First Edition.
Edited by Ahmad Fauzi Ismail and Takeshi Matsuura.

oxygen demand (BOD) value and discharged into the effluent ponds. Anaerobic ponds situated near the latex concentrate factory take the bulk of the effluent discharged from the skim latex processing activities. Microbial breakdown takes place below the scum layer of the anaerobic ponds, whereby sulfate ion originating from sulfuric acid is converted to poisonous hydrogen sulfide (H_2S gas). This pungent smelling H_2S gas is released once the scum layer is disturbed due to a heavy downpour, thus affecting the environment of the surrounding area and leading to a chain of unpleasant events such as malodor and health-related issues [2].

10.1.2 NR Field Latex Concentration by Membrane Separation

Recent study [3, 4] showed the use of membrane separation technology, in particular, ultrafiltration (UF), which could become an alternative method of concentrating NRL. By using UF, preserved NRL could be concentrated to produce concentrated latex where the DRC could be increased from 28% to 50% and to yield a latex-free serum as a by-product. UF could also eliminate the production of NRSL as a by-product. Concentration by centrifugation is the main source of the environmental-related problems currently faced by many latex concentrate factories [2, 4, 5].

10.1.3 Recovery of Skim Latex by UF

An immediate measure to solve the environmental problems faced by the latex concentrate factories is to recover the NRSL by concentration it via UF. Malaysian Rubber Board (MRB) has successfully carried out R&D work to concentrate NRSL that was obtained as a by-product via the centrifugation process, where the DRC was increased from 5% to about 30% (same as the DRC value of field latex) [6]. Latex free, clear serum which was not contaminated by any acid is obtained from this UF process. The serum has a great potential to be turned into fertilizers as a low-end product. On the higher end, it can be used to extract value-added biochemicals such as industrial protein and quebrachitol [7, 8]. The concentrated skim latex is being currently evaluated as a new value-added raw material. Therefore, UF has prevented the NRSL being turned into a waste latex and scrap rubber and in addition prevented the discharge of acid-tainted effluent into the anaerobic ponds. Instead, it has produced two new valuable raw materials [6].

10.1.4 Lack of Effective Membrane Cleaning Procedures for Good Flux Recovery

Recently there has been a great degree of interest in the development and utilization of membrane separation technology for improving indigenous processes such as NRL processing, waste NRL recovery, palm oil, and raw rubber processing mill effluent [4, 6]. Porous membranes, like UF and microfiltration (MF) membranes, offer alternative operational advantages such as lower driving force and smaller space requirement due to high packing density and cleaner processing tool which could be used to prevent wastage, recovery, and recycling waste raw materials [6–8]. However, membrane fouling is one of the most serious of all the problems in employing this technology. Decrease in performance of membrane separation due to fouling has hindered its widespread application to replace the current inefficient method [4, 5]. In fact, fouling has many adverse effects on the membrane system including flux decline, significant requirement for increase in transmembrane pressure (TMP), biodegradation of membrane materials, and system failure. Several studies had been carried out to enhance membrane separation in various industrial processes such as in the dairy, brewery, and food processing sectors to prevent or reduce membrane fouling as well as establishing suitable membrane cleaning procedures to bring about good flux recoveries so as to increase the economic life of membrane [9–15]. As for NRL processing and recovery of waste latex, very little studies were done so far in establishing a suitable membrane cleaning procedure to bring about good flux recovery after every concentration run. This is necessary to establish UF as a suitable alternative process for processing NRSL.

This study was focused on establishing an effective membrane cleaning procedure to bring about a high permeate flux recovery for the polyvinylidene difluoride (PVDF) membrane fouled during the NRSL concentration process. The cleaning procedure incorporated chemical (alkaline and acidic solutions) and physical (ultrasonic) applications, which were conducted at room temperature as well as at an elevated temperature of 50°C.

10.2 MATERIAL/THEORY

10.2.1 Original Source of Feed: NRL

The NRL is predominantly *cis*-1,4-polyisoprene. The rubber particles are usually spherical with diameter ranging from 0.02 to 3 μm. Only 5% of the particles have diameters larger than 0.4 μm. The weight average molecular weight (M_w) of natural rubber ranges from 300,000 to about 10 million Dalton. A random blend would have an M_w of about 2×10^6 and a number average weight (M_n) of 5×10^5 [16]. Composition of NRL: DRC 30–35%; proteinous substance 1.4%; neutral lipids 1.0%; phospholipids 0.6%; ash 0.5%; inositol and carbohydrates 1.6%; other nitrogenous compounds 0.3%; water 60% [17].

10.2.2 Feed Material—NRSL

Large volumes of NRSL are produced as a by-product in the concentration process of NRL by centrifugation; 10–15% of the rubber which enters the centrifuge effluxes as skim latex. The DRC of skim varies from 2.5 to 10%. The ratio of nonrubber solids to rubber is very high, around 1/1. The nonrubber constituents are mainly proteinous and other nitrogenous substances [15, 16]. Typical composition of skim latex is rubber hydrocarbon: 4.5–5.5%; protein: 3.0–3.5%; carbohydrates: 2.5–3.0%; other nitrogenous substances: 1–1.5%; metallic ions: <1%, and the rest is water. NRSL has a narrow particle size distribution ranging from 0.2 to 0.6 μm with the mean value of 0.5 μm. The smaller rubber particles of NRSL could contribute to membrane fouling together with NR proteins which have a molecular weight ranging from 5 to 50 kD [4, 6] and could be the other major membrane fouling component.

10.2.3 Preservation System

NRSL obtained from the centrifugation process carries along the remnants of the preservation chemicals from the previous process. Chemical analysis is needed to be done to determine its chemical composition before a topping up of preservation chemicals before commencement of the NRSL concentration [4, 6].

10.2.4 UF System

The UF system developed to concentrate NRSL is as shown in Figure 10.1. This is an in-house designed system which was incorporated with a tubular cross-flow membrane of 200 kD molecular weight cutoff (MWCO) of PVDF type, manufactured by PCI Membranes, UK. It consisted of two tubular membranes connected in series with each tube measuring 0.4 m in length and 10 mm of internal diameter with an effective membrane area of 0.024 m^2. The membranes were mounted onto a stainless steel module (PCI, Model MICRO

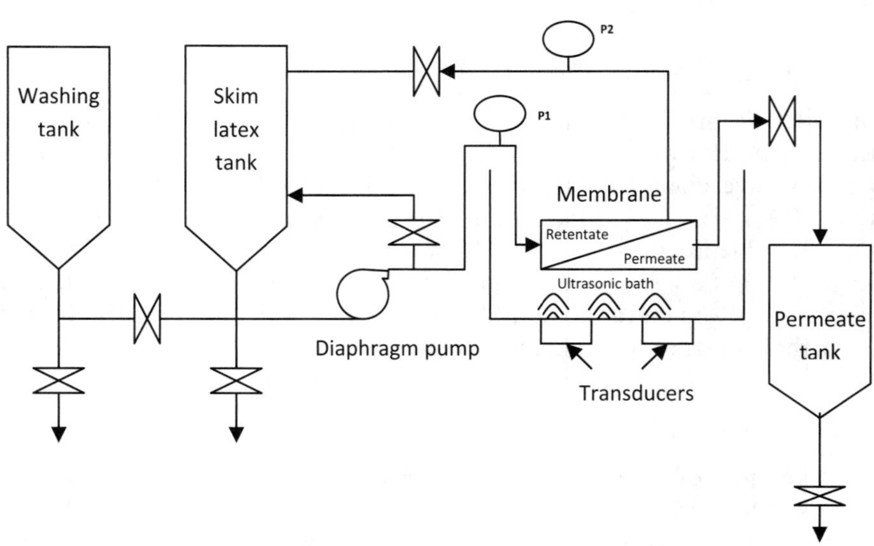

Figure 10.1. Schematic diagram of experimental rig.

240). According to manufacturer's specification, the membranes need to be utilized at temperatures below 80°C and a TMP not exceeding 10 bar and ideally at a pH of between 1.5 and 12. This UF system has an ultrasonic bath (Model SON SC—420, Sonicor) incorporated where the membrane module with the membranes intact would be completely immersed and could hold seven gallons of water to enhance the effect of ultrasonication on the module. The ultrasonic bath is built with a piezoelectric ultrasonic transducer at nominal power of 400 W output for an emission of 45 kHz frequency ultrasound.

10.2.5 NRSL Concentration Run

Before the PVDF membrane was used for the concentration of NRSL, the preservation chemicals coated on the inner surface of the membrane was cleaned by following the membrane manufacture's set procedure. Deionized water flux test (DIWFT) was carried out to the new membrane, and the intrinsic membrane resistance (R_m) was determined [4]. Twenty kilograms of preserved NRSL was fed into the feed tank after obtaining its initial DRC and total solid content (TSC) values. The feed pump was turned on to circulate the feed for about 10 minutes. The TMP value was set at 2 bar. The system was left to stabilize for 5 minutes before commencing of the experiment. From then on, the reading of permeate was taken every hour with the TMP maintained at the set value to determine the value of permeate flux with the last reading taken at the end of 10th hour when the membrane deemed as fully fouled. It has become customary after repeated runs with the fixed set of hydrodynamic conditions the 200 kD tubular PVDF membrane would become fully fouled after 10 hours of concentration run. Then the feed pump was stopped and the retentate (NRSL concentrate) was drained into another container. Three hundred milliliters of the concentrated NRSL was taken to conduct DRC and TSC determinations accurately.

10.2.6 Membrane Cleaning Cycles

The following cleaning cycles were performed to evaluate the cleaning efficiency and to determine flux recovery: (1) Deionized (DI) water replaced the NRSL, and the applied pressure was released to remove CP. The fouled membrane was then rinsed (the flow rate was increased) with DI water so that the gel layer (highly concentrated latex layer) was removed from the membrane surface. Rinsing was performed until rinsed DI water became clear. (2) Filtration with 0.1 M NaOH and (3) filtration with 0.1 M HNO_3. (4) These three steps were repeated at an elevated temperature of 50°C and (5) finally with an elevated temperature of 50°C with ultra sonication.

We used 0.1 M NaOH, which has a pH of 12 and is known to possess the maximum cleaning efficiency to remove the organic foulants, especially proteins found in the feed, which would easily foul the membrane during the concentration process. Similarly 0.1 M HNO_3 is ideal enough to remove scales and inorganic contaminants that foul the membrane [9–15].

All the cleaning filtrations were performed for 10 minutes each with the system operating with no TMP, TMP of 1 bar, and finally with 2 bar taking a total time of 30 minutes for each type of cleaning. The retentate and permeate were collected separately and analyzed for total organic carbon (TOC) and metallic elements. DIWFTs were carried out during the beginning and also at the end of each mode of cleaning process to determine its flux recovery. Subsequently, DIWFT was also carried out at the end of completing all the modes of cleaning cycle to determine the overall flux recovery.

10.2.7 Fouled Membrane and Autopsy Procedure

A replicate (second set of membranes) of this process was carried out using a different set of membranes. Rinsing was performed until rinsed DI water was clear. This set of membranes was deemed fully fouled and were dismantled from the module for carrying out the membrane autopsy procedures. Membrane autopsy is a destructive procedure requiring a sacrificial membrane element to be removed from the plant for destructive analyses. Membranes are cut open and their surfaces are scientifically examined by conducting chemical and physical analyses to determine the type of foulant present and its effect on the membrane performance. Analytical techniques such as Fourier transform infrared attenuated total reflection (ATR/FTIR) spectroscopy and scanning electron microscope were used to enhance the membrane autopsy procedure.

10.2.8 Microscopic Observation of Fouled Membrane

PVDF tubular membrane was sliced through to reveal the inner filtering matrix of the asymmetric membrane. Segments of the membranes were cut off at various places to be observed through an scanning electron microscope. Scanning electron microscope low vacuum model: FEI Quanta 200 field emission scanning electron microscopy (FESEM) with back-scattered electron (BSE) detector was used to study the surfaces of the clean and the fouled membrane. There was no sample preparation needed for this system, just the correct magnification and the spot on the membrane to do the scanning.

10.2.9 TOC, Atomic Absorption Spectrometer, and ATR/FTIR Analyses

TOC analysis was carried out on the chemically and physically extracted solution in each washing using TOC Analyzer (TOC-VCSH/CSN Shimadzu, Japan) and the metallic compositions by an AAS (Analytikjena Contra AA 300). The ATR/FTIR analyses were carried out using Thermo Nicolet Instrument Corporation (Madison, WI). The clean and fouled membranes were dried at room temperature for 3 days before ATR/FTIR analyses were performed with a resolution of 2.0 cm^{-1} and the results averaging 100 scans. ATR/FTIR analyses were used to determine the functional groups of foulants on the membrane surface as well to determine the membrane cleaning efficiency.

10.2.10 DRC and TSC Determinations

The DRC is defined as the percentage by weight of latex, which is coagulated by acetic acid under closely defined conditions. TSC is the percentage by weight of latex that is nonvolatile at a definite temperature. The difference between percentage TSC and percentage DRC gives a measure of the percentage nonrubber content (protein and carbohydrate). The NRSL concentration efficiency is determined by the gain in DRC and TSC with processing time. For this study, TSC value determination is used as the basis due to its ease of determination [4–6].

10.3 RESULTS AND DISCUSSION

10.3.1 PVDF Membrane Fouling Characteristics

A typical flux decline profile measured during NRSL concentration is shown in Figure 10.2. The average starting flux was 36.25 L/m^2/h during the first hour of the concentration process and from then on there was a gradual declination (except for the initial abrupt drop of 3.75 L/m^2/h) until the seventh hour. The declination became more gradual until it reached the lowest flux value of 7.5 L/m^2/h at the end of the 10th hour. TSC of NRSL rose from 5.02% to 27.8%. The concentration process was carried out at a constant TMP of 2 bar well below the recommended critical TMP of 2.5 bar [4–6]. This was important

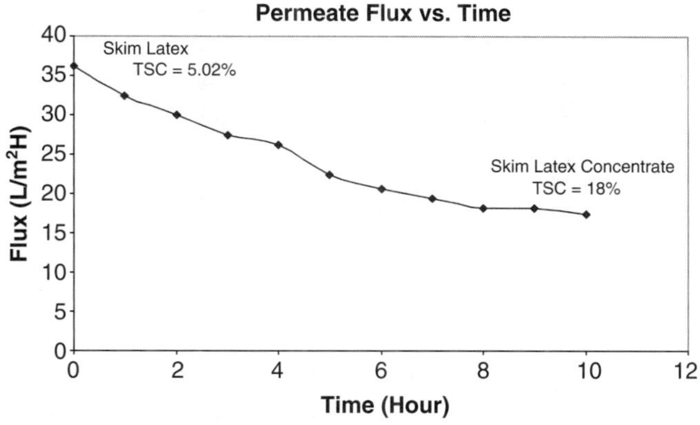

Figure 10.2. Variation in flux during NRSL concentration process at constant TMP.

because the low TMP gave a higher feed flow rate (>200 mL/s) and caused more shearing action on the membrane surface to prevent the formation of concentration polarization (CP). For NR latex filtration, a high TMP is not recommended because membrane compaction coupled with high CP could cause a faster decline in flux [5, 6].

10.3.2 Cleaning Cycles on the Fouled Membrane (After 10 Hours of NRSL Concentration)

The membrane cleaning cycle was designed based on the work by Shorrock and Bird [14, 15]: the combination 0.1 M of NaOH followed by 0.1 M of HNO_3 after rinsing the system with DI water once the NRSL concentration was completed. It was reported that fouling resistance was reduced more rapidly at temperatures of 50°C and 60°C [14, 15]. It is important to do the alkaline cleaning first after the rinsing of the system with DI water once NRSL filtration had been completed. The alkaline solution could prevent the fouled rubber particle from getting coagulated in the pores of the membrane, as would have happened if the acid cleaning was carried out first. The coagulated rubber particles in the membrane would have enhanced the permanent fouling of the membrane.

The organic foulants in the fouled membrane refers to the protein and rubber particles that had fouled the membrane and were extracted by different modes of membrane cleaning. After comparing all the different modes, NaOH had the highest extraction efficiency in all the three modes as shown in Figure 10.3. The TOC extraction by 0.1 M NaOH at room temperature was 52%; at 50°C it was 79%; at 50°C with sonication it was 78%, respectively. PVDF

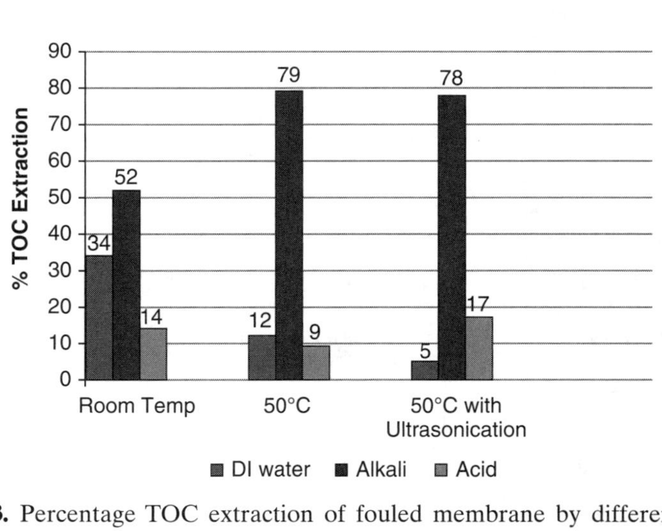

Figure 10.3. Percentage TOC extraction of fouled membrane by different cleaning methods.

TABLE 10.1. Composition of Chemically Extracted Solution from Fouled Membrane

Type of Cation	Room Temperature			50°C			50°C with Sonication		
	DI	NaOH	HNO_3	DI	NaOH	HNO_3	DI	NaOH	HNO_3
Fe^{3+} (mg/L)	0.7	0.8	8.6	0.7	0.6	2.2	1.8	2.1	10.2
Mg^{2+} (mg/L)	ud	ud	ud	ud	ud	ud	ud	ud	ud
Zn^{2+} (mg/L)	ud	6.0	9.8	ud	ud	2.3	0.1	ud	3.3

ud, undetectable.

membrane is highly hydrophobic, but it was still preferred for this study based on its high chemical resistance. PVDF membrane would be fouled less by protein compared to polyethersulfone (PES) and poly sulfone (PS) membranes as reported by Cheryan [18]. The elevated temperature improved TOC extraction with NaOH by 27% compared to its efficiency at the room temperature. NaOH still maintained its high TOC extraction efficiency with sonication at 50°C, although considerable amount of TOC foulant had already been extracted after many cycles of cleaning. The TOC extraction by HNO_3 at 50°C with sonication showed an increase of 8% compared to without sonication. This is because at higher temperatures, the acidic cleaning solution enhanced the hydrolysis of the fouled rubber protein from the membrane by converting back to amino acids, which would be still loosely attached to the membrane. With the application of ultrasonication, they were flushed out and got extracted [17, 19, 20].

Table 10.1 shows the AAS readings of metallic ions by different modes of washings (only Fe^{3+}, Zn^{2+}, and Mg^2 known to be present in NRSL). HNO_3 enabled the extraction of more metallic ions compared to NaOH, and DI water performed poorly in all the modes of washings. Of the total cations extracted, acid washing accounted for 36.4 mg/L, alkaline washing, 9.5 mg/L, and for DI water washing, it was only 3.3 mg/L. This was because Fe^{3+}, Mg^{2+}, and Zn^{2+} ions are more soluble in HNO_3 compared to NaOH. Utilization of sonication enabled the extraction of a total of 17.5 mg/L of cations from the three washing media and modes, which was 35% of the total cations extracted. The foulants containing Zn^{2+} ions had originated from the original NR latex preservation chemicals of ammonia and zinc oxide/tetra methyl thiuram disulfide (ZnO/TMTD). The Fe^{3+} ions might have originated from the gradual erosion of the centrifuges by the highly alkaline latex and also by the water contamination by rust, due to poor housekeeping. The Mg^{2+} ions might have originated from the groundwater used by the factory. Except for Zn^{2+} and Mg^{2+} (slightly soluble), Fe^{3+} formed insoluble complexes with ammonia and NaOH and might have got rejected by the membrane and subsequently might have been fouled by the membrane.

TABLE 10.2. Sequence of Membrane Cleaning Process

Sequence	Type of Cleaning Procedures
1	Cleaning with DI water at room temperature
2	Cleaning with 0.1 M NaOH water at room temperature
3	Cleaning with 0.1 M HNO_3 water at room temperature
4	Cleaning with DI water at 50°C
5	Cleaning with 0.1 M NaOH at 50°C
6	Cleaning with 0.1 M HNO_3 at 50°C
7	Cleaning with DI water at 50°C with ultrasonication
8	Cleaning with 0.1 M NaOH at 50°C with ultrasonication
9	Cleaning with 0.1 M HNO_3 at 50°C with ultrasonication

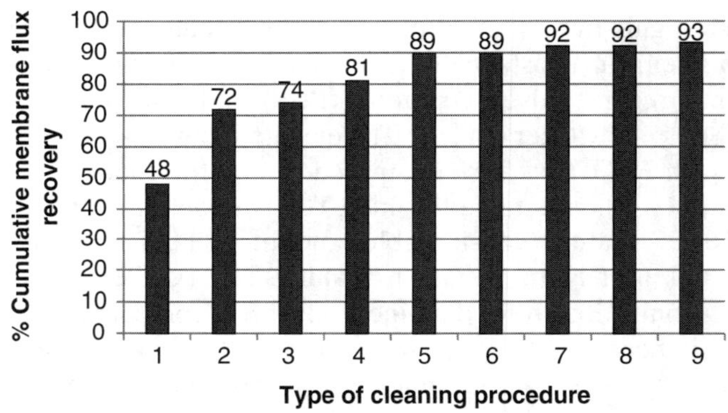

Figure 10.4. Percentage membrane flux recovery after different cleaning processes.

10.3.3 Flux Recoveries after the Cleaning Cycles

The flux of the cleaned membrane was measured through filtration with DI water after each membrane cleaning process. The various cleaning processes were conducted as shown in Table 10.2. The ratio of the flux (L/m^2/h) at room temperature of the cleaned membrane to the new membrane flux (L/m^2/h) was used to evaluate the flux recovery, and the results are as shown in Figure 10.4. The flux recovery obtained after cleaning with DI water at room temperature of the fouled membrane was only 48%. The hydraulic cleaning was found capable of removing the outmost and loose fouling layer of the fouled membrane surface. The cumulative flux recoveries obtained after cleaning with 0.1 M NaOH and 0.1 M HNO_3 solution were 72 and 74%, respectively. Alkaline cleaning helped to obtain an increase of 24% of flux recovery from the initial DI water cleaning, as compared to that of the acid cleaning, which brought an additional increase of only 2%. Alkaline cleaning enabled the removal of organic foulants such as the NR particles, protein, and other nitrogenous compounds. Almost all the foulants were organic in nature, which could be deduced from the mass balance. The inorganic foulant that could be removed by acid washings include metallic ion complexes such as Zn and Fe, which would have originated from the feed as contaminants. The only organic foulant, which could be cleaned sparingly by acid washings, include foulants which were derived from latex proteins. With the elevated temperature cleaning at 50°C, using DI water and 0.1 NaOH brought about additional flux recoveries of 7 and 8%, respectively. There was no increase in membrane flux recovery after cleaning with 0.1 M HNO_3 at 50°C. Nevertheless, there were increases of 3 and 1% after cleaning with DI water and 0.1 M HNO_3, respectively, but there was no further increase in flux obtained after cleaning with 0.1 NaOH at 50°C with ultrasonication. This brought a cumulative flux recovery of 93% at the end of the full membrane cleaning procedures. The elevated temperature opened up the fouled membrane pores, and the sonication provided the additional force to flush out probably the weak irreversible foulants (weak adsorbents), which caused an increase of 4% in cumulative flux recoveries due to the application of ultrasonication [17–21].

10.3.4 Spectral Analysis of UF Fouled Membrane

Figure 10.5 shows the superimposition of the ATR/FTIR spectrums of PVDF membranes: the fouled (top), fouled and cleaned (middle), and new (bottom). When a membrane was fouled, the FTIR peaks of the clean membrane were

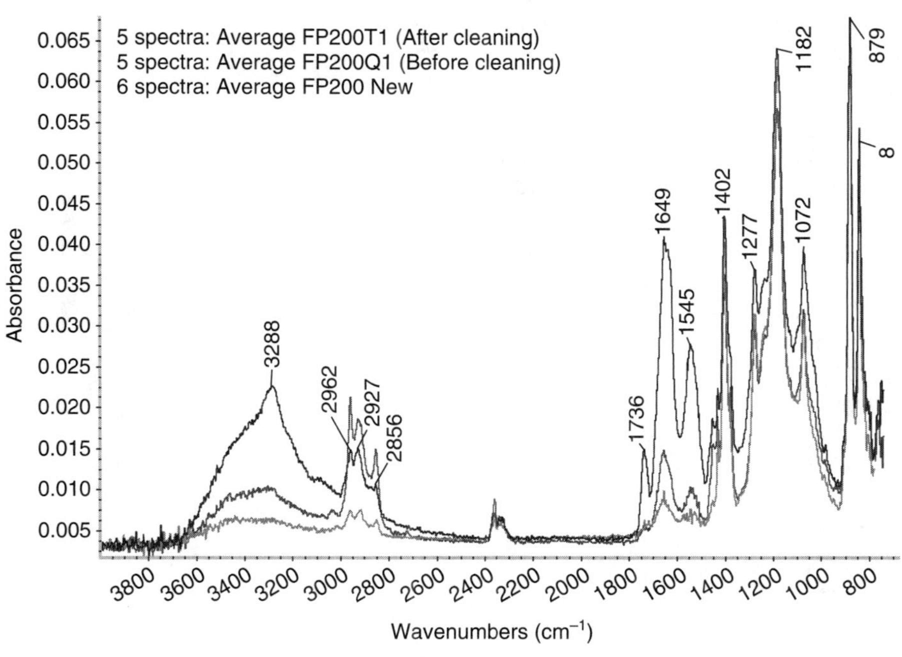

Figure 10.5. ATR/FTIR spectra of new membrane; fouled but not cleaned membrane; fouled and cleaned.

changed in absorbance intensity, indicating the covering of the original clean surface by functional groups of foulants. The fouled membrane had distinct peaks at wavenumbers 3288, 1649, and 1545. The wavenumber 3288 indicates the presence of the following: free-OH aromatics; bonded—OH; free NH; amino acids; bonded NH; NH_2. The wave numbers 1619 and 1515 indicates the presence of amino acids and amine. Majority of these foulants were derivatives of rubber proteins and amino acids. These proteins were derived from latex proteins and have their M_w ranging from 5 kD to 50 kD [4, 19, 20]. The presence of NR as foulants can be seen from the high peaks of the fouled membrane spectrum at wave numbers from 800 to 950. The almost identical peak heights of the spectrums of the fouled and cleaned and the new unused cleaned membrane showed that the cleaning methods were effective enough to remove most of the foulants.

Figure 10.6 is the ATR/FTIR spectrum of components of NRSL that fouled a PVDF membrane (the PVDF membrane spectrum has been digitally subtracted; the negative peaks from wave number 700 to 1450 were due to oversubtraction). Wave numbers of 1647, 1540, 1447, and 1357 were the foulant peaks of the amino acids and amine from rubber protein. These peaks were very marginal in the subtraction spectra of (fouled and cleaned membrane—new membrane [top]) compared to (fouled membrane—new membrane [bottom]). This is an indication of the efficiency of the membrane cleaning cycles which had eradicated most of the protein and NR foulants. This is confirmed by the absence of NR at its fingerprint from 800 to 950 cm^{-1} which were at the oversubtracted regions [20–23].

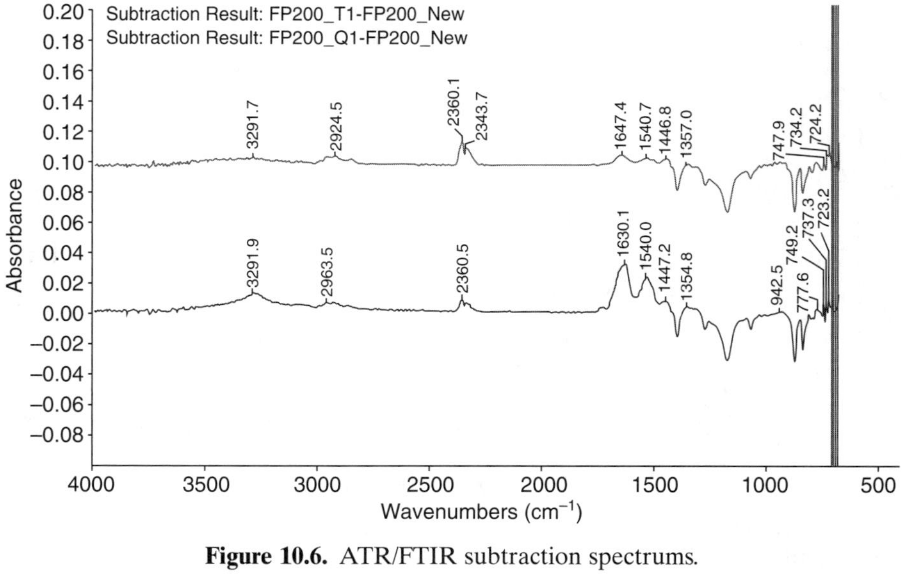

Figure 10.6. ATR/FTIR subtraction spectrums.

10.3.5 Scanning Electron Microscope Analysis of the Membrane

Figure 10.7a–c were scanning electron microscope images of inner surface of 200 kD PVDF tubular membranes of unused, fouled, and fouled and cleaned membrane with 15,000 magnifications, respectively. From the scanning electron microscope image of unused membrane surface, the unblocked pores could be seen very clearly compared to the fouled membrane surface which was completely covered with scales of cake layers seen with some white spots. The scales could be the rejected rubber particles and rubber proteins formed on the PVDF membrane which constitutes the formation of the gel layer. Figure 10.7c is the scanning electron microscope photograph of the membrane which had gone through the full cycles of membrane cleaning procedure (DI water, alkaline, and acid) at room temperature and at 50°C, coupled with

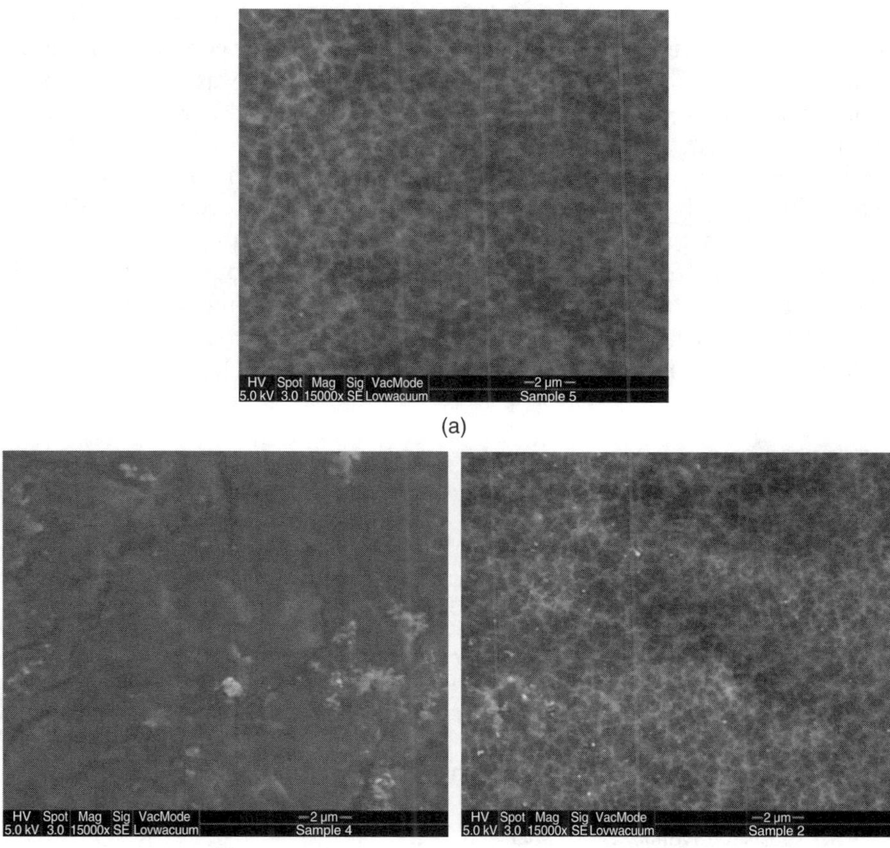

Figure 10.7. SEM micrograph image of (a) new, (b) fouled, and (c) fouled and cleaned PVDF membrane with 16,000 times magnification.

ultrasonication at 50°C. These membrane cleaning procedures were able to remove most of the scales and pore blocks but could not remove completely all the scales, which can be seen as scattered white spots. This could be part of the irreversible foulants which could probably account for the remaining 7% of the unrecovered membrane flux recovery.

10.4 CONCLUSION

The fouled 200 kD PVDF membrane from 10 hours of concentration run with NRSL concentration was made to undergo three cycles of membrane cleaning procedures (DI water, alkali, and acid) at room temperature, at 50°C, and finally ultrasonication at temperature 50°C. The membrane autopsy revealed that the fouling on the interior surface of the membrane was mainly due to latex particles and the protein derivatives of NRSL. Major portions of these two types of foulants, which make up the organic foulants, were effectively removed by alkaline cleaning (sodium hydroxide) whereas acid cleaning (nitric acid) removed inorganic fouling caused by cationic contaminants in the feed as well as from water used in the factory (the feed was obtained from a commercial source). Three modes of membrane cleaning (DI water, alkali, and acid) at room temperature brought a flux recovery of only 74%. Subsequent cleaning at 50°C followed by cleaning with the application of ultrasonication at 50°C brought about a total cumulative membrane flux recovery of 93%. The foulants still intact in the membrane pores could be mainly of the strong irreversible type. The remnants of weak and major portions of strong irreversible foulants could be further removed by cleaning using a higher concentration of NaOH and longer durations of ultrasonic application at 50°C or slightly higher (60°C) temperature.*

REFERENCES

[1] Cheng S. F. Types, composition, properties, storage and handling of natural rubber latex concentrates. Notes on Natural Rubber Examination Glove Manufacture, Rubber Research Institute of Malaysia. Limited Circulation, 1988, pp. 1–12.

[2] Ahmad I., S. Sethu, M. K. Zin, and I. Zaid. Anaerobic/facultative ponding system for treatment of latex concentrate effluent. Proceedings of Rubber Research Institute Malaysia Planters Conference, Kuala Lumpur, 1979.

[3] Devaraj V. and M. N. Zairossani. The use of membrane separation technology to achieve environment friendliness in natural rubber processing. Proceedings of International Rubber Conference, Yokohama, Japan, October 24–28th, 2005.

[4] Devaraj V. Concentration of natural rubber field latex by ultrafiltration, M. Eng. Sc. Thesis. University of Malaya, 2004.

[5] Devaraj V., N. S. Meriam, J. Nambiar, and A. Yusof. Environmentally friendly natural rubber latex concentration via membrane separation technology. Proceed-

* Similar data has appeared in *Desalination* 2011 (in press) during the production of this book.

ings of The 5th International Membrane Science and Technology Conference, University of New South Wales, Sydney, Australia, November 10–14, 2003.

[6] Devaraj V. and M. N. Zairossani. Alternative and environment friendly skim latex processing for value-added products recovery. Proceedings of Malaysian Rubber Board Latex Seminar, Kuala Lumpur, Malaysia, May 18, 2006, 2007.

[7] Lau C. M. Quebrachitol—an additional role for Hevea latex, rubber developments. Malaysian Rubber Research and Development Board, Kuala Lumpur, vol. 49. 1996, pp. 11–13.

[8] Lau C. M. Concentration of natural rubber serum by reverse osmosis. *Journal of Natural Rubber Research* 9(4) (1994) 226–241.

[9] Kobayashi K., X. Chai, and N. Fujii. Ultrasound enhanced cross-flow membrane filtration. *Separation and Purification Technology* 17 (1999) 31–40.

[10] Muthukumaran S., S. Kentish, S. Lalchandani, M. Ashokkumar, R. Mawson, G. W. Stevens, and F. Gieser. The optimization of ultrasonic cleaning procedures for dairy fouled ultrafiltration membranes. *Ultrasonics Sonochemistry* 12 (2005) 29–35.

[11] Muthukumaran S., K. Yang, A. Seuren, S. Kentish, M. Asokkumar, G. W. Stevens, and F. Gieser. The use of ultrasonic cleaning for ultrafiltration membranes, in the dairy industries. *Separation and Purification Technology* 39 (2004) 99–107.

[12] Mo L. and X. Huang. Fouling characteristics and cleaning strategies in a coagulation-microfiltration combination process for water purification. *Desalination* 159 (2003) 1–9.

[13] Li J., R. D. Sanderson, and E. P. Jacobs. Ultrasonic cleaning of nylon microfiltration membranes fouled by Kraft paper mill effluent. *Trans I. Chem. E.* Vol. 6 and Part C: 30–38.

[14] Shorrock C. J. and M. R. Bird. Membrane cleaning: chemically enhanced removal of deposits formed during yeast cell harvesting. *Trans I. Chem. E.* Vol. 6 and Part C: 30–38.

[15] Bird M. R. and M. Bartlett. Measuring and modeling flux recovery during the chemical cleaning of MF membranes for the processing of whey protein concentrate. *Journal of Food Processing Engineering* 53 (2002) 143–152.

[16] Blackley D. C. *Polymer Lattices: Science and Technology Volume 1: Fundamental Principles*, 2nd ed. Chapman and Hall, London, 1997.

[17] Blackley D. C. *Polymer Lattices: Science and Technology Volume 2: Types of Lattices*, 2nd ed. Chapman and Hall, London, 1997.

[18] Cheryan M. *Ultrafiltration and Microfiltration Hand Book*, 2nd ed. Technomic Publishing Co., Lancaster, CA, 1998.

[19] Zeman L. J. and A. L. Zydney. *Microfiltration and Ultrafiltration Principles and Applications*. Marcel Dekker Inc., New York, 1996.

[20] Devaraj V., A. F. Ismail, and M. N. Zairossani. Fouling characteristics and autopsy of a PVDF tubular ultrafiltration membrane in natural rubber skim latex (waste latex) recovery. Proceedings of The 6th International Membrane Science and Technology Conference, University of New South Wales, Sydney, Australia, November 5–9, 2007.

[21] Devaraj V., A. F. Ismail, and M. N. Zairossani. ATR/FTIR spectrometry study on fouling of a PVDF tubular ultrafiltration membrane in natural rubber skim latex recovery. Proceedings of The Regional Membrane Science and Technology Conference, Phuket, Thailand, August 13–15, 2008.

[22] Cho C., G. Amy, and J. Pellegrino. Membrane filtration of natural organic matter: factors and mechanism affecting rejection and flux decline with charged ultrafiltration (UF) membranes. *Journal of Membrane Science* 164 (2000) 89–110.

[23] Sakinah A. M. M., A. F. Ismail, M. I. Rosli, and O. Hasan. Fouling characteristics and autopsy of PES ultrafiltration membrane in cyclodextrins separation. *Desalination* 207 (2007) 227–242.

11

REMOVAL OF DIETHANOLAMINE (DEA) FROM WASTEWATER USING MEMBRANE SEPARATION PROCESSES

Binyam Seyoum, Hilmi Mukhtar, and Kok Keong Lau

11.1 INTRODUCTION

The preservation of our environment is a very important and pressing topic, particularly when we deal with energy issues. There are a few types of fossil fuels that are widely used for energy production, including petroleum, coal, and natural gas. Natural gas, the most flexible of all primary fossil fuels, is the fastest growing energy source in the world [1]. It is a mixture of hydrocarbon (HC) gases consisting predominantly of methane (CH_4). Although methane is always the main component, natural gas may also include ethane, ethylene, propane, butane, pentane, and higher HCs as well as some impurities particularly hydrogen sulfide (H_2S) and carbon dioxide (CO_2) [2]. Natural gas is considered as an environmentally friendly clean fuel, offering important environmental benefits when compared to other fossil fuels. The superior environmental qualities over coal or oil are that commercialized natural gas is practically sulfur free and thus, it produces virtually no sulfur dioxide (SO_2) or that the level of nitrous oxide (NO_x) and CO_2 emissions is lower [1]. Moreover, combustion of natural gas releases virtually no ash or particulate matter. Thus, increased use of natural gas could help to reduce or combat problems of acid rain, ozone layer, or greenhouse gases. However, natural gas must be

Sustainable Membrane Technology for Energy, Water, and Environment, First Edition.
Edited by Ahmad Fauzi Ismail and Takeshi Matsuura.

purified in order to remove impurities, particularly CO_2 and H_2S down to pipeline quality before it can be used for domestic and industrial purposes.

There are many acid gas-treating processes available for removal of H_2S and CO_2 from natural gas. These processes include absorption processes using chemical or physical solvents including amines, potassium carbonate, and selexol [3]; hybrid solution using mixed physical and chemical solvents such as sulfinol [4]; adsorption processes using activated carbon; and physical separation using membrane separation process [5]. Among the available technologies, amine-based sweetening processes are the most prominent and have been the process of choice for removal of H_2S and CO_2 from sour natural gas for meeting the pipeline quality [6, 7]. Amine solutions provide flexibility, low cost, and high reliability to industries seeking a proven acid gas removal technology. Amine absorption is also the technology of choice for low pressure acid gas removal applications. However, during amine absorption–desorption process, a small amount of amine is carried over and discharged into the effluent stream. This carryover is usually caused by foaming or excessive gas velocities in the absorption tower, leakage due to spills or corrosion and during process turnaround. Others are due to vaporization from the absorber and the overhead condenser [8].

Amines containing wastewater is generally characterized by high chemical oxygen demand (COD) typically about 17,000 mg/L [9]. Thus, the wastewater from amine sweetening plants cannot be directly discharged to the surrounding water; otherwise, it can deplete dissolved oxygen in receiving waters, stimulate aquatic plant growth, exhibit toxicity toward aquatic life, present a public health hazard, and affect the suitability of wastewater for reuse purposes [10]. Treatment of amine wastewater using existing wastewater treatment plant (WWTP) without any dilution is very challenging since it can affect the performance of the activated sludge. However, dilution increases the volume of the wastewater and requires extension of the existing WWTP. In addition, the slow degradation rate and disposal requirement of excess sludge are the other drawbacks of existing WWTP.

Several studies, including adsorption, advanced oxidation, and biological treatment methods have been conducted in order to remove or mineralize amines in wastewater. Fox [11] has proposed the use of activated carbon as an efficient adsorbent for removal of nitrogen containing compounds such as ammonia and amine. The findings of the study showed that activated carbon had poor adsorptivity for removal of amines from wastewater [12]. In addition, the adsorption decreases with increase in pH and resulted in very poor adsorption capacity at pH > 9.0 [13]. Fritz and Schluender [14] have proposed silica gel adsorbent to overcome the drawback of poor adsorptivity of activated carbon or active clay. However, when a highly concentrated amine containing wastewater at a level of 20,000 ppm was treated with silica gel, the removal ratio was at a level of 57%. Fox [11] also proposed silica-titania and silica-titania-magnesia gel as highly activated adsorbents. However, it was found that removal of only 58% at best against a 100 ppm amine containing wastewater was achieved.

Advanced oxidation processes (AOPs), especially the photo-Fenton process, are considered to be highly promising for the remediation of highly contaminated wastewater and are consequently proposed for degradation of amines [15]. As compared to other conventional oxidant species, the photo-Fenton process is capable of completely oxidizing (mineralize) even the less reactive pollutants. However, organic compound oxidation by photo-Fenton process would be inhibited by the presence of inorganic species such as phosphate, sulfate, chloride, and carbonate ions [16]. Moreover, the main disadvantage of AOPs is the operational cost associated with their high electrical energy input (UV radiation generation) and expensive chemical demand (H_2O_2, O_3, etc.). In fact, studies show that only wastewaters with relatively small concentration (COD $\leq$ 5000 ppm) can be economically treated with these technologies [15].

Biological systems are effective methods to economically remove large quantities of biodegradable wastewater organics, converting them to water and CO_2. However, the use of biological processes to treat amine contaminated wastewater is limited to low concentration [17]. High COD amine wastewater can harm the bacteria populations and slow down their activity, hence may take several weeks for complete degradation. Aerobic degradation of methyldiethanolamine in a batch system was studied [18]. The findings showed that the degradation did not succeed within standardized batch experiments, whereas experiments with flow through reactors resulted in a removal of about 96% within 28 days. Moreover, all biological processes produce excess sludge that may have to be further reduced and stabilized before disposal [10]. Therefore, the development of a technology which is suitable for effective removal of amines in the high COD amine wastewater is required.

Membranes, on the other hand, provide several advantages that make them attractive for wastewater treatment. They are widely used in water purification industry or WWTP to remove suspended and dissolved solids, heavy metals, and other kinds of impurities from water streams. The uses of reverse osmosis (RO) and nanofiltration (NF) have also been expanded to include a wide variety of wastewater treatment applications such as natural and synthetic organic pollutants, petrochemicals, and pesticides in the process of drinking water production [19]. Therefore, it is the objective of the current study to evaluate the flux and rejection characteristics of aqueous diethanolamine (DEA) solutions across commercialized AFC99 (RO), AFC40 (NF), and CA202 (ultrafiltration [UF]) membranes. Thus, to fulfill this objective, membrane separation processes have been proposed as an alternative for the conventional WWTP under the present study.

11.2 MATERIALS AND METHODS

11.2.1 Materials

The membranes used in the experimental study covered three classes of membrane processes including RO, NF, and UF. They are tubular membranes and commercially known as AFC99, AFC40, and CA202, respectively, each having an internal diameter of 12.5 mm, length of 1.2 m, and effective membrane surface area of 0.05 m^2. AFC99 and AFC40 membranes are thin film composite membranes with a polyamide skin layer, whereas CA202 is made from cellulose acetate. Deionized water and DEA were used for the preparation of the artificial wastewater under the present. The range of concentrations of the feed synthetic wastewater was varied between 5000 and 15,000 mg/L. In addition, hydrochloric acid (36%) was used in order to adjust the pH of the artificial wastewater and consequently investigate the effect of pH on the membrane performance. Hach UV-Spectrophotometer was used in order to measure the concentration of the feed, retentate, and permeate samples from each membrane module.

11.2.2 Methods

The experiment was carried out using a membrane test unit, which is capable of testing four different tubular membranes simultaneously. Before the water flux study was carried out, the membranes were stabilized at 25 bar in order to offset membrane compaction, which may lead to reduction in permeability during the experiment.

The pure water permeation was studied using deionized water under different operating pressures. The range of operating pressure was varied between 4 and 24 bars for each membrane under the study. The experiments to study the permeate fluxes were carried out at different operating conditions: (1) feed concentrations (5000, 10,000, and 15,000 mg/L); (2) cross-flow velocities (1.5, 3, 4.5, and 6 L/min); (3) feed pHs (3 and 8); and (4) operating pressures (4, 8, 12, 16, 20, and 24 bar). The feed temperature was maintained constant at 25±1°C for all experiments. The system was operated in batch circulation mode. Consequently, both permeate and retentate streams were returned back to the feed vessel (except during sample collection for analyses) in order to maintain constant bulk concentration. Sufficient time was given (usually an hour) for the process to reach steady state before collecting the permeate sample for every change in the operating conditions. The volume of permeate collected versus time was recorded online from each membrane module, simultaneously for data analysis.

Both the permeate flux and the observed rejection R_o of the membranes were calculated, respectively, as given by the equations below:

$$J_v = \frac{Q_p}{A_m} \tag{11.1}$$

$$R_o = \frac{C_b - C_p}{C_b} \tag{11.2}$$

TABLE 11.1. Characteristics of Artificial Amines Wastewater in Terms of COD and pH

Concentration (mg/L)	1500	3500	6000	9000	12,000	15,000
COD(mg/L)	2355	5425	9215	13,825	18,430	23,440
pH	10.62	10.80	10.91	10.99	11.05	11.09

where J_v (L/m^2·h) is the permeate flux, Q_P (L/h) is the total flow rate for solvent (water) and solute through the membrane, A_m (m^2) is the surface area of the membrane, C_b (mg/L) and C_p (mg/L) are the bulk and permeate concentrations, respectively.

11.3 RESULTS AND DISCUSSION

11.3.1 Wastewater Characterization

Table 11.1 shows the characteristic of artificial amine wastewater in terms of COD and pH. Both properties form the critical characteristic of the artificial amine wastewater. Results showed that as the concentration of amines increased, the COD has also increased as shown in Table 11.1.

This can be ascribed to the presence of more pollutants in the artificial wastewater [17, 18]. The higher amount of pollutants in the wastewater results in consumption of more dissolved oxygen during the oxidation reaction. Hence, the oxygen demand required to oxidize the wastewater increases with increase in concentration.

11.3.2 Water Flux Study

Figure 11.1 shows the water flux against operating pressure for the three membranes under study. Results show that the water flux increases linearly with respect to the operating pressure. It is understood that the increment of the operating pressure has increased the driving force, hence increasing the permeate flux. These results also show that there are no signs of membrane compaction that occurred within the range of operating pressure.

It was found that the water flux of CA202 (UF) membrane was the highest, followed by AFC40 (NF) and AFC99 (RO) membranes, respectively. This is attributed to the difference in the pore sizes of the membranes. Studies show that membranes with lowest pore size like RO would give the highest resistance for the solvent (water) transportation across the membrane [20]. Under the present study, CA202 has the largest pore size as compared with very tight pore size for AFC40 or nonpore for AFC99 membranes. Hence, CA202 exhibited the highest amount of water flux.

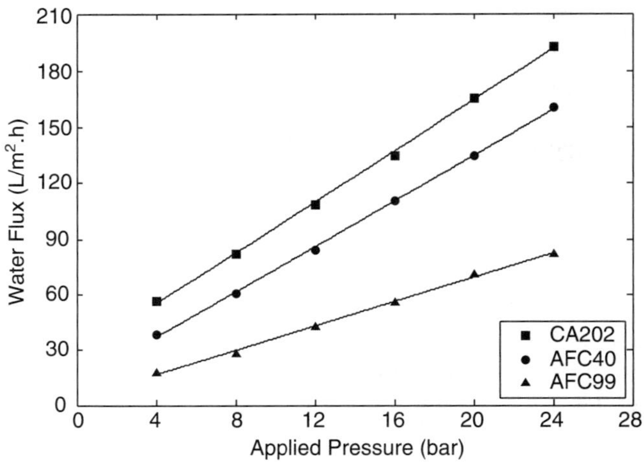

Figure 11.1. Effect of operating pressure on pure water flux across different types of membranes.

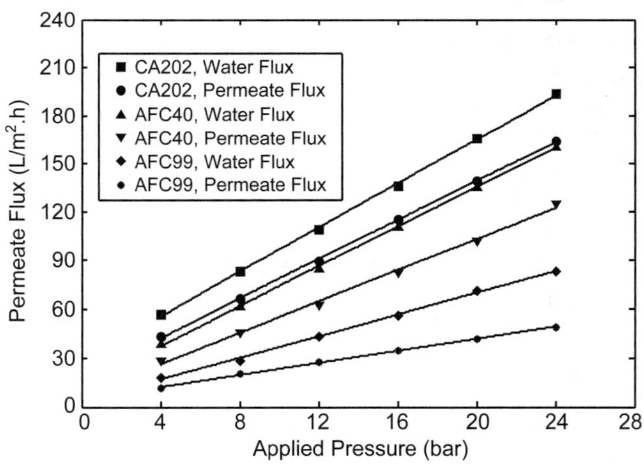

Figure 11.2. Effect of operating pressure on DEA solution and water flux across various membranes (C_b = 5000 mg/L, u = 6 L/min, and pH = 8).

11.3.3 Permeate Flux Study

11.3.3.1 Effect of Operating Pressure on Permeate Flux. Figure 11.2 shows the permeate flux for AFC99 (RO), AFC40 (NF), and CA202 (UF) membranes under different operating pressure conditions for DEA solution. As expected, the figures show that the permeate flux increases linearly with increases in operating pressure for all the membranes. As explained earlier, increasing the operating pressure would increase the net pressure as well, hence increasing the permeate flux.

As discussed earlier in Section 3.2, the same sequential trend was also found for the permeate flux of the three membranes. It was also found that the permeate flux of the amine solution for all membranes was lower as compared with pure water flux. This is due to the presence of osmotic pressure in the amine solution which reduces the net driving pressure, hence reducing the permeate flux. In addition, the presence of the solute on the membrane surface increases the overall membrane resistance due to concentration polarization, lowering the permeate flux [20].

11.3.3.2 Effect of Cross-Flow Velocity on Permeate Flux. Figures 11.3–11.5 show the effect of cross-flow velocity on permeate flux of DEA solution across AFC99 (RO), AFC40 (NF), and CA202 (UF) membranes, respectively. The results show that the permeate flux increases with an increase in cross-flow velocity in the range of operating conditions. The results were contributed by the increment of the boundary layer mass transfer under higher tangential velocity, which sweeps away the retained solutes on the membrane surface. Thus, the performance of the membrane can be improved [21, 22].

According to the present study, the AFC99 and AFC40 membranes reject more solutes; as a result, they can develop significant concentration polarization layer as compared to CA202 membrane [21]. Hence, the increase in the cross-flow velocity can give considerable flux improvement for AFC99 followed by AFC40 and CA202 membranes, respectively, as shown on Figures 11.3–11.5. Under higher cross-flow velocity, the permeate fluxes of DEA solutions increase by 29.65%, 20.61%, and 9.66% across the AFC99, AFC40, and CA202 membranes, respectively.

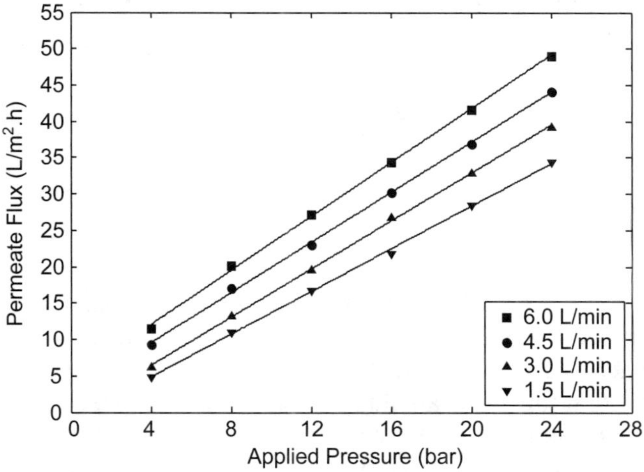

Figure 11.3. Effect of cross-flow velocity on the permeate flux across AFC99 membrane (C_b = 5000 mg/L and pH = 8).

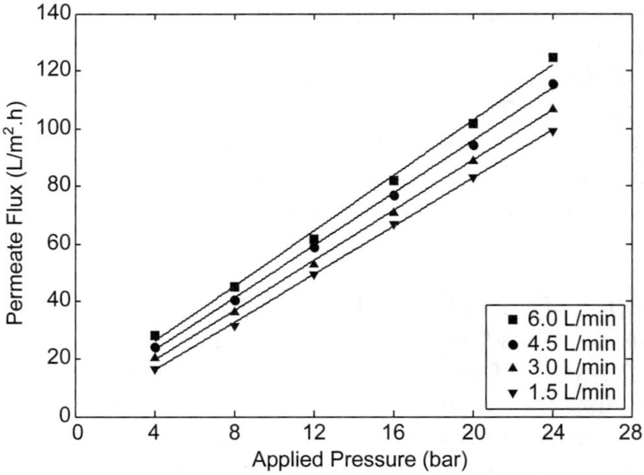

Figure 11.4. Effect of cross-flow velocity on the permeate flux across AFC40 membrane ($C_b = 5000$ mg/L and pH = 8).

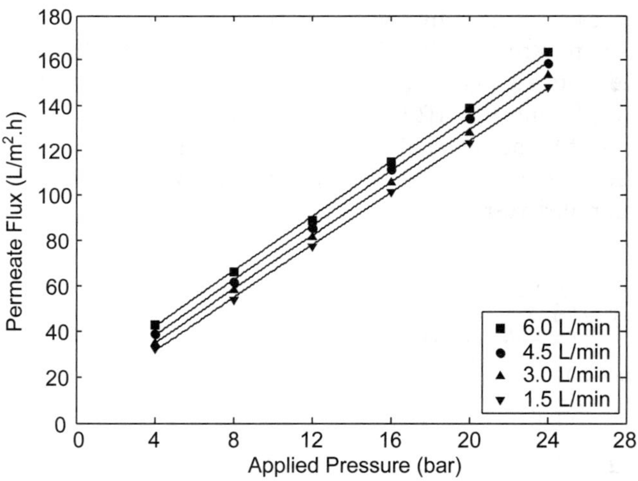

Figure 11.5. Effect of cross-flow velocity on the permeate flux across CA202 membrane ($C_b = 5000$ mg/L and pH = 8).

11.3.3.3 Effect of Feed Concentration on Permeate Flux. Figure 11.6 shows the effect of feed concentration on permeate flux under different pressure differences across AFC99 (RO), AFC40 (NF), and CA202 (UF) membranes. Results show that the permeate flux decreases as the concentration of the feed increases. Increasing the feed concentration can effectively increase the osmotic pressure in the solution as well as the overall membrane resistance. As a result, it reduces the net driving pressure, resulting in reduction of the permeate flux [20].

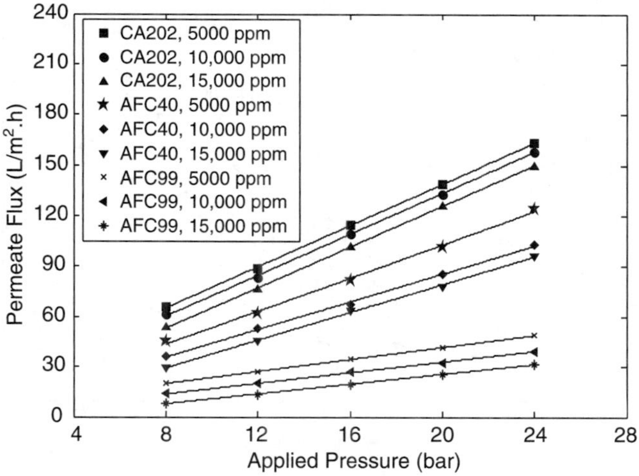

Figure 11.6. Effect of feed concentration on the permeate flux across various membranes (u = 6 L/min and pH = 8).

TABLE 11.2. Percentage Reduction of Permeate Fluxes of the Amines

	% Reduction of DEA Permeate Flux		
Change in Concentration	Across AFC99	Across AFC40	Across CA202
Water as compared to 5000 mg/L feed solution	40.53	22.28	15.29
Water as compared to 15,000 mg/L feed solution	61.18	39.86	22.34
5000 mg/L as compared to 15,000 mg/L feed solution	34.72	22.61	8.32

Table 11.2 summarizes the percentage reduction of permeate flux of DEA solutions against concentration and type of membrane, in comparison with water flux. The table shows that the permeate flux reduction is highest for AFC99 (reverses osmosis) followed by AFC40 (NF) and CA202 (UF), respectively.

11.3.3.4 Effect of pH on Permeate Flux. Figure 11.7 shows permeate flux of DEA solution versus operating pressure at pH 3 and 8. It is observed that the permeate flux increased for AFC99 membrane, whereas it decreased for AFC40 and CA202 membranes as the pH decreased from 8 to 3.

The effect of pH on membrane performance can be explained by the surface chemistry of the membranes such as the presence of dissociable functional groups, the degree of their dissociability, and the orientation of the

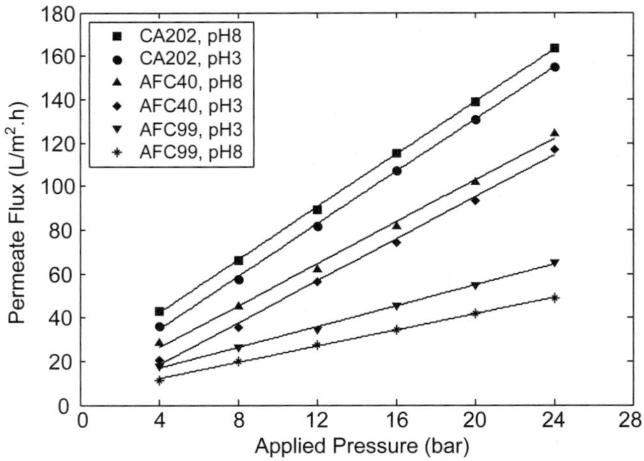

Figure 11.7. Effect of feed pH on the permeate flux across various membranes (C_b = 5000 mg/L and u = 6 L/min).

functional groups [23]. Several researchers discussed the presence of excess carboxylic and amine functional groups at the surface of thin film composite polyamide membranes, which give them either positive or negative surface charge depending on the pH of the surrounding medium [24, 25]. The positive surface charge is due to the protonation of the amine functional groups in strongly acidic medium, and the negative charge is due to deprotonation of the carboxylic groups in alkaline medium. In both cases, the electrostatic repulsion between the charged groups would cause an increase in pore size or free volumes between the polymer chains of the membranes, resulting in permeate flux increase [24]. However, the orientation of the functional groups affected the response of the membranes when the pH varies as can be seen from Figure 11.7. This difference in permeate flux characteristics of AFC99 and AFC40 membranes could be attributed to the different types of monomers and solvents used during fabrication of the membranes [26]. Ghosh et al. [27] also discussed that selecting the organic solvent is also critical for morphology of the thin film since it governs the solubility and diffusivity of the amine monomer in the reaction zone during the interfacial polymerization. In addition, AFC99 and AFC40 membranes also have different hydrophilicities, which could result in different permeability behavior of the membranes as provided by the manufacturer [28].

11.3.4 Rejection Study

11.3.4.1 Effect of Operating Pressure on Observed Rejection. Figure 11.8 shows the effect of operating pressure on observed rejection of DEA across AFC99 (RO), AFC40 (NF), and CA202 (UF) membranes. The observed

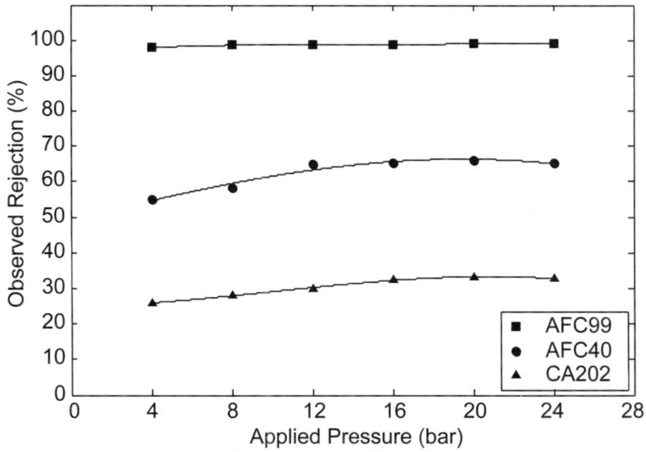

Figure 11.8. Effect of operating pressure on observed rejection of DEA across various membranes (C_b = 5000 mg/L, u = 6 L/min and pH = 8).

rejection of the amine across the membranes was found to increase as the operating pressure increases.

Generally, in pressure-driven membrane processes, the water flux is proportional to the operating pressure, but the solute flux is independent of operating pressure [20]. Thus, as the operating pressure increases, the solute passage is increasingly overcome as solvent (water) is pushed through the membrane at a faster rate than solute can be transported. This means that the membrane becomes more selective as the pressure increases, hence the observed rejection increases.

The findings show that AFC99 membrane is the most effective membrane followed by AFC40 and CA202 membranes, respectively. Generally, AFC99 membrane has achieved the highest rejection efficiency (above 96.5%) for DEA under the present study. The observed rejection of the amine across AFC40 and CA202 membranes was also found to increase with operating pressure and reaches a plateau region at higher operating pressure. The AFC40 and CA202 membranes exhibited the highest rejection efficiency up to 65.9 and 33.3%, respectively. The variation in rejection of the membranes is due to the difference in the membranes pore size. AFC99 membrane has a tight or poreless structure and hence, the mechanism of solute transport is only due to diffusion. AFC40 and CA202, on the other hand, have relatively lose and porous structure, hence there would also be convection transport of solutes through the membrane pores, which results in higher permeate concentration and lower rejection efficiency [20, 22].

11.3.4.2 Effect of Cross-Flow Velocity on Observed Rejection. Figures 11.9–11.11 show the effect of cross-flow velocity on observed rejection for DEA solutions across AFC99 (RO), AFC40 (NF), and CA202 (UF)

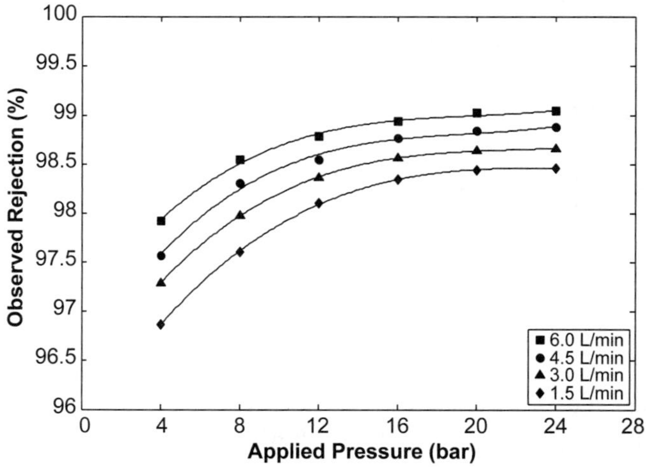

Figure 11.9. Effect of cross-flow velocity on observed rejection of DEA across AFC99 membrane (C_b = 5000 ppm and pH = 8).

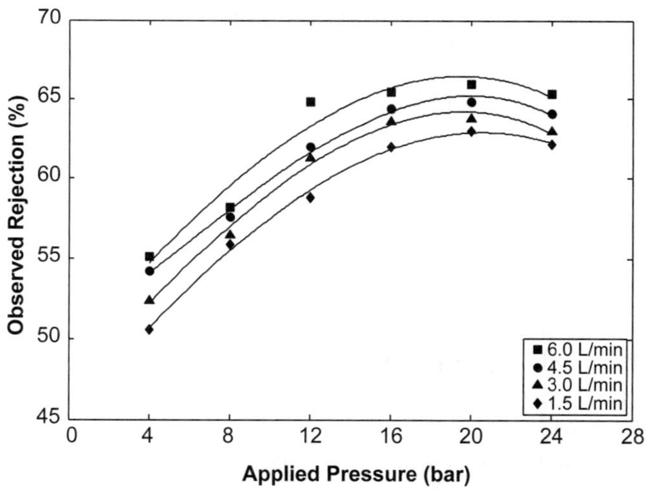

Figure 11.10. Effect of cross-flow velocity on observed rejection of DEA across AFC40 membrane (C_b = 5000 ppm and pH = 8).

membranes. The results show that the observed rejection increases with an increase in cross-flow velocity for all the membranes under the study. Cross-flow velocity is one of the hydrodynamic factors that affect the accumulation of solutes on the membrane surface and the subsequent formation of concentration polarization which affects the membrane performance negatively. However, a higher cross-flow velocity increases the mass transfer coefficient and subsequently minimizes the effect of concentration polarization, resulting in higher observed rejection of the solutes by the membranes [22].

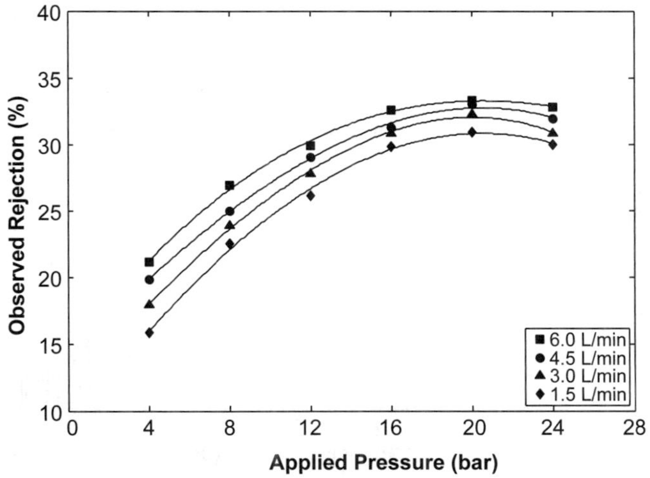

Figure 11.11. Effect of cross-flow velocity on observed rejection of DEA across CA202 membrane (C_b = 5000 ppm and pH = 8).

11.3.4.3 Effect of Feed Concentration on Observed Rejection. Figures 11.12–11.14 show the effect of feed concentration on the observed rejection DEA solution across AFC99, AFC40, and CA202 membranes. The findings show that the observed rejection of the amines decreases as the feed concentration increases. This is because solute flux across the membranes increases with increases in feed concentration due to the increment of concentration gradient of the solutes across the membranes. It was discussed by many researchers that observed rejection increases with increasing water flux due to the dilution effect, which occurs when water flux is increased without a subsequent increase in solute flux [20, 29]. However, as feed concentration increases, so does osmotic pressure which offsets the feed driving pressure. Therefore, if feed pressure remains constant, higher feed concentration results in lower membrane water flux. As the water flux declines, the increase in solute passage through the membrane due to increased feed concentration decreases the observed rejection.

11.3.4.4 Effect of pH on Observed Rejection. Figures 11.15–11.17 show the effect of feed pH on observed rejection of DEA solutions across AFC99, AFC40, and CA202 membranes. The results show that the observed rejection increases as the pH of the feed decreases. The chemistry of amines is dominated by the lone pair electrons on nitrogen atom. Because of this lone pair, amines are basic and can be readily protonated by strong acids like HCl. Thus, at low pH, DEA gets protonated and acts as a positively charged solute resulting in electrostatic repulsion with the positively charged membrane at pH 3 [30, 31].On the other hand, at pH 8, the membrane is negatively charged, but DEA will have no charge. This is because amines are weak bases and

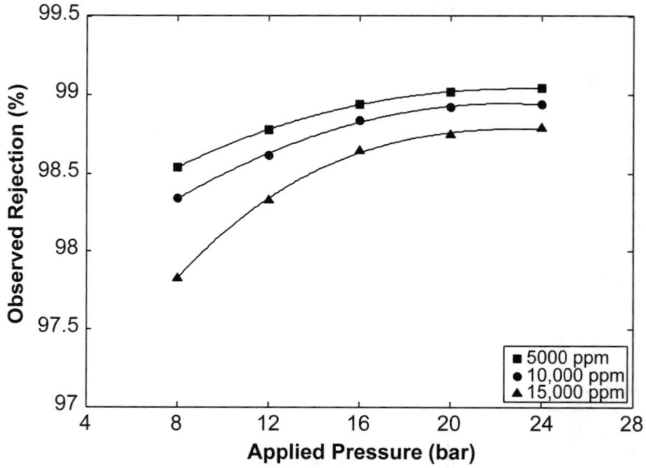

Figure 11.12. Effect of feed concentration on observed rejection of DEA across AFC99 membrane (C_b = 5000 ppm and pH = 8).

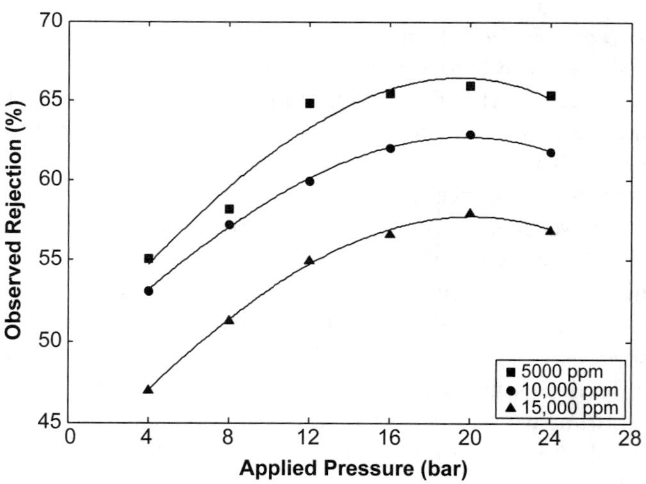

Figure 11.13. Effect of cross-flow velocity on observed rejection of DEA across AFC40 membrane (C_b = 5000 ppm and pH = 8).

consequently do not readily protonate at high pH values [32]. Hence, there is no electrostatic repulsion working between DEA and the membrane.

The finding shows that the rejection effect of the pH is more significant for AFC40 membrane compared to AFC99 and CA202 membranes. The investigation shows that the observed rejection of the membranes increases on average by 0.50%, 21%, and 11% for AFC99, AFC40, and CA202 membranes, respectively, when the pH decreases from 8 to 3. This shows that for AFC40 and CA202 membranes, the rejection mechanism depends not only on sieving

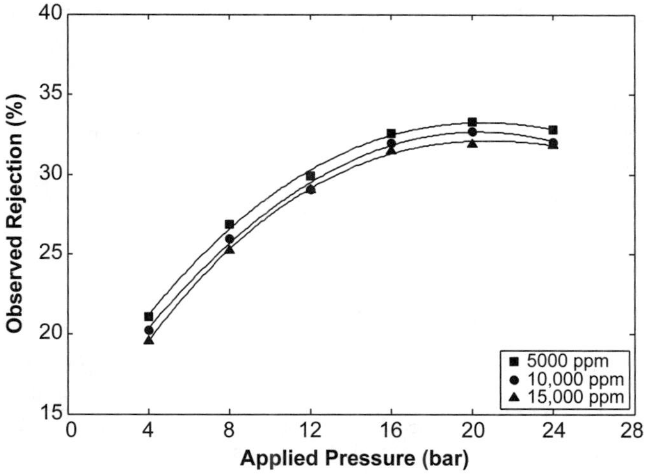

Figure 11.14. Effect of cross-flow velocity on observed rejection of DEA across CA202 membrane (C_b = 5000 ppm and pH = 8).

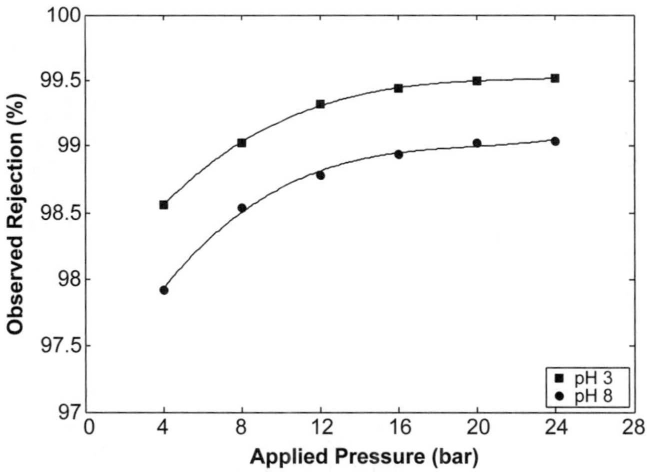

Figure 11.15. Effect of feed pH on observed rejection of DEA across AFC99 membrane (v = 6 L/min and C_b = 5000 ppm).

but also on the charge interaction between the feed solution and the membranes, whereas for AFC99 membrane, the rejection was almost due to sieving.

11.4 CONCLUSION

The flux study shows that the permeate flux of the membranes increased linearly with an increase in operating pressure. In addition, the permeate flux of CA202 (UF) membrane was found to be the highest followed by AFC40 (NF)

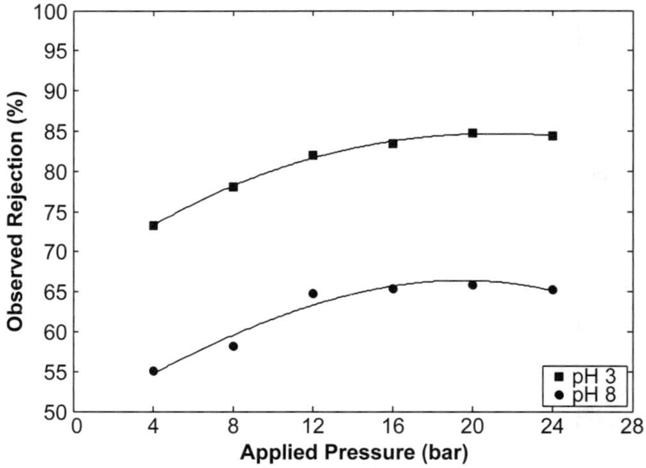

Figure 11.16. Effect of feed pH on observed rejection of DEA across AFC40 membrane (v = 6 L/min and C_b = 5000 ppm).

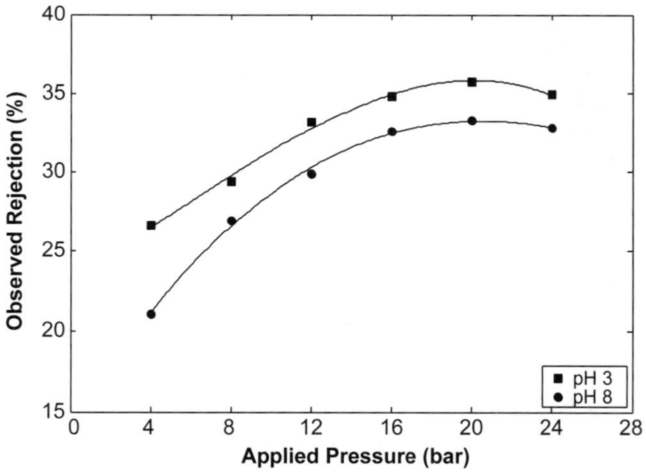

Figure 11.17. Effect of feed pH on observed rejection of DEA across CA202 membrane (v = 6 L/min and C_b = 5000 ppm).

and AFC99 (RO) membranes, respectively. Moreover, the permeate flux was found to be decreasing as the feed concentration increased due to the effect of concentration polarization and osmotic pressure. On the other hand, increasing cross-flow velocity could reduce the concentration polarization effect and subsequently increase the permeate flux. The rejection study by the membranes shows that the observed rejection of the amine has generally increased with an increase in operating pressure and cross-flow velocity, whereas it decreased with an increase in feed concentration. It was found that the rejection efficiency of AFC99 membrane was the highest followed by AFC40 and CA202 membranes, respectively. The experimental results showed that AFC99 membrane was able to reject more than 96.5% of the DEA under the present study. The highest rejection of DEA by AFC40 and CA202 was found to be 65.9% and 33.3%, respectively. The study also showed the permeate flux and observed rejection were affected by feed pH. When the feed pH decreased from 8 to 3, the permeate fluxes of AFC40 and CA202 membranes decreased, whereas the permeate flux of AFC99 membrane increased. On the other hand, the findings showed that the observed rejection of DEA increased when the feed pH decreased from 8 to 3 for all the three membranes under the present study. It was found that for AFC40 and CA202 membranes, the rejection mechanism depended not only on molecular sieving but also on the charge interaction between the feed solution and the membranes.

The overall results show that membrane separation process, especially AFC99 membrane, has excellent observed rejection behavior for removal of DEA from artificial wastewater. Similarly, NF and UF membranes such as AFC40 and CA202 can also be selected for the same purpose but both are more suitable to be employed for complementing the existing biological treatment.

REFERENCES

[1] Chandra V. *Fundamental of Natural Gas: An International Perspective*. PennWell, Tulsa, OK, 2006.

[2] William C. L. and J. P. Gary. *Standard Handbook of Petroleum and Natural Gas Engineering*, 2nd ed. Gulf Professional Publishing, New York, 2005, chapter 6, p. 371.

[3] Granite E. J. and T. O'Brien. Review of novel methods for carbon dioxide separation from flue and fuel gases. *Fuel Processing Technology* 86 (2005) 1423–1434.

[4] Kidnay A. J. and W. R. Parrish. *Fundamentals of Natural Gas Processing*. Taylor and Francis Group, CRC Press, Boca Raton, FL, 2006, pp. 91–133.

[5] Kohl A. and R. Nielsen. *Gas Purification*, 5th ed. Gulf Publishing, Houston, TX, 1997.

[6] Jou F. Y., F. D. Otto, and A. E. Mather. The solubility of mixtures of H_2S and CO_2 in an MDEA solution. *The Canadian Journal of Chemical Engineering* 75 (1997) 1138–1141.

[7] Sohbi B., M. Meakaff, M. Emtir, and M. Elgarni. The using of mixing amines in an industrial gas sweetening plant. *Proceedings of World Academy of Science. Engineering & Technology* 25 (2007) 1307–6884.

[8] Abry R. G. F. and R. S. DuPart. *Hydrocarbon Processing: Amine Plant Troubleshooting and Optimization*. GAS/SPEC Technology Group, Gulf Publishing Co., Houston, TX, 1995, pp. 1–11.

[9] Omar A. A., R. M. Ramil, and N. F. Putri. Fenton oxidation of natural gas plant wastewater. *Canadian Journal on Chemical Engineering & Technology* 1 (2010) 1–6.

[10] Hawthorne S. B., A. Kubätovä, J. R. Gallagher, J. A. Sorensen, and D. J. Miller. Persistence and biodegradation of monoethanolamine and 2-propanolamine at an abandoned industrial site. *Environmental Science & Technology* 39 (2005) 3639–3645.

[11] Fox C. R. Industrial wastewater control and recovery of organic chemicals, in F. L. Slejko (ed.), *Adsorption Technology*. Marcel Dekker, New York, 1985, pp. 167–185.

[12] Mitarai K., M. Fujii, I. Inoue, and S. Kumoi. Method for treating an amine containing wastewater. United States Patent 5039424, 1999, 1–11.

[13] Casey T. J. *Unit Treatment Processes in Water and Wastewater Engineering*. John Wiley and Sons Ltd., England, 1997, pp. 113–114.

[14] Fritz W. and E. U. Schluender. Simultaneous adsorption equilibria of organic solutes in dilute aqueous solutions on activated carbon. *Chemical Engineering Science* 29 (1974) 1279–1282.

[15] Pignatello J., E. Oliveros, and A. Mackay. Advanced oxidation processes for organic contaminant destruction based on the Fenton reaction and related chemistry. *Critical Reviews in Environmental Science & Technology* 36 (2006) 1–84.

[16] DeLaat J., L. G. Truong, and B. Legube. A comparative study of the effects of chloride, sulfate and nitrate ions on the rates of decomposition of H_2O_2 and organic compounds by Fe (II)/H_2O_2 and Fe (III)/H_2O_2. *Chemosphere* 55 (2004) 715–723.

[17] Isa A. A., B. Hassan, and A. Shafawi. Treatment of amine contaminated effluent water via membrane application. In Proceedings of the 18th Symposium of Malaysian Chemical Engineers, 2005, pp. 368–380.

[18] Furhacker M., A. Pressl, and R. Allabashi. Aerobic biodegradability of methyldiethanolamine (MDEA) used in natural gas sweetening plants in batch test and continuous flow experiments. *Chemosphere* 52 (2003) 1743–1748.

[19] Ozaki H. and H. Li. Rejection of organic compounds by ultra-low pressure reverse osmosis membrane. *Water Research* 36 (2002) 123–130.

[20] Baker R. W. *Membrane Technology and Applications*, 2nd ed. Membrane Technology and Research, Inc., Menlo Park, CA, 2004, pp. 15–66.

[21] Damak K., A. Ayadi, B. Zeghmati, and P. Schmitz. Concentration polarization in tubular membranes—a numerical approach. *Desalination* 171 (2004) 139–153.

[22] Jude S. and B. Jefferson. *Membrane for Industrial Wastewater Recovery and Re-Use*. Elsevier Science Ltd, Oxford, UK, 2003.

[23] Zydney A. L. and L. J. Zeman. *Microfiltration and Ultrafiltration: Principles and Applications*. CRC, New York, 1996.

[24] Manttari M., A. Pihlajamaki, and M. Nystrom. Effect of pH on hydrophilicity and charge and their effect on the filtration efficiency of NF membranes at different pH. *Journal of Membrane Science* 280 (2006) 311–320.

[25] Hurwitz G., G. R. Guillen, and E. M. V. Hoek. Probing polyamide membrane surface charge, zeta potential, wettability, and hydrophilicity with contact angle measurements. *Journal of Membrane Science* 349 (2010) 349–357.

[26] Freger V., A. Pihlajamaki, Y. Shabtai, and J. Gilron. Distribution of fixed charge functional groups in the polyamide composite membranes. In: Supplementary Book of Abstracts, ICOM'02, Toulouse, France, 2002, 144.

[27] Ghosh A. K., B. H. Jeong, X. Huang, and E. M. V. Hoek. Impacts of reaction and curing conditions on polyamide composite reverse osmosis membrane properties. *Journal of Membrane Science* 311 (2008) 34–45.

[28] *PCI Membranes, Membranes and Modules: A User Manual.* ITT PCI Membrane Ltd, Hampshire, UK, 2000.

[29] Murthy Z. V. P. and L. B. Chaudhari. Application of nanofiltration for the rejection of nickel ions from aqueous solutions and estimation of membrane transport parameters. *Journal of Hazardous Materials* 160 (2008) 70–77.

[30] Smith J. G. *Organic Chemistry*, 1st ed. McGraw-Hill, New York, 2006, pp. 892–941.

[31] Schafer A. I., A. G. Fane, and T. D. Waite. *Nanofiltration—Principles and Applications*. Elsevier Ltd., Oxford, UK, 2005, pp. 96–169.

[32] McMurny J. *Organic Chemistry*, 6th ed. Thomson Learning, Belmont, CA, 2004, pp. 892–941.

12

THE EFFECT OF CHITOSAN MEMBRANE PREPARATION PARAMETERS ON REMOVAL OF COPPER IONS

AZADEH GHAEE, MOJTABA SHARIATY-NIASSAR, AND JALAL BARZIN

12.1 INTRODUCTION

Heavy metal contamination exists in aqueous waste streams of many industries. The soils surrounding these industries are very polluted and pose a risk of metal contamination of groundwater and surface water, which causes a worldwide problem. Due to potentially unacceptable ecological risks to plants, animals, and microorganism, such contamination is hazardous and can be carcinogenic. On the other hand, many metals such as cadmium (Cd), chromium (Cr), lead (Pb), mercury (Hg), nickel (Ni), and copper (Cu) are playing important roles in most industries [1].

Among them copper is used in industries as engine moving parts, brake linings, metal plating, mining operations, electroplating, pharmaceuticals, fungicides, insecticides, etc. This metal is also considered as a key constituent to many enzymes that are responsible for catalytic oxidation–reduction reactions. It has an essential role in animal metabolism and enables the body to regulate its level of hemostatic ally, though large acute doses can be harmful and even lead to fatal effects. Short-term exposure to copper ions for an amount of about 1.3 mg/L causes stomach and intestinal problems, while long-time exposure can lead to a severe damage to liver, kidney, and brain, and further to

Sustainable Membrane Technology for Energy, Water, and Environment, First Edition.
Edited by Ahmad Fauzi Ismail and Takeshi Matsuura.

vomiting, cramps, convulsions, and even death. Moreover, it can also cause damages to a variety of aquatic fauna (fish and invertebrates) [1, 2].

It is understood that the heavy metals are not biodegradable and tend to accumulate in living organisms, leading to severe problems. Therefore, the treatments of such polluted streams are of considerable importance. Currently, the most popular processes for the treatment of metal contaminated waste streams are chemical precipitation, membrane processes, reverse-osmosis, and ion exchange. Cost-effective alternative technologies or sorbents for treatment of metal contaminated waste streams are also sought for. Toxic metal ions in trace quantities are difficult to remove from aqueous solutions. Adsorption is one of the few alternatives available for such situations [3].

When materials of biological origin, specifically living or dead microorganisms are involved, biosorption applies. Biosorbents are available either as large quantities in nature or as certain waste products of industrial units [3]. They are normally inexpensive materials. Among them, chitosan and chitin are both natural polysaccharide biopolymer sorbents. They represent renewable resources and are friendlier with environment than other materials.

Chitosan is used in the biomedical, cosmetic, food, and textile industries [4]. Its useful features include hydrophilicity, biocompatibility, biodegradability, and high affinity for biomacromolecules. It is worth noting that among many biopolymers, chitosan has the highest adsorption capacity for metals [5]. The growing need for new sources of low-cost adsorbent, the increased problems of waste disposal, the high cost of synthetic resins undoubtedly make chitosan as one of the most attractive known biopolymer material sorbents for wastewater treatment [6]. It is obtained by industrial alkaline deacetylation of chitin which is widely found in the exoskeleton of shellfish and crustaceans [6]. The chemical structure of chitosan is shown in Figure 12.1.

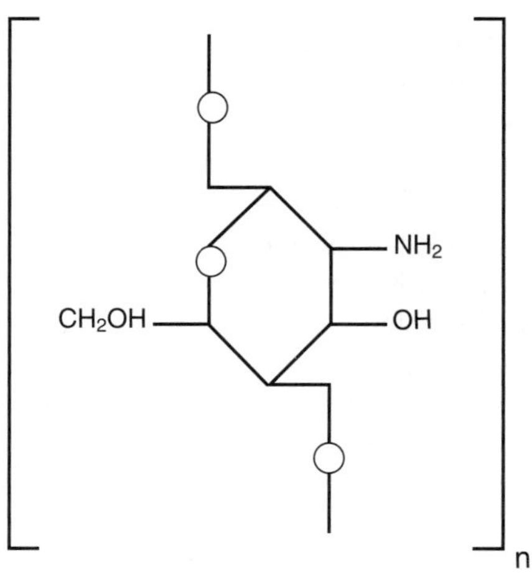

Figure 12.1. Chitosan structure.

Both nitrogen and oxygen atoms have lone pair(s) of electrons that can bind a proton or a metal ion through an electron pair sharing to form a complex. Because of the stronger attraction of the electrons to the nucleus in an oxygen atom than in a nitrogen atom, the nitrogen atoms would have a greater tendency to donate electrons for sharing with a metal ion to form a metal complex than the oxygen atoms [7]. For this reason, the amine groups are therefore easily protonated in acidic solutions. Hence, the protonation may cause electrostatic attraction of anionic compounds, including metal anions (resulting from metal chelation by chloride, anionic legends, etc.) [8]. Therefore, amino groups are mostly responsible for the uptake of metal ions such as copper. The mechanisms involved in chelation are

$$R-NH_2+H^+ \rightarrow R-NH_3^+. \tag{12.1}$$

$$R-NH_2+Cu^{2+} \rightarrow R-NH_2Cu^{2+}. \tag{12.2}$$

$$R-NH_3^+ +Cu^{2+} \rightarrow R-NH_2Cu^{2+}+H^+. \tag{12.3}$$

The industrial scale application of chitosan is difficult due to its handling. The use of its raw powders or flakes has drawbacks such as unsatisfied mechanical properties and solubility in acidic conditions [9]. Also characteristics like density, shape, and size can usually cause strong hydrodynamic limitations (i.e., head loss and column clogging) which may restrict the application of these systems in large-scale columns. In addition, high capital and energy costs of adsorption columns make the membrane processes more favorable for removal of metal ions [10].

Most of the studies dedicated to metal ion sorption on chitosan membranes are focused on proposing an equilibrium model to investigate sorption behavior of metal ions on chitosan membrane [11–13]. Other researches considered the influence of some adsorption process parameters like metal type, metal concentration, temperature, and pH on metal sorption in batch and continuous systems [10–17]. Kamiński et al. used chitosan membrane in a continuous process (cross flow and dead end) and investigated the influence of pressure on the membrane permeation flux and performance [14, 15].

However, only a few papers have discussed the influence of membrane preparation parameters on metal removal. Matsuyama et al. [18] investigated the effect of solvent type on membrane performance and found that the membrane prepared with acetic acid had the highest permeability compared to other solvents. Modrzejewska et al. [19] also studied such effects as well as that of rheological changes in the membrane-forming solution on the separation properties. For membrane preparation, most researchers used evaporation method [10–13, 16–18] except a few who prepared chitosan membrane by immersion-precipitation method [14, 15, 20].

The effects of membrane preparation conditions and the preparation method on metal sorption are discussed in this work. Depending on the membrane preparation parameters such as concentrations of polymer, concentrations of chemicals required in solvent and nonsolvent, the morphology of membrane changes and, consequently, adsorption rate, varies too.

12.2 EXPERIMENTS

12.2.1 Materials

Chitosan was purchased from Chitotech with deacetylation degree of 90. All other chemicals (glutaraldehyde, acetic acid, sodium hydroxide, cupric sulfate) were of analytical grade used without further purification. The solutions were prepared using distilled water.

12.2.2 Membrane Preparation

Membranes were prepared by dissolving various amounts of chitosan (i.e., 2.5% and 7% w/w) in acetic acid solutions of different concentrations as solvent (i.e., 5% and 8% v/v). Two methods were applied:

For the immersion-precipitation method, the solution was poured onto a flat surface after deaeration. A proper membrane thickness was formed by using a 2-mm slot applicator. Immediately after the solution was poured (without solvent evaporation), the process of coagulation was carried out in aqueous solutions of different sodium hydroxide (i.e., 1 and 1.5 M) as nonsolvent for 24 hours, after which the membranes were washed with distilled water.

For the evaporation method, the solution was poured on a petri dish to the same thickness as above, and was dried at room temperature until the weight of the membrane became constant. Then the membrane was immersed in the aqueous solution of sodium hydroxide (either 1 or 1.5 M) for 6 hours, followed by washing with distilled water. Other drying methods were further attempted. For this purpose, 7% w/w chitosan solution (5% v/v acetic acid solution) was poured on a petri dish, and the membrane was prepared by one of the following methods:

1. Dried at 50°C until the weight became constant.
2. Dried in an oven with gradual temperature increase between 20 and 50°C until the weight became constant.
3. Dried until 50% reduction from the initial weight at 50°C.

Then the membranes were immersed for about 6 hours in an aqueous solution of sodium hydroxide (1, 1.5 M), followed by washing with distilled water.

12.2.3 Membrane Cross-Linking

Since chitosan is soluble in acidic media, crosslinking treatment is required to prevent polymer degradation. Chitosan cross-linking is performed using specific bifunctional chemicals (glutaraldehyde and epichlorohydrin) that react with specific groups of biopolymers [11]. Natural chitosan membranes were heterogeneously cross-linked by immersion in 0.25% (w/w) aqueous glutaraldehyde solution (3.0 g of wet chitosan membrane in 50 mL of glutaraldehyde solution) at a temperature of 25°C for 1 hour, followed by washing with distilled water to remove the unreacted glutaraldehyde residues. The membranes were finally dried between two filter papers.

12.2.4 Adsorption Experiments

Copper solutions were prepared by dissolving hydrated copper sulfate ($CuSO_4 \cdot 5H_2O$) in water to the initial concentration of 1000 mg/L. Batch adsorption experiments were carried out by soaking 3×3 cm^2 (wet base) of a cross-linked chitosan membrane in 100 mL of copper solution for 18 hours at 20°C with stirring by a magnetic stirrer. The copper concentration in the solution was measured by atomic absorption spectrophotometer.

The amount of adsorbate in the membrane phase (q (mg/g)) is calculated by the following expression:

$$q = (C_0 - C_{eq}) \times \frac{V}{m} \tag{12.4}$$

Where C_0 and C_{eq} are the initial and equilibrium concentrations of metal ion in the liquid phase (mg/L) respectively, V is the volume of the solution (L), and m is the weight of wet chitosan membrane (g).

12.2.5 Scanning Electron Microscopy (SEM)

For the cross-sectional analysis, membranes were fractured in liquid nitrogen before their cross sections were covered with a thin layer of gold using a sputter coater (SCDOOS—Baltec, Switzerland) and then the samples were observed with a scanning electron microscope, SEM (XL30-Philips-Netherlands).

12.2.6 Atomic Force Microscopy (AFM)

Top surface morphology of the chitosan membranes (with polymer concentration of 7% w/w) prepared with different phase inversion method were characterized with AFM (Dualscope/Rasterscope C26, DME, Herlev, Denmark) in noncontact mode. Various roughness parameters can be measured by AFM. Some of these are reported in this work; the definitions of these parameters are as follows [20, 21]:

S_z is the average height of the five highest local maximums plus the average height of the five lowest local minimums.

S_a is the average roughness evaluated over the complete surface and is defined as

$$S_a = \iint_a |Z(x, y)| \cdot dx \cdot dy \tag{12.5}$$

where Z is the height function of the area.

S_q is the root mean square (rms) roughness expressed as follows:

$$S_q = \sqrt{\iint_a |Z(x, y)|^2 \cdot dx \cdot dy} \tag{12.6}$$

The average surface roughness and rms roughness data indicate significant deviations in texture characteristics.

S_{dq} (rms gradient) defines the slope for each point of area excluding points on the edge and is calculated as below:

$$S_{dq} = \sqrt{\frac{1}{A}\int_0^{l_x}\int_0^{l_y}\left(\left(\frac{\partial Z(x, y)}{\partial x}\right)^2 + \left(\frac{\partial Z(x, y)}{\partial y}\right)^2\right) \cdot dx \cdot dy} \tag{12.7}$$

where A is the cross-sectional area.

Pore sizes were measured by visual inspection of line profiles of different areas of the same membrane. These values were arranged in ascending order and were assigned median ranks [22, 23].

12.2.7 Atomic Absorption Experiments

The concentrations of metal ions were analyzed using a flame atomic absorption spectrophotometer (Varian AA 240).

12.2.8 Tensile

The membrane tensile strength was measured at room temperature according to ASTM D 882 by the universal testing machine STM-150, Santam Co., with crosshead speed of 5 mm/min.

12.3 RESULTS AND DISCUSSION

In Table 12.1 the chitosan membrane preparation conditions are shown along with their copper sorption data.

12.3.1 The Effect of Polymer Concentration on Morphology and Metal Sorption

As the sorption mechanism is chelation with amine groups of polymer, it is obvious that by increasing polymer concentration, metal sorption must also increase. That is clearly seen in comparing M1, M2, M3 sorption results with those of M4, M5, M6, respectively. The comparison of resulting images of the two dense membranes in Figure 12.2 indicates that the thickness of M4 (Fig. 12.2d) is about 2.5 times thicker than that of M1 (Fig. 12.2a).

TABLE 12.1. Membrane Preparation Conditions and Their Metal Sorption

Membrane Code	Chitosan Solution Conc. (% w/w)	Acetic Acid Solution Conc. (% v/v)	NaOH Conc. (M)	Preparation Method	Adsorption Amount (mg/g)
M1	2.5	5	1	Evaporation	9.87
M2	2.5	8	1	Evaporation	9.61
M3	2.5	5	1.5	Evaporation	10.36
M4	7	5	1	Evaporation	18.23
M5	7	8	1	Evaporation	18.04
M6	7	5	1.5	Evaporation	19.15
M7	7	5	1	Immersion-Precipitation	15.74
M8	7	8	1	Immersion-Precipitation	15.69
M9	7	5	1.5	Immersion-Precipitation	15.81
M10	7	5	1	Evaporation at 50°C	18.54
M11	7	5	1	Evaporation with temperature ramp	18.43
M12	7	5	1	Evaporation until 50% reduction	19.78

12.3.2 The Effect of Acetic Acid Concentration on Morphology and Metal Sorption

By comparing sorption results of M1, M4, M7 with M2, M5, M8, it can be seen that increasing acetic acid concentration causes a slight decrease in sorption. The effect of acetic acid concentration on morphology is shown in Figure 12.2. By comparing M1 (Fig. 12.2a) and M4 (Fig. 12.2d) with M2 (Fig. 12.2b) and M5 (Fig. 12.2e), respectively, it can be concluded that the porosity increases with acetic acid concentration.

12.3.3 The Effect of NaOH Concentration on Morphology and Metal Sorption

When NaOH concentration in the nonsolvent coagulation bath is increased, exchange of solvent and nonsolvent occurs faster, the membrane porosity increases leading to an increase in adsorption. That can be observed by comparing M1, M4, M7 sorption results with M3, M6, M9 results. Figure 12.2 shows that M3 (Fig. 12.2c) and M6 (Fig. 12.2f) are more porous than M1 (Fig. 12.2a) and M4 (Fig. 12.2d), respectively.

12.3.4 The Effect of Phase Inversion Method on Morphology and Metal Sorption

Membranes prepared by solvent evaporation method have asymmetric structure with a thick dense layer as shown in Figure 12.2a–g. Due to the exchange of solvent and nonsolvent in the membranes prepared by immersion precipitation method, the membranes are porous with sponge-like structure. The

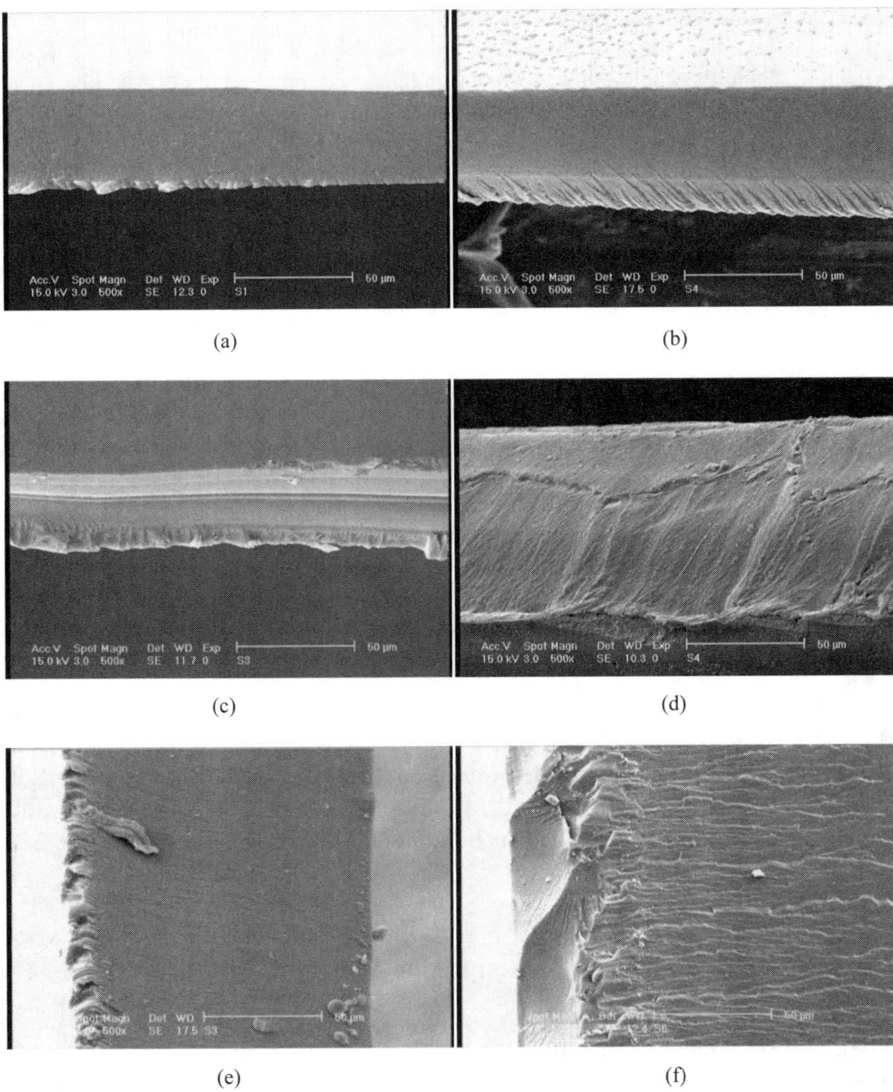

Figure 12.2. SEM images of different chitosan membranes: (a) M1, (b) M2, (c) M3, (d) M4, (e) M5, (f) M6, (g) M12, (h) M8 with ×500, 2000, 7500 magnifications, and (i) M9 with ×500, 2000, 7500 magnifications.

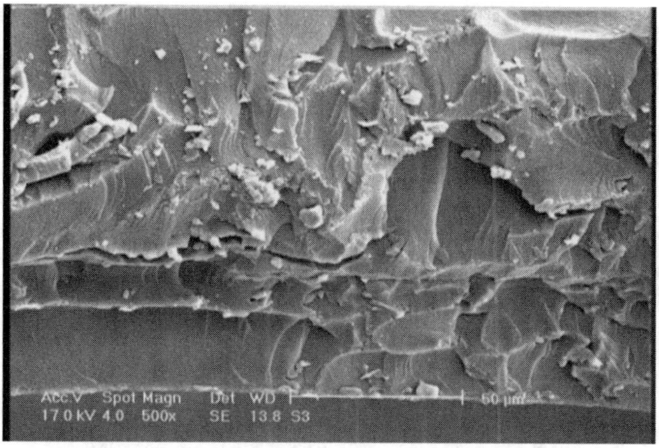

(g)

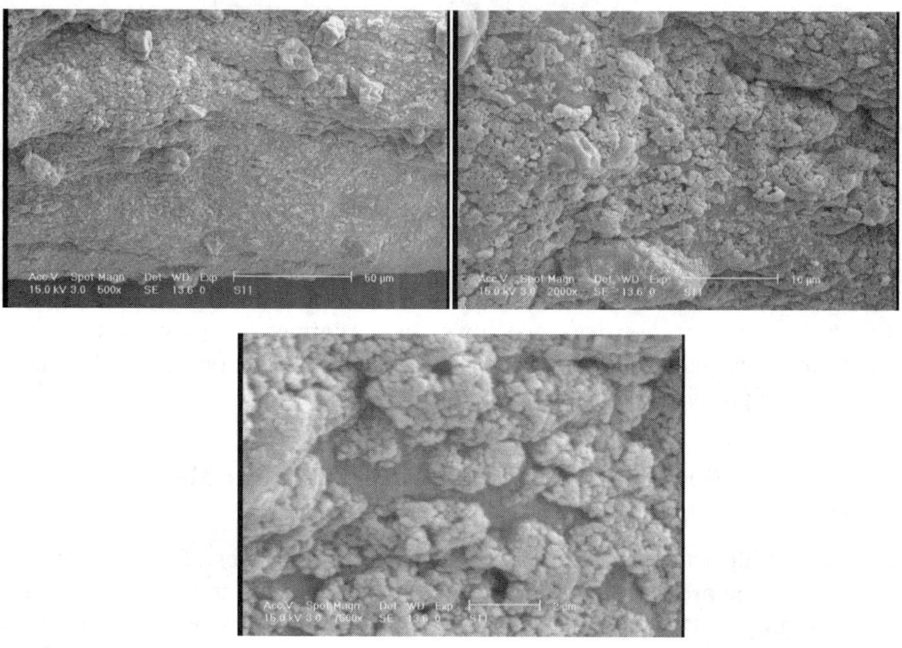

(h)

Figure 12.2. (*Continued*)

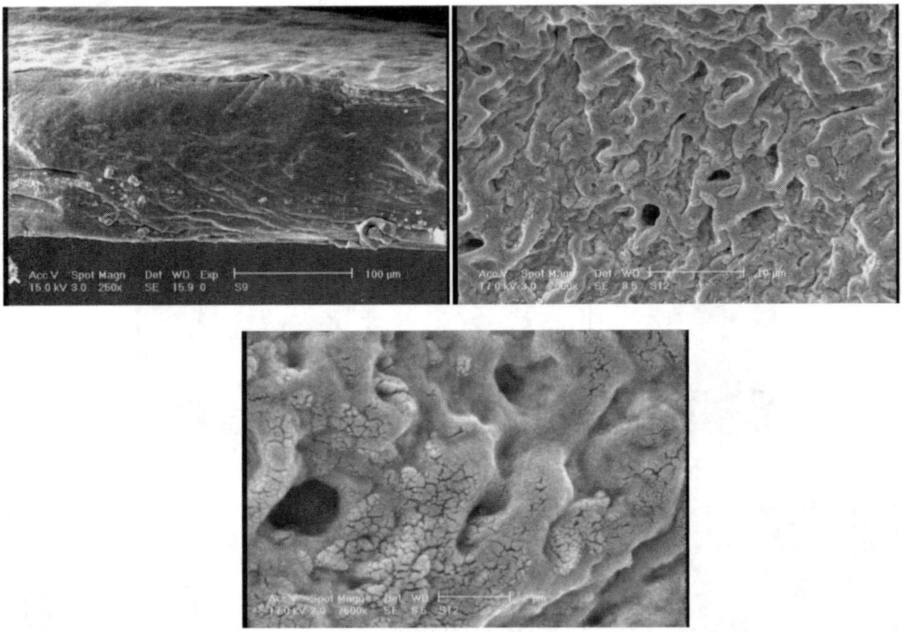

(i)

Figure 12.2. (*Continued*)

adsorption amount, however, was lower in the immersion-precipitation case due to the presence of sodium hydroxide in the membrane pores, which caused copper precipitation.

12.3.5 The Effect of Phase Inversion Method on Surface Morphology and Pore Sizes

Figure 12.3 illustrates two-dimensional topography images of chitosan membranes' surface prepared by the different phase inversion methods. Table 12.2 is a summary of the roughness parameters. For M7 (prepared by immersion precipitation), these parameters are greater than the others (prepared by evaporation) due to the exchange of solvent and nonsolvent. Chitosan membranes prepared by evaporation method shrank more during the solvent evaporation and showed lower porosity and roughness. Among them, M4 (dried at room temperature) showed the lowest roughness. That is due to lower evaporation rate and slower phase inversion. The difference in the copper sorption capacity of the membranes is due to their surface roughness and functional group accessibility.

By using the AFM images in conjunction with digitally stored line profiles, it is possible to measure the diameter of individual pores [10, 15]. Figure 12.4

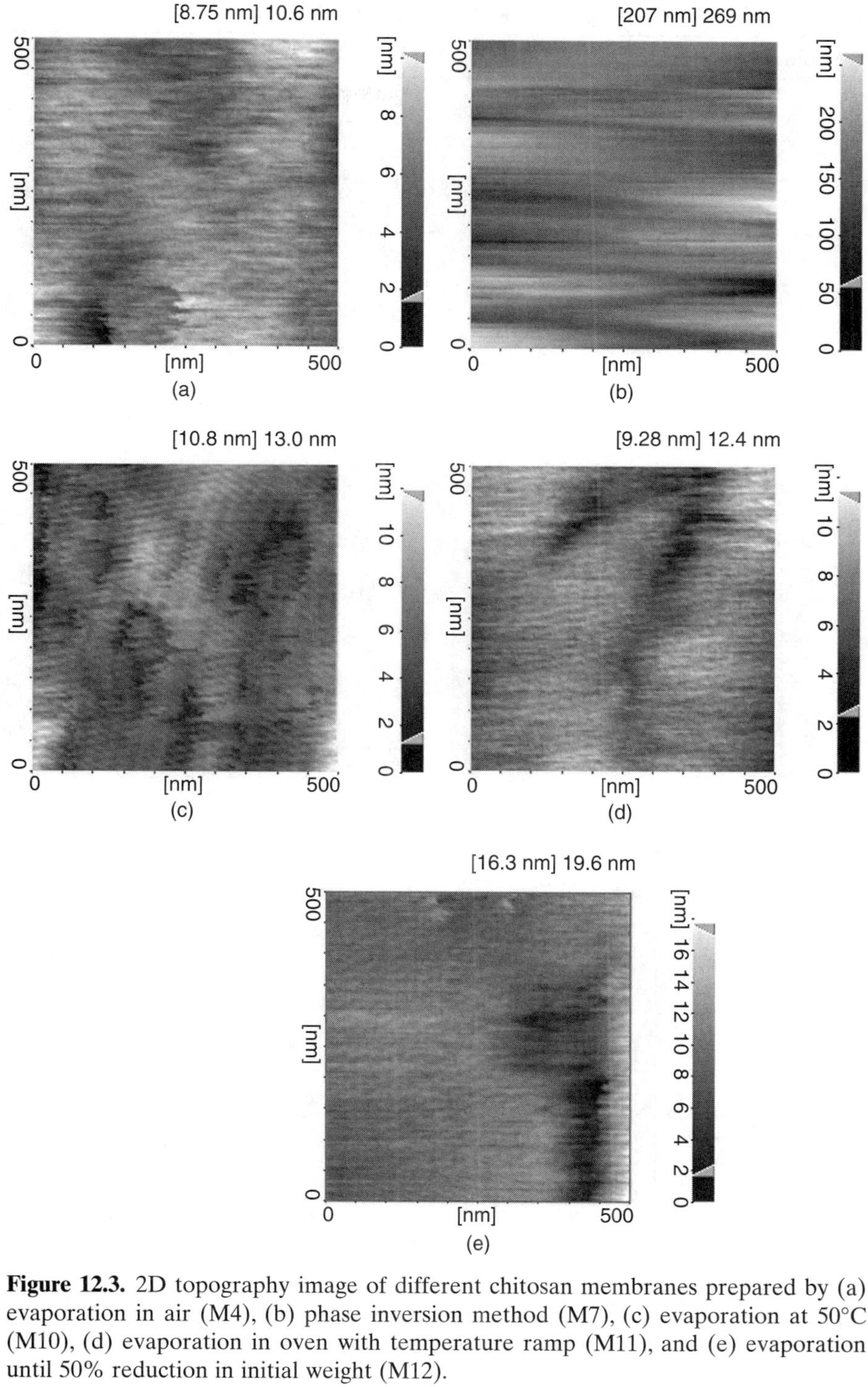

Figure 12.3. 2D topography image of different chitosan membranes prepared by (a) evaporation in air (M4), (b) phase inversion method (M7), (c) evaporation at 50°C (M10), (d) evaporation in oven with temperature ramp (M11), and (e) evaporation until 50% reduction in initial weight (M12).

TABLE 12.2. Surface Roughness Parameters of Membranes

Membrane	S_z (nm)	S_a (nm)	S_q (nm)	S_{dq}	Mean Pore Size (nm)
M4	9.69	1.01	1.29	0.27	3.40
M7	18.3	21.2	28.1	3.86	15.87
M10	11.5	1.32	1.68	0.33	5.12
M11	10.7	1.18	1.48	0.30	4.26
M12	17	1.64	2.21	0.35	6.61

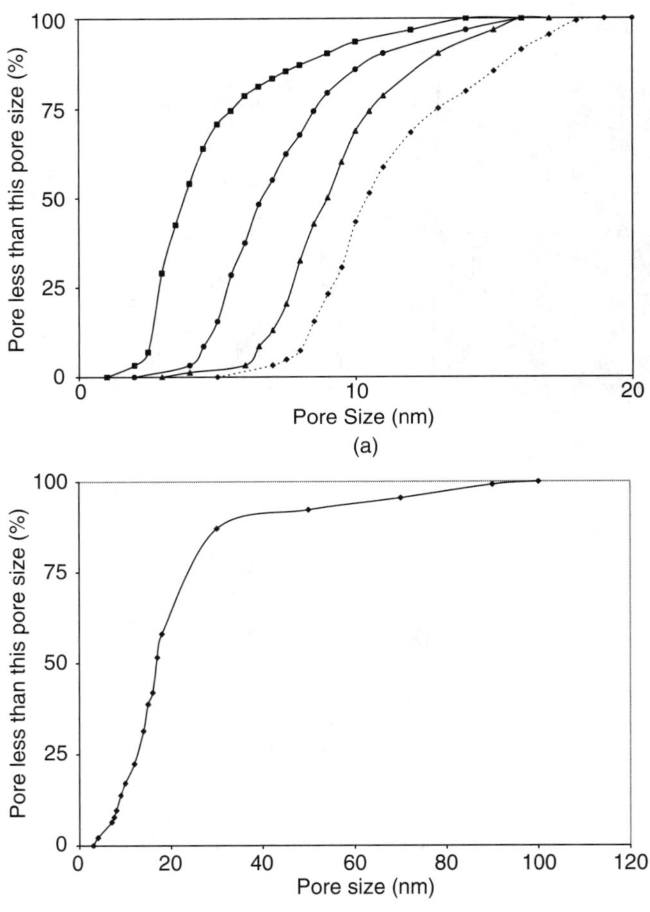

Figure 12.4. Pore size distribution of different chitosan membrane prepared by (a) evaporation method solid line (■) M4, (◆) M11, (▲) M10, and dash line (—) for M12, (b) phase inversion method (M7).

shows the pore size distribution of 5 chitosan membranes (i.e., M4, M7, M10, M11, and M12) obtained for each membrane from the images of 50 pores. The pore size distribution of the membranes is shifted to the right due as phase inversion becomes faster. The mean pore size of the above membranes is reported in Table 12.2 [15].

12.3.6 Membrane Stability

The membranes produced by immersion-precipitation method were brittle, while those prepared by evaporation method could resist a pressure of 8.4 MPa.

12.4 CONCLUSION

Based on the present work, one can conclude that the chitosan membrane becomes more porous by an increase in acetic acid concentration in the solvent and sodium hydroxide concentration in the nonsolvent solution. The membrane also becomes more porous by immersion precipitation method than solvent evaporation method. In batch experiments, the lateral surface is more involved in adsorption process, so the availability of functional groups at the surface together with the mass transfer resistance controls the sorption amount. An increase of polymer concentration in the casting solution formed a thicker membrane with higher sorption capacity. The membrane porosity was increased with an increase in acetic acid concentration, but the copper sorption was decreased. An increase in sodium hydroxide concentration in nonsolvent lead to larger pores and caused copper sorption to increase, due to the lower mass transfer resistance. Membranes prepared by immersion-precipitation had sponge-like structure with higher roughness and larger pores than the membranes prepared by solvent evaporation. It can be concluded that faster phase inversion can cause higher porosity, roughness, and larger pore size.

REFERENCES

[1] Ho Y. S. and G. Mckay. Sorption of copper (II) from aqueous solution by peat. *Water Air Soil Pollut.* 158 (2004) 77–97.

[2] Boddu V. M., K. Abburi, A. J. Randolph, and E. D. Smith. Removal of copper (II) and nickel (II) ions from aqueous solutions by a composite chitosan biosorbent. *Sep. Sci. Technol.* 43 (2008) 1365–1381.

[3] Shafaei A., F. Zokaee Ashtiani, and T. Kaghazchi. Equilibrium studies of the sorption of Hg(II) ions onto chitosan. *Chem. Eng. J.* 133 (2007) 311–316.

[4] Vieira R. S. and M. M. Beppu. Interaction of natural and crosslinked chitosan membranes with Hg(II) ions. *Colloids Surf. A* 279 (2006) 196–207.

[5] Wan M. W., C. C. Kan, B. D. Rogel, and M. L. P. Dalida. Adsorption of copper (II) and lead (II) ions from aqueous solution on chitosan-coated sand. *Carbohyd. Polym.* 80 (2010) 891–899.

[6] Babel S. and T. A. Kurniawan. Low-cost adsorbents for heavy metals uptake from contaminated water: a review. *J. Hazard. Mater. B* 97 (2003) 219–243.

[7] Jin L. and R. Bai. Mechanisms of lead adsorption on chitosan/PVA hydrogel beads. *Langmuir* 18 (2002) 9765–9770.

[8] Vold I. M. N., K. M. Varum, E. Guibal, and O. Smidsrod. Binding of ions to chitosan—selectivity studies. *Carbohyd. Polym.* 54 (2003) 471–477.

[9] Francisco C. F., F. C. F. Barros, F. W. Sousa, R. M. Cavalcante, T. V. Carvalho, F. S. Dias, D. C. Queiroz, L. C. G. Vasconcellos, and R. F. Nascimento. Removal of copper, nickel and zinc ions from aqueous solution by chitosan-8- hydroxyquinoline beads. *Clean* 36 (2008) 292–298.

[10] Vieira R. S. and M. M. Beppu. Dynamic and static adsorption and desorption of Hg(II) ions on chitosan membranes and spheres. *Water Res.* 40 (2006) 1726–1734.

[11] Vieira R. S., E. Guibal, E. A. Silva, and M. M. Beppu. Adsorption and desorption of binary mixtures of copper and mercury ions on natural and crosslinked chitosan membranes. *Colloid Surf. A: Physicochem. Eng. Aspects* 279 (2006) 196–207.

[12] Lopes E. C. N., F. S. C. Anjos, E. F. S. Vieira, and A. R. Cestari. An alternative avrami equation to evaluate kinetic parameters of the interaction of Hg(II) with thin chitosan membranes. *J. Colloid Interface Sci.* 263 (2003) 542–547.

[13] Cestari A. R., E. F. S. Vieira, J. D. S. Matos, and F. S. C. Anjos. Determination of kinetic parameters of Cu(II) interaction with chemically modified thin chitosan membranes. *J. Colloid Interface Sci.* 285 (2005) 288–295.

[14] Kamiński W. and Z. Modrzejewska. Application of chitosan membranes in separation of heavy metal ions. *Sep. Purif. Technol.* 32(16) (1997) 2659–2668.

[15] Kamiński W. and Z. Modrzejewska. Separation of Cr(VI) on chitosan membranes. *Ind. Eng. Chem. Res.* 38 (1999) 4946–4950.

[16] Beppu M. M., E. J. Arruda, R. S. Vieira, and N. N. Santos. Adsorption of Cu (II) on porous chitosan membranes functionalized with histidine. *J. Membrane Sci.* 240 (2004) 227–235.

[17] Wang X. P. and Z. Q. Shen. Studies on the effects of copper salts on the separation performance of chitosan membranes. *Polym Int.* 49 (2000) 1426–1433.

[18] Matsuyama H., Y. Kitamura, and Y. Naramura. Diffusive permeability of ionic solutes in charged chitosan membrane. *J. Appl. Poly. Sci.* 72 (1998) 397–404.

[19] Modrzejewska Z., I. Korus, and P. Owczarz. The effect of seasoning a membrane-forming solution on the separation properties of chitosan membranes. *J. Membrane sci.* 181 (2001) 229–239.

[20] Barzin J., C. Feng, K. C. Khulbe, T. Matsuura, S. S. Madaeni, and H. Mirzadeh. Characterization of polyethersulfone hemodialysis membrane by ultrafiltration and atomic force microscopy. *J. Membrane Sci.* 237 (2004) 77–85.

[21] Khulbe K. C., C. Feng, and T. Matsuura. Synthetic Polymeric Membranes: Characterizationby AtomicF orceM icroscopy,vol. 1. Springer, Berlin, 2008,pp . 38–43.

[22] Singh S., K. C. Khulbe, T. Matsuura, and P. Ramamurthy. Membrane characterization by solute transport and atomic force microscopy. *J. Memb. Sci.* 142 (1998) 111–127.

[23] Bowen W. R. and T. A. Doneva. Atomic force microscopy studies of nanofiltration membranes: surface morphology, pore size distribution and adhesion. *Desalination* 129 (2000) 163–172.

13

ANALYSIS OF FOULING AND FLUX BEHAVIOR IN CROSS-FLOW MICROFILTRATION OF NONALCOHOLIC BEER BY CERAMIC MEMBRANE

MEHDI YAZDANSHENAS, SEYYED ALI REZA TABATABAEI NEJAD, MOHAMMAD SOLTANIEH, AND LUC FILLAUDEAU

13.1 INTRODUCTION

Beer is the fifth most widely consumed beverage in the world behind tea, carbonates (sodas), milk, and coffee, and it continues to be a popular drink with an average consumption of 23 L/person per year. The consumption of nonalcoholic beers is significantly increasing due to its fewer calories, the health consideration (against alcohol disadvantages), and prohibitions in some cultures and religions. This has led to the development of a healthy market and most major brewers have included alcohol-free beer in their product portfolios [1–3]. Brewers are very concerned that the techniques they use are the best in terms of product quality, cost-effectiveness, and environmental impact. Consequently, energy consumption, water use, and wastewater generation constitute real economic opportunities for improvements in the existing process. Beer clarification is probably one of the most important operations, when rough beer is filtered in order to eliminate yeast and colloidal particles responsible for haze. In addition, this operation should also ensure

Sustainable Membrane Technology for Energy, Water, and Environment, First Edition.
Edited by Ahmad Fauzi Ismail and Takeshi Matsuura.

the biological stability of the beer. It should comply with the haze specification of a lager beer in order to produce a clear bright beer. Standard filtration consists of the retention of solid particles (yeast cells, macrocolloids, suspended matter) during dead-end filtration with filter aids. The variety of compounds (chemical diversity, large size range) to be retained makes this operation one of the most difficult to control [3, 4]. Recently, membrane technology has been successfully used in clarification operation wherein kieselguhr filtration is replaced by cross-flow microfiltration (MF). Since cross-flow MF has been introduced, it was found that there could be additional benefits in quality, environments, health, safety, simplicity, flexibility, and cost. At the time being, several industrial membrane filtration plants are being operated in the big companies for producing famous brands [5–8].

One of the most important disadvantages of using MF in clarification of beer is the fouling and flux decline during the process [6, 9]. The fouling formation can be subdivided into three major types: (1) the cake layer consisting of a compact deposit of yeast cells, debris, and coagulated materials; (2) discrete particles blocking the pore; and (3) soluble macromolecules adsorbed onto the membrane surface and pore walls [3, 6, 10, 11]. Since 1995, several works have tried to identify fouling factors and to quantify their contribution regarding fouling mechanisms [12–14]. Analyzing flux decline data using the standard blocking laws, Blanpain et al. [13, 15, 16] have shown that fouling occurred initially by the deposition of material within the pores followed by cake filtration [13, 15–17]. Gan et al. [12] have identified the in-pore membrane fouling as the most predominant flux reduction factor. At present, scientific literature confirms that proteins, carbohydrates (β-glucan, pentosan with molecular weight [MW] > 100 kDa), and starch molecules or particles are identifiable as key membrane foulants [6, 10, 18].

The mechanisms of membrane fouling can be analyzed by one of the classical fouling models of pore blockage, pore constriction, or cake filtration which can all be conveniently written in a common mathematical form as:

$$\frac{d^2t}{dV^2} = k\left(\frac{dt}{dV}\right)^n \tag{13.1}$$

where t (s) is the filtration time, V (m^3) is the total filtered volume, and parameters of n and k are characterized by the fouling models. For complete pore blockage mechanism $n = 2$, for standard blocking (or pore constriction) $n = 1.5$, for intermediate blocking $n = 1$, and for cake filtration $n = 0$ [19, 20]. In the complete pore blockage model, it is assumed that each particle arriving to the membrane blocks pores with no superposition of particles, and the number of blocked pores is assumed to be proportional to the volume of solution filtered through the membrane and particle's concentration in feed. On the basis of this, the filtrate flux decreases exponentially with time. In the standard blocking or pore constriction model, it is assumed that each particle arriving to the membrane is deposited onto the internal pore walls leading to a decrease in the pore volume. The intermediate pore blockage model is similar to the complete pore blockage model and accounts the possibility of settling particles on other particle previously arrived and already blocking some pores. The cake filtration model (Eq. 13.2) presumes that a uniform cake layer of particles forms over the entire membrane surface and this fouling layer is permeable to fluid flow. The rate of change of cake layer resistance is directly proportional to the convective transport of particles to the membrane surface:

$$J = J_0\left(1 + f'R'\frac{2\Delta P}{\mu R_m^2}C_p t\right)^{-0.5} = J_0(1 + Ct)^{-0.5} \qquad (13.2)$$

where J (m^3/m^2 s) is the permeate flux, J_0 (m^3/m^2 s) is the initial flux, f' is the fraction of the aggregates contributing to the growth of the cake layer, R' (m/kg) is the specific resistance of the cake layer, ΔP (Pa) is the TMP, μ (kg/m s) is the permeate viscosity, Rm (m^{-1}) is the intrinsic resistance of the membrane, C_p (kg/m^3) is the particle concentration, t (s) is the time of operation, and C (1/s) is the cake filtration coefficient.

Prevalently, a nonalcoholic beer is known to contain at last 0.5% v/v alcohol which is produced either by the removal of alcohol in a postfermentation treatment (alcohol removal) or by the restriction of alcohol production during the brewing process. The fermentation by-pass is the most accepted method widely used in the Muslim countries where using alcohol is prohibited by the religion and people prefer nonfermented and "Halal" products [2]. Fermentation bypass differs from lager beer process not only in passing over the fermentation but also in other operations. These operational discrepancies make the filtration feed to be different in comparison with the lager beer. Other than the alcohol content, the yeast content of rough nonalcoholic beer (RNAB) is limited, it does not contain any carbon dioxide, and as there is not any biological modification, the molecular and macromolecular composition of filtration feed (like proteins, amino acids, and acetyls) is dissimilar. These make the physicochemical properties of this type of beer deviated from lager beer and needs accurate consideration in the process design.

Despite the long lasting history of applied research and development efforts in membrane clarification of lager beer and the significant difference between the nonalcoholic and lager beer properties, no study has been performed on the nonalcoholic beer clarification yet. This research aims to analyze the flux decline during cross-flow MF and to get insight into fouling mechanisms.

13.2 MATERIALS AND METHODS

13.2.1 Experimental Setup and Instrumentation

Experiments were carried out using a pilot plant composed of two major parts: feed (retentate) and permeate loops (Fig. 13.1). The feed temperature and the

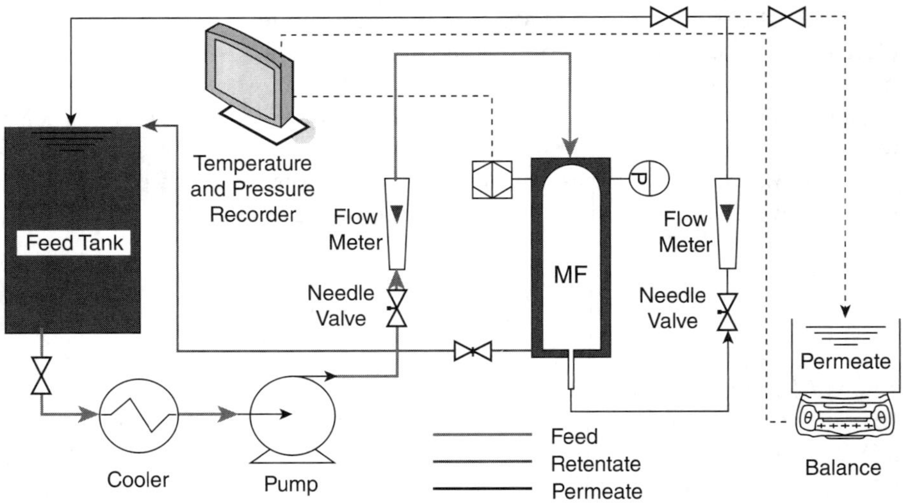

Figure 13.1. Schematic flow diagram of the microfiltration setup.

transmembrane pressure (TMP) have been measured with the accuracy of ±2 mbar and ±0.02°C, respectively, by an acquisition device (RPU353 Haffmans®). A rotameter (precision ±8 L/h) has been used to determine the feed flow rate, and the filtrate flow rate was measured by timed collection using a digital balance (Sartorius® GP 5202, accuracy ±0.1 g) or by a rotameter (precision ±2 L/h) in the filtrate line. The feed was circulated through the module by a rotary vane pump. A constant temperature was provided for feed stream by passing it through a fixed temperature bath. A tubular mono-channel aluminum–silicon oxide membrane (Atlas Filtri®, L = 0.208 m, Øint/ext = 0.032/0.05 m, S = 0.029 m^2) was used for clarification process with the nominal pore diameter of 0.45 μm (manufacturer specification). The pore size distribution of the membrane has been determined by mercury intrusion test (ISO 15901-1:2005), and the bubble point test was performed for the evaluation of the maximum pore diameter.

The subjected nonalcoholic beer in this research was produced in the fermentation bypassed process from pure malt. The RNAB was prepared by taking out from the upper part of the conditioning vessel and adjustment to the desired concentration. Rough beers are real products; consequently, a large and inevitable variability of product quality is noticeable [4]. Therefore, a bulk of malt extract has been used for dilution and preparing the feed for the situations where constant feed properties were needed. The soluble solids concentration of feeds was adjusted to 3.8 ± 0.1 Brix. The particle size distribution of the RNAB and diluted malt extract (DME) was determined by laser light scattering (LLS) method (CILAS® 1064 Liquid). The concentration of suspended particles of the RNAB and/or DME has been evaluated by filtration of a distinct volume of the samples with a preweighted filter paper and measuring its weight after drying. The soluble solids concentration has been determined by digital refractometer (ATTAGO® Model PR-32α). The viscosity was measured with a controlled rotational speed concentric cylinders viscometer (Fungilab® Model Visco Elit). The haze has been determined in NTU scales by HACH® (2100 P) haze meter.

13.2.2 Experimental Procedure and Operating Conditions

Intrinsic membrane resistances were determined by measuring the permeation of distilled water before each experiment. The MF process started after adjusting the feed concentration and temperature to the desired values of 3.8 ± 0.1°Brix and 3.5 ± 0.5°C. The feed flow rate and TMP were controlled by adjusting the feed and retentate valves. About 19 ± 0.5 L of RNAB (or DME) were used in the feed tank, and the filtration was carried out for about 4–8 hours in a constant pressure of 0.1–2.5 bar and a feed flow rate of 100–350 L/h. With changing the internal tube (exactly fitted to the inside of the module) and using baffles around the membrane, the cross-flow velocity has been adjusted to 0.02–2.0 m/s (the feed flow was on the outer surface of the membrane). About 12 membranes were used, and filtrations were done in the feed and bleed mode. After filtration, the loop was emptied and rinsed with water. Then, membranes were backwashed by 2% NaOH at 70–75°C for about 5 minutes and afterward the soda solution was recirculated along feed side of the membrane. The effectiveness of cleaning treatment has been characterized by comparing resistance of the cleaned membrane with the initial intrinsic resistance. For the situation which the membrane resistance increment exceeded 10% of initial intrinsic resistance, the cleaning procedure was repeated or the membrane was changed.

13.3 RESULTS AND DISCUSSIONS

13.3.1 Membrane and RNAB Characterization

Despite the nominal pore size of the membrane indicated by its manufacturer, 0.45 μm, pore size distribution diagram showed that the average pore diameter was about 2.4 μm. According to the volume distribution diagram, the pores with the diameter smaller than 1.35 μm and 2.58 μm represent, respectively, 10 and 90% of the pores' relative volume. The bubble point test also confirmed that the maximum pore size of the membrane was about 2.85 μm, which was in good agreement with the pore size distribution curve.

The volumetric distribution of particles size of the DME and RNAB are exhibited in Figure 13.2. According to volumetric distribution curves, the particles in feed exhibit a wide range of sizes, and the average particle sizes of RNAB and DME are 115 μm and 76 μm, respectively. Comparison of this information with the corresponding researches [11] confirms the big difference between nonalcoholic and alcoholic rough beer.

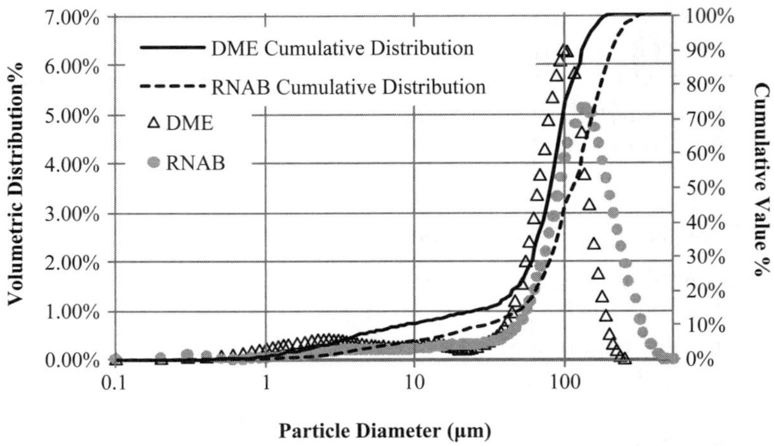

Figure 13.2. Particle size distribution of a typical rough nonalcoholic beer (RNAB) with pH = 4.4 and haze = 161 NTU and a typical diluted malt extract (DME) with pH = 5.2 and haze = 596 NTU.

Suspended particle concentration exhibited a very good linear proportional relationship with the measured turbidity ($R^2 = 99.7\%$) as

$$C_p = 4.45 \times 10^{-4} H \tag{13.3}$$

where C_p (kg/m^3) is the concentration of particles and H (NTU) is the turbidity.

The formation of "chill haze" is a temperature-dependent and reversible process, and the large colloidal aggregates may re-dissolve into the beer when temperature rises [6, 10]. Therefore, in order to normalize the fluctuation in the haze measurement due to temperature, the ratio of the haze in 20°C ($H_{T=20°C}$) to the hazes in different temperatures has been correlated with temperature ($R^2 = 95.54\%$) as

$$\frac{H_{T=20°C}}{H} = 0.0123T - 2.5468 \tag{13.4}$$

This relation predicts the haze of a RNAB sample at 20°C ($H_{T=20°C}$) with the haze of H (NTU) at temperature of T (°K).

In the beginning of MF, flux declined drastically and continued to reach a quasi-steady-state flux. Figure 13.3 shows this phenomenon in a typical MF of the RNAB and DME. As Figure 13.3 revealed, the rate of flux decline in MF of DME is much greater than the rate of flux decline in RNAB MF, but both of them converge finally to almost the same steady-state flux.

Figure 13.4 represents the variation of log(d^2t/dV^2) versus log(dt/dV) in different TMPs for DME. According to Figure 13.4, during early stage of filtration corresponding to small dt/dV, there is a linear relationship on the log–log

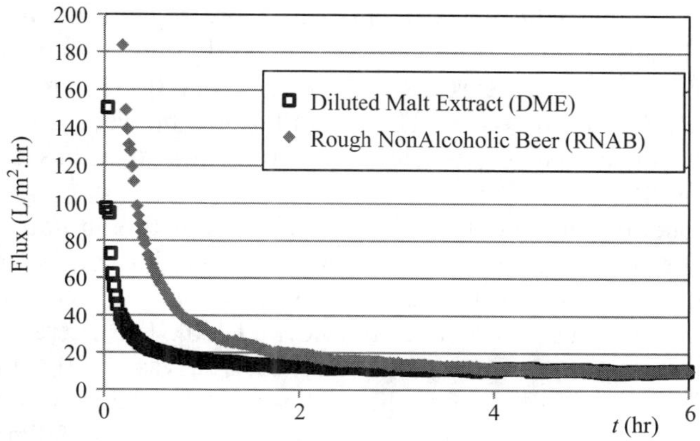

Figure 13.3. Flux decline during microfiltration at TMP = 0.9 bar, T = 7°C, v = 0.9 m/s, Brix = 3.8.

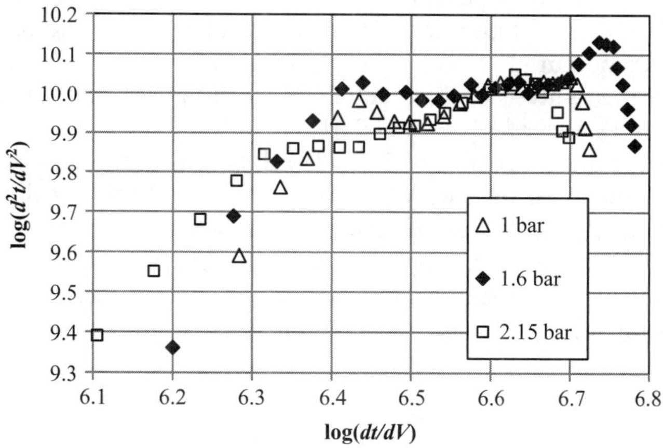

Figure 13.4. Analysis of MF flux decline for DME at different TMP, v = 0.9 m/s, T = 3.5°C and Brix = 3.8.

plot with the slope of 2 ± 0.2 and correlation coefficient of $R^2 > 98\%$, which is close to the value predicted by classical pore blockage model. In higher *dt/dV*, the data deviated from the straight line and came to a constant value representing the cake filtration mechanism for fouling. The transition from pore blocking to cake filtration shifts to a higher *dt/dV* (lower flux) with decreasing pressure implicating that the cake formation started sooner in higher pressures. Note that the data obtained at the higher TMPs start at lower values of *dt/dV* due to the higher initial flux.

These consecutive steps in the whole process of membrane blocking can be explained in terms of the successive or simultaneous stages. Initially, the protein aggregates and cell debris block the entrance of the pores and then a cake layer or deposit begins to form over those regions of the membrane that have already been "blocked" by the initial deposit. This phenomenon is in accordance with the researches of Bowen et al. [20], Ho and Zydney [21], and Van Reis and Zydney [17]. The big difference between membrane pore size distribution and feed particle size distribution diagrams confirms this phenomenon [22].

The secondary increment of the $\log(d^2t/dV^2)$ with $\log(dt/dV)$ after the plateau in the characteristic curves should correspond to long small pore tails in pore size distributions. It means after construction of the cake layer on the bigger pores, the small pores became blocked [20]. The curves were finally followed by a fast decrease of the second derivatives d^2t/dV^2, implying the existence of a nonzero limit flow [20]. These results are in accordance with the conclusions of the researches by Blanpain et al. and Fillaudeau et al. on the fouling in clarification of beer [4, 11, 15].

According to the Figure 13.4 and the similar sets of data, the blockage of pores dominates the fouling mechanism in the first 15 minutes of operation, and the cake formation governs the flux decline for the remaining of MF process (after about 20 minutes). The analysis of second zone of flux decline with cake filtration model shows a very good correlation. Table 13.1 represents the result of the analysis at different TMPs with the correlation factors.

Linear regression of C versus TMP ($R^2 = 92.8\%$) is in accordance with the cake filtration model (Eq. 13.2); and with consideration of viscosity, membrane resistances, and feed concentration, the $f'R'$ factor is calculated to be 8.1×10^{14} m/kg. J_0 can be described with resistance in series model. For the $t = 0$, other than the membrane resistance (R_m), there is a resistance of adsorbed materials and initially blocked particles on the intact surfaces of the membrane occurring after soaking the membrane by feed.

$$J_0 = \frac{\Delta P}{\mu(R_m + R_0)} \tag{13.5}$$

where R_0 can be defined as the initial resistance of the deposit. It is interesting that the R_0 increases proportionally with pressure, that is, $R_0 = 1.30 \times 10^6 \Delta P$ ($R^2 = 96.6\%$).

TABLE 13.1. The Results of Cake Filtration Correlation with the Flux Data in the Second Zone of Flux Decline in Constant Concentration, Cross-Flow Velocity, and Shear Rate (C_p = 145 mg/L, v = 0.65 m/s, and γ = 1576/s)

TMP (bar)	0.52	1.0	1.56	2.15	2.70
C (1/s)	0.23	0.65	0.76	0.89	1.14
J_0 ($m^3/m^2.s$)	1.35×10^{-4}	2.13×10^{-4}	2.25×10^{-4}	2.64×10^{-4}	2.94×10^{-4}
R^2	98.40%	98.10%	98.70%	98.20%	98.20%

According to the theoretical basis of cake filtration model, with a constant concentration and TMP, C (in Eq. 13.2) is independent of shear rate. Therefore, for correlation of cake filtration model with the flux data in different shear rate, the C factor is calculated by the operation parameters, and J_0 is calculated by minimizing the sum of the squared residuals between the experimental data and the model calculations. It has been found that the variation of J_0 with shear rate is linear, which is in agreement with the theory of MF of micron-sized and larger particles [22]. Alternatively, at a constant TMP, initial resistance of the deposit diminishes with shear rate ($R^2 = 92.75\%$) as $R_0 = 4.48 \times 10^{13}\gamma^{-0.81}$.

The initial flux did not change considerably with concentration; therefore, the equation of cake filtration model for MF of RNAB has been generalized as below:

$$J = \frac{\Delta P}{\mu\left(R_m + 4.57\times10^{8}\Delta P\gamma^{-0.81}\right)}\left(1 + 8.1\times10^{14}\frac{2\Delta P}{\mu R_m^2}C_p t\right)^{-0.5} \qquad (13.6)$$

Application of Equation 13.6 for other experiments describes their flux behavior meaningfully with correlation factor greater than 85%.

13.4 CONCLUSION

The clarification of rough beer representing one of the most important operations in the brewing process has been recently performed successfully by membrane technology instead of the former technology of kieselguhr filtration. The application of cross-flow MF in clarification of RNAB and DME has been investigated. A pilot plant including a ceramic membrane with nominal mean pore diameter of 0.45 μm and actual mean pore diameter of 2.4 μm has been utilized. The particle size distribution of the feed is wide, and the particles with about 50–120 μm diameter have the most volumetric concentration. The concentration of the particles has been correlated to the haze proportionally, and a linear regression has been found between reverse of haze and temperature.

The flux of MF declined drastically with time due to the fouling. A characterizing curve used to describe the fouling phenomenon during MF process shows that at the beginning of MF, the pore blocking mechanism governs the initial flux reduction, and the subsequent flux decline is due to the cake filtration mechanism. In other words, initially the big particles block the entrance of the pores and then a cake layer or deposit begins to form over the blocked entrance of the membrane pores. Analyzing these zones of flux decline with the appropriate models presented a good agreement. The model's parameters has been evaluated in different conditions, and they have been generalized for using in the MF of RNAB.

On the basis of this work, it can be concluded that the formation of the cake and gel layer on outer surface of the membrane is the main reason of flux reduction. Therefore, it can be predicted that for enhancement of flux, hydrodynamic techniques may play a significant role.

ACKNOWLEDGMENT

This work is supported by Applied Research Project of Ministry of Industries and Mines of Islamic Republic of Iran, and the authors would like to thank the management of Iran Behnoush Co. for its support and providing the RNAB and DME.

REFERENCES

[1] Euromonitor. The world market for beer. Euromonitor, 2006.

[2] Briggs D. E., C. A. Boulton, P. A. Brookes, and R. Stevens. *Brewing: Science and Practice*. Woodhead Publishing Limited, Cambridge, 2004.

[3] EBC. Beer filtration, stabilization and sterilization, manual of good practice, 1999.

[4] Fillaudeau L., B. Boissier, A. Moreau, P. Blanpain-Avet, S. Ermolaev, N. Jitariouk, and A. Gourdon. Investigation of rotating and vibrating filtration for clarification of rough beer. *Journal of Food Engineering* 80 (2007) 206–217.

[5] Fillaudeau L., P. Blanpain-Avet, and G. Daufin. Water, wastewater and waste management in brewing industries. *Journal of Cleaner Production* 14 (2006) 463–471.

[6] Gan Q., J. A. Howell, R. W. Field, R. England, M. R. Bird, C. L. O'Shaughnessy, and M. T. MeKechinie. Beer clarification by microfiltration—product quality control and fractionation of particles and macromolecules. *Journal of Membrane Science* 194 (2001) 185–196.

[7] Yazdanshenas M., A. R. Tabatabaeenezhad, R. Roostaazad, and A. B. Khoshfetrat. Full scale analysis of apple juice ultrafiltration and optimization of diafiltration. *Separation and Purification Technology* 47 (2005) 52–57.

[8] Schuurman R., L. Broens, A. Mepschen, and D. Meijer. Full scale results of (kieselguhr free) beer membrane filtration and latest developments on combined membrane filtration and stabilisation. In: 10th brewing convention, Pilanesberg, South Africa, 2005.

[9] Reed R. Current and future roles of membranes in brewing. *Membrane Technology* 101 (1998) 5–8.

[10] Gan Q., J. A. Howell, R. W. Field, R. England, M. R. Bird, and M. T. McKechnie. Synergetic cleaning procedure for a ceramic membrane fouled by beer microfiltration. *Journal of Membrane Science* 155 (1999) 277–289.

[11] Fillaudeau L. and H. Carrère. Yeast cells, beer composition and mean pore diameter impacts on fouling and retention during cross-flow filtration of beer with ceramic membranes. *Journal of Membrane Science* 196 (2002) 39–57.

[12] Gan Q., R. W. Field, M. R. Bird, R. England, J. A. Howell, M. T. McKetchnie, and C. L. O'Shaughnessy. Beer clarification by cross-flow microfiltration: fouling mechanisms and flux enhancement. *Transactions of the Institution of Chemical Engineers* 75 (1997) 3–8.

[13] Blanpain-Avet P. and M. Lalande. Investigation of fouling mechanisms governing permeate flux in the crossflow microfiltration of beer. *Filtration and Separation* 34 (1997) 1065–1069.

[14] Czekaj P., F. López, and C. Güell. Membrane fouling during microfiltration of fermented beverages. *Journal of Membrane Science* 166 (2000) 199–212.

[15] Blanpain-Avet P., N. Doubrovine, C. Lafforgue, and M. Lalande. The effect of oscillatory flow on crossflow microfiltration of beer in a tubular mineral membrane system—Membrane fouling resistance decrease and energetic considerations. *Journal of Membrane Science* 152 (1999) 151–174.

[16] Blanpain-Avet P., L. Fillaudeau, and M. Lalande. Investigation of mechanisms governing membrane fouling and protein rejection in the sterile microfiltration of beer with an organic membrane. *TransIChemE* 77 (1999) 75–89.

[17] Van Reis R. and A. Zydney. Bioprocess membrane technology. *Journal of Membrane Science* 297 (2007) 16–50.

[18] Gan Q. Beer clarification by cross-flow microfiltration—effect of surface hydrodynamics and reversed membrane morphology. *Chemical Engineering and Processing* 40 (2001) 413–419.

[19] Chandler M. and A. Zydney. Effects of membrane pore geometry on fouling behavior during yeast cell microfiltration. *Journal of Membrane Science* 285 (2006) 334–342.

[20] Bowen W. R., J. I. Calvo, and A. Hernández. Steps of membrane blocking in flux decline during protein microfiltration. *Journal of Membrane Science* 101 (1995) 153–165.

[21] Ho C. C. and A. L. Zydney. A combined pore blockage and cake filtration model for protein fouling during microfiltration. *Journal of Colloid and Interface Science* 232 (2000) 389–399.

[22] Belfort G., R. H. Davis, and A. L. Zydney. The behavior of suspensions and macromolecular solutions in crossflow microfiltration. *Journal of Membrane Science* 96 (1994) 1–58.

14

COMPARISON AND UPGRADING OF WASTEWATER TREATMENT PLANTS FOR WASTEWATER RECLAMATION AND REUSE BY MEANS OF MEMBRANE BIOREACTOR (MBR) TECHNOLOGY

MAHDI KHOSRAVI, GAGIK BADALIANS GHOLIKANDI, AND HAMID REZA TASHAOUEI

14.1 INTRODUCTION

Domestic wastewater is usually treated by conventional activated sludge processes (CASPs), which involve the natural biodegradation of pollutants by heterotrophic bacteria (i.e., activated sludge) in aerated bioreactors [1]. Activated sludge could be separated by gravitational settling. The treatment efficiency is usually limited by the difficulties in separating suspended solids (SS) [2].

Effluent recycle and reuse are developing for irrigation, reclamation, and agriculture, as well as for both indirect and direct potable water supply. Up to now, the approach has consisted of providing an advanced treatment to a secondary treated effluent to meet the standards for reuse. For irrigation and reclamation, this treatment may be limited to filtration and disinfection, as illustrated in Figure 14.1a [3, 4].

A simplified treatment scheme (Fig. 14.1b) consists of replacing the biological treatment with microfiltration (MF) or ultrafiltration (UF). This process has been used in a number of areas, including industrial and municipal treatment, domestic wastewater reuse and reclamation.

Sustainable Membrane Technology for Energy, Water, and Environment, First Edition.
Edited by Ahmad Fauzi Ismail and Takeshi Matsuura.

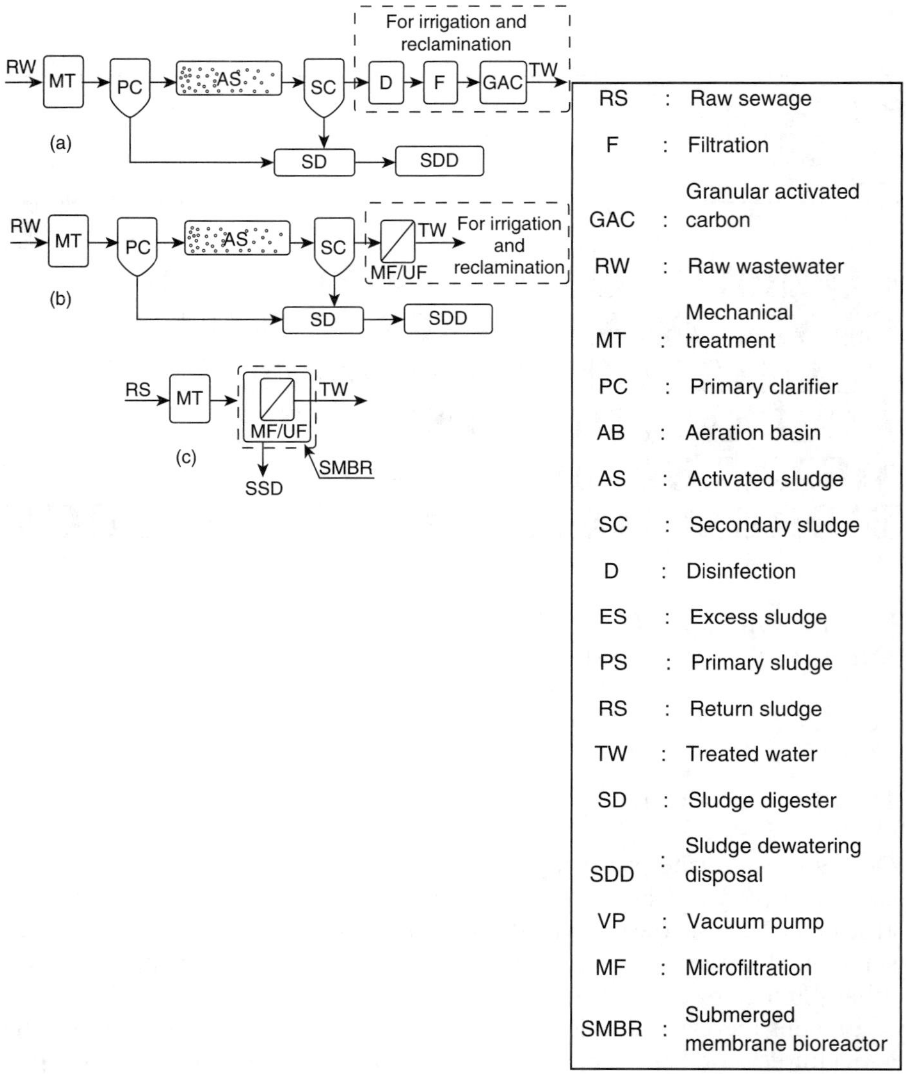

Figure 14.1. Evolution of a water reclamation treatment scheme.

The simple treatment scheme shown in Figure 14.1c provides the same level of treated quality water as the two schemes described above (Fig. 14.1a,b). The traditional secondary clarifiers for the separation of treated water from the mixed solution in the bioreactor are replaced by a membrane unit (MF or UF). It is based on the use of a process, where activated sludge and membrane separation are integrated into a single treatment step. This configuration of membrane bioreactor (MBR) is known as a submerged membrane bioreactor (SMBR).

SMBR technology has the following advantages: high sludge concentration [5], high quality of effluent, long contact time between activated sludge and organic pollutants [6], and complete separation of the hydraulic retention time (HRT) and sludge retention time (SRT) [7–9]. Moreover, effluent from SMBR is free from bacteria and has the potential for municipal and industrial reuse [10].

Problems, such as nonsettling, foam, and bulking sludge, are very widespread operational problems in treatment plants, which exist in South Wastewater Treatment Plant Isfahan (SWWTPI). It is important for the activated sludge plant manager to be aware of the possible causes and control strategies. The MBR can provide stable effluent and excellent performance independent of filamentous bulking or other phenomena affecting sludge settling characteristics [6, 11–14].

The MBR process is an emerging advanced wastewater treatment technology that has been successfully applied at an ever increasing number of locations around the world. This chapter gives an overview about some relevant preconditions and basic upgrading concepts in SWWTPI with using MBR technology.

14.2 MATERIALS AND METHODS

14.2.1 CASP

SWWTPI was used to compare the SMBR with the CASP (Fig. 14.2). The SWWTP is located approximately 10 km of Isfahan and receives wastewater from several zones and an industrial area; it is estimated that the 85% of the receiving wastewater is domestic and 15% industrial. The current plant consisted of main screening units, grit chamber, primary settling tank, activated sludge (aeration) tank, final settling tank, sludge thickening, digestion tank, and dried beds. The main characteristics of the wastewater treatment plant (WWTP) are shown in Table 14.1.

14.2.2 SMBR

The MBR was operated in parallel to a conventional WWTP in Isfahan, Iran. MBR was inoculated with activated sludge, fed with the raw wastewater of the SWWTPI after the primary treatment. MBR was fitted with Kubota membranes (Osaka, Japan). The configuration is shown in Figure 14.3. The pilot plant consisted of one compartment with a total volume of 3 m^3.

The experimental and operational conditions are shown in Table 14.2. An air diffuser was placed under the membrane module to prevent or reduce the fouling of the membrane and to maintain an aerobic environment for the normal growth of the activated sludge. A study of the global parameters is shown in Table 14.3.

Figure 14.2. South wastewater treatment plants, Isfahan.

TABLE 14.1. The Main Characteristics of CAS of SWWTPI

Type of treatment	Primary and secondary
Type of influent	Mixed (urban [85%]/industrial [15%])
Inflow (m^3/day)	60,000
Medium inflow (m^3/h)	2000
Inhabitants equivalents	277,000
COD (ppm)	
Influent	586
Influent after primary treatment	400
Effluent	89
BOD_5 (ppm)	
Influent	363
Influent after primary treatment	241
Effluent	50

COD, chemical oxygen demand; BOD, biological oxygen demand.

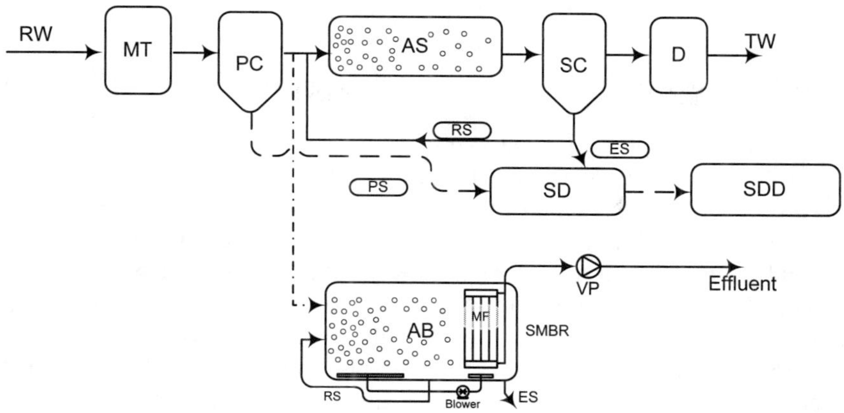

Figure 14.3. Conventional process and submerged bioreactor.

TABLE 14.2. Characteristics of MBR

Average inflow (L/m^{2}h)	10–20
Hydraulic retention time (h)	10–20
Mixed liquor suspended solids (MLSS) (g/L)	7–10
Type of membrane	Kubota (plate and frame)
Type of filtration	Microfiltration
Pore size (μm)	0.4
Effective membrane area (m^2)	40
Aerobic reactor volume (m^3)	3

TABLE 14.3. Global Parameters during the Studied Period (Standard Deviation)

	SWWPTI (*n* = 25)	CASP (*n* = 25)		SMBR (*n* = 29)	
Parameter	Inf. CASP	Effluent	Efficiency	Effluent	Efficiency
COD (mg/L)	586 (50)	89 (12.6)	84.8	9.9 (2.1)	98.3
BOD (mg/L)	363 (42)	50 (10)	86.2	3.6 (1.2)	99
TP (mg/L)	15 (4.1)	7.9 (2.8)	47.3	2.4 (1.1)	84
PO_4^{-3}-P (mg/L)	10.5 (1.5)	7.1 (2.1)	32.4	2.1 (1.2)	80
SS (mg/L)	115 (11)	41.8 (8.9)	63.6	0	100
Turbidity (NTU)	40 (5)	15.8 (3)	60.5	0.64 (1.0)	98.4
TKN (mg/L)	48.3 (2.1)	30.2 (6)	37.5	3.4 (2)	93
NH_4^+-N (mg/L)	35 (3)	20 (9.3)	42.9	1.9 (1.0)	94.6
NO_3^--N (mg/L)	0.94 (0.1)	3 (1)	–	13.5 (3)	–

14.3 ANALYTICAL ITEMS AND METHODS

Two samples of WWTP influent wastewater samples (the effluent from the primary treatment), WWTP effluent, and MBR effluent wastewater samples were taken as composite samples (integrated samples for 24 hours). The flux, temperature and TMP were monitored by daily measurements. Turbidity was measured by a turbidity meter (Model 965-10, Orbeco Analytical System Inc., Farmingdale, NY). The parameters of the SMBR and the CASP, such as the chemical oxygen demand (COD), total Kjeldahl nitrogen (TKN), ammonium nitrogen ($NH4^+$-N), nitrate (NO_3-N), total phosphorus (TP), orthophosphate (Orto-P) and SS of the influent and the filtrate, and mixed liquor suspended solids (MLSS) and mixed liquor volatile suspended solids (MLVSS) in the bioreactor were analyzed according to standard methods [15].

14.4 RESULTS AND DISCUSSION

From a technical point of view, reclamation of urban wastewater by SMBR is applicable because the technology offers several advantages such as reliability, compactness, and excellent treated water quality, over conventional processes used to date. Treated water quality of SMBR indicated that the removal of organic matter and SS were relatively successful (Table 14.3).

In the SMBR and CASP, the average removal of biological oxygen demand (BOD) was 99%, 86.2%, and the average removal of COD was approximately 98.3%, 84.8%, respectively. While the average COD concentration in the effluent of the SMBR was 9.9 mg/L, the average COD in the CASP was 89 mg/L. The COD removal increased with an increase in organic load, which was removed by >98% at organic loads in range of 0.6–0.8 kg COD/m^3.d. This indicates that the SMBR has a potential in treating high-strength urban wastewater. It can be concluded that the removal of organic pollutants by the SMBR was high in terms of COD, and therefore, a good quality effluent could be achieved during thelong -termoper ation.

TP was less than 1 mg/L at the period in which the bioreactor was continuously aerated, and the dissolved oxygen concentration varied between 1.90 mg/L and 4 mg/L. It was shown that the filtrate TP and Orto-P concentrations increased dramatically during the periods of intermittent aeration because the phosphorus was released again under anoxic conditions. Average TP and Orto-P in the municipal wastewater were 15 mg/L and 10.5 mg/L, respectively. In the intermittent aeration conditions, the average TP and Orto-P in the filtrate from the SMBR increased up to 2.4 mg/L and 2.1 mg/L, respectively. However, the average TP and Orto-P in the effluent of CASP were 7.9 and 7.1 mg/L, respectively. While the average TP and Orto-P removal in the SMBR were 84% and 80%, respectively, TP and Orto-P removal in the CASP were only 47.3% and 32.4, respectively.

The average TKN and $NH4^+$-N in the municipal wastewater were 48.3 mg/L and 35 mg/L, respectively, and nitrate was 0.94 mg/L. The average TKN and

$NH4^+$-N in the filtrate of the SMBR were 3.4 mg/L and 1.9 mg/L, respectively, and the average nitrate concentration increased up to 13.5 mg/L. Therefore, the average removal of TKN and $NH4^+$-N were 93% and 94.6%, respectively, indicating a successful nitrification process. The autotrophic nitrifiers, which have long generation times, were forced to proliferate speedily because the nitrifying population was completely confined within the bioreactor. The nitrifiers in the bioreactor were not negatively influenced by the heterotrophic bacteria because the sludge production was low in the SMBR.

The nitrate concentration in the filtrate decreased during interval aeration conditions. The nitrate concentration in the filtrate decreased to 2.4 mg/L during the period in which the aeration occurred, that is, 60 minutes on and 90 minutes off, indicating that the nitrate which occurs through the nitrification process has been strongly denitrified in the SMBR. Both nitrification and denitrification could be significantly decreased through intermittent aeration. The average TKN and $NH4^+$-N in the effluent of the CASP were 30.2 and 20 mg/L, respectively, that is, the nitrification efficiency was 42.9%, and the denitrification value was not available. While the average SS in the influent was 115 mg/L, the SS concentration in the filtrate of the SMBR was not detected because of the pore size of the MF membrane module.

However, the effluent SS in the CASP was 41.8 mg/L. the SS removal efficiency of the SMBR remained high, for example, 100%, which demonstrates the better separation effect of the MF membrane module than that of the secondary clarifier in the CASP.

Average turbidity in the influent was 40 NTU. The turbidity in the filtrate of SMBR was less than 1 NTU during the 90-day experiment, while the average turbidity of the CASP effluent was close to 15.8 NTU. However, total coliform bacteria were not detected in the filtrate, so disinfection of the wastewater can be assumed to have taken place. The membrane module should also be able to combine further tertiary treatment steps normally achieved by additional process units such as filtration, disinfection, and nitrification/denitrification.

The results of all the researchers clearly indicate that SMBRs can be used for good quality reclamation and reuse of domestic wastewater.

The SMBR plant for sewage treatment is simple in its construction and operation. With minimum process equipment, that is, no clarifiers, no sludge return pumps, which require periodic sludge control, and no sludge digestion. Thus, the process is easy to operate and maintain. The entire operation of such a plant is now based on MLSS control instead of the sludge volume index (SVI), sludge return ratios, and strict SRTs.

14.5 CONCLUSIONS

The SMBR appears technically more feasible for upgrade and retrofit of south wastewater treatment plants. It provides a high degree of treatment in terms of SS, turbidity, nitrogen, phosphorus, and organic removal. The phosphorus removal during continuous aeration was better than that under intermittent aeration conditions. Also, the phosphorus removal can be achieved at a high level with the use of a coagulant. Nitrification was quite successful, and therefore, the denitrification process was achieved in the same unit, providing the anoxic conditions with intervals of aeration. It also provides a high level of physical disinfection, filtration, and separation.

Problems, such as nonsettling, foam, and bulking sludge, do not affect the performance of SMBRs. The SMBR allows the use of long sludge ages, with the benefits of reduced sludge production, which is about 50%, when compared to a CASP, and a simplification of the sludge treatment line. The mass loading rates of SMBR fell within the CASP range, but the volumetric loading rates were two to four times those of CASP. This demonstrated that the volume of the bioreactor of the SMBR could be reduced in comparison with that of a CASP, if the same wastewater was treated. Therefore, a large amount of space and investment could be saved.

In addition, the SMBR effluent can be applied to irrigation of crops, flowers, and landscaped areas. Also, it may be used as process water such as cooling tower and metal processing. Furthermore, purified by reverse osmosis (RO) and ion exchange units, SMBR effluent could be used for water manufacture and, moreover, in consideration of the high price of ultrapure water (normally US\$10–20/m^3 in Iran). Upgrading and retrofitting of SWWTPI, Iran and also the reuse potential of SMBR effluent in this case could be promising.

REFERENCES

[1] Metcalf & Eddy. *Wastewater Engineering: Treatment, Disposal and Reuse*, 3rd ed. McGraw-Hill, New York, 1991.

[2] Stephenson T., S. Judd, B. Jefferson, and K. Brindle. *Membrane Bioreactors for Wastewater Treatment*. IWA Publishing, London, 2000.

[3] Dufresne R., H. Lavalle, R. Lebrun, and S. Lo. Comparison of performance between membrane bioreactor and activated sludge system for treatment of pulping process wastewater. *TAPPI Journal* 81 (1998) 131–135.

[4] Clech P. L., B. Jefferson, and S. Judd. Impact of aeration, solids concentration and membrane characteristics on the hydraulic performance of a membrane bioreactor. *Journal of Membrane Science* 218 (2003) 117–129.

[5] Halil H., K. Cumali, and U. Ayhan. Viability of microbial mass in a submerged membrane bioreactor. *Desalination* 150 (2002) 263–268.

[6] Brindle K. and T. Stephenson. Application of membrane biological reactors for the treatment of wastewaters. *Biotechnology and Bioengineering* 49 (1996) 601–610.

[7] Ueda T., K. Hata, and Y. Kikuoka. Treatment of domestic sewage from rural settlements by a membrane bioreactor. *Water Science and Technology* 34 (1996) 189–196.

[8] Devies W. J., M. S. Le, and C. R. Heath. Intensified activated sludge process with submerged membrane microfiltration. *Water Science and Technology* 38 (1998) 21–428.

[9] Guender B. and K. Krauth. Replacement of secondary clarification by membrane separation: results with plate and hollow fiber modules. *Water Science and Technology* 38 (1998) 383–393.

[10] Xing H., X. Wen, Y. Qian, and E. Tardieu. Microfiltration membrane-coupled bioreactor for urban wastewater reclamation. *Desalination* 141 (2000) 63–73.

[11] Gui P., X. Huang, Y. Chen, and Y. Qian. Effect of operational parameters on sludge accumulation on membrane surfaces in a submerged membrane bioreactor. *Desalination* 150 (2002) 185–194.

[12] Bouhbila H., B. Aimr, and H. Buisson. Fouling characterization in membrane bioreactors. *Separate Purification Technology* 1(22–23) (2001) 123–132.

[13] Choo K. H. and C. H. Lee. Membrane fouling mechanisms in the membrane-coupled anaerobic bioreactor. *Water Research* 30(8) (1996) 1771–1780.

[14] Tardieu E., A. Grasmick, V. Geaugey, and J. Manem. Fouling mechanisms in membrane bioreactors applied to wastewater treatment, in *Proceedings of the Seventh World Filtration Congress*, Budapest, Hungary, 1996, pp. 571–575.

[15] AWWA. Using reclaimed water to augment potable water resources. Washington, D.C., 1998.

PART IV

APPLICATIONS IN ENVIRONMENT

15

SURFACE TREATMENT AND CHARACTERIZATION OF POLYPROPYLENE HOLLOW FIBERS BY SOL-GEL METHOD FOR LIQUID PHASE MICROEXTRACTION

Mohd Marsin Sanagi, Yanuardi Raharjo, Wan Aini Wan Ibrahim, Ahmedy Abu Naim, Syairah Salleh, and Mazidatulakmam Miskam

15.1 INTRODUCTION

The term polymers is used for large molecules where simple repeating units of identical structure are joined by covalent bonds. Polymers comprise a large class of natural and synthetic materials with a variety of properties and purposes [1]. According to Hiemenz [2], polymers are exceedingly long molecules that consist of repeating molecular units called monomers. Monomers have relatively simple structures consisting of anywhere from 4 to 100 atoms per molecule. When chained together, they can form polymers consisting of hundreds of thousands of atoms per molecules.

Polymer degradation occurs when the properties of the polymer undergo several changes for instance tensile strength, color, and shape. Degradation could happen under the influence of one or more environmental factors such as heat or high temperature (thermal degradation), light (photo degradation), electromagnetic (γ ray, UV ray), moisture (hydrolysis), or chemical agents [1].

Sustainable Membrane Technology for Energy, Water, and Environment, First Edition.
Edited by Ahmad Fauzi Ismail and Takeshi Matsuura.

Such changes are primarily based on the chemical composition of the polymer. Polymers that contain functional groups such as polyester, polyamide, and polyurethane in their chains are much more subject to hydrolytic and biodegradation compared to polymers containing carbon–carbon backbones.

Sol-gel coating technique offers many advantages compared to the conventional coating techniques, such as strong adhesion between the coating and the bare fiber surface. This is due to the chemical bonding that is important to enhance operation stabilities involving temperature and solvents. In addition, the porous surface structure of sol-gel coating can provide high surface area that allows high extraction efficiency and ease of composition change for the coating and thus different selectivity [3].

In general, the degradation effect on the polymer should be minimized or prevented by understanding the structure of a polymer in order to recycle or reuse the polymer. This presents a problem in polymer required for short-term use and which are then discarded [4]. Thus, it is of interest to investigate the coating of polypropylene hollow fiber with poly(dimethylsiloxane) PDMS by sol-gel technology. In this work, the coated hollow fiber was applied in liquid phase microextraction system for the analysis of selected organophosphorus pesticides from environmental water samples.

15.2 METHODS

15.2.1 Reagent and Materials

Accurel Q3/2 polypropylene hollow fiber membranes (600 μmi.d., 200 μm wall thickness, and 0.2 μm pore size) were purchased from Membrane (Wuppertal, Germany). Hydrogen peroxide (H_2O_2) was obtained from QReC (Steinheim, Germany). Ferrosulfate ($FeSO_4$) was obtained from Fluka (Buchs, Switzerland). The sol-gel substances, methyltrimethoxysilane (MTMOS) 95%, hydroxyl-terminated polydimethylsiloxane (OH-TPDMS), trimethylmethoxysilane (TMMS) 99%, poly(methylhydroxysilane) (PMHS), and trifluoroacetic acid (TFA) 98% were purchased from Sigma-Aldrich (St. Louis, MO).

15.2.2 Instrumentations

The coated hollow fiber membranes were characterized to study their chemical properties and morphologies. Fourier transform infrared (FTIR) spectrophotometer spectra of the PDMS composite were recorded using a Perkin Elmer Spectrum One spectrometer (Ueberlingen, Germany) in the range 400–4000/cm. The thickness and surface texture of PDMS fiber coating was determined using JEOL JSM-6701 field emission scanning electron microscopy (FESEM) equipped with energy dispersive X-ray (EDX) analyzer JED-2300 series. The high performance liquid chromatography (HPLC) system used was a Perkin Elmer Series 600 LC interfaced with Series 200 vacuum degasser and UV detector. The column was a Nucleosil 100-5 C18 with particle size 5 μm, length 250 mm, and internal diameter 4.6 mm. The isocratic elution was carried out with acetonitrile-water (60:40) as the mobile phase at a flow rate of 1.0 mL/min. Data was interpreted using Perkin Elmer software Total-Chrom Navigator Series 200. The detection of analytes was carried out with the UV detector at 200 nm.

15.2.3 Preparation of PDMS-Coated Hollow Fiber Membranes

Initially, the polypropylene hollow fiber membranes were cut manually (1.5 cm length) and sealed at one edge by a sealer machine. The membranes were washed with acetone for several minutes to remove any contaminants and allowed to dry completely. The membranes were transferred into a small vial and subjected to Fenton's reaction in order to activate the surface of the fiber. The Fenton's reaction was performed by adding iron sulfate ($FeSO_4$) and hydrogen peroxide (H_2O_2) step-by-step in the vial containing the membranes. After Fenton's reaction was completed, the membranes were irradiated with a UV lamp for 1 hour at 365 nm. The fiber was dried at the room temperature prior to coating.

The PDMS sol-gel phase was prepared by mixing 300 μL MTMOS, 400 μL OH-TPDMS, 20 μL PMHS, and 200 μL 95% TFA in a plastic tube. The mixture was vortexed for 5 minutes before centrifuging at 12,000 rpm for 10 minutes to produce two layers. The top sol solution (clear solution) was transferred into another vial. The hollow fiber membranes were dipped vertically into the sol solution for 2 minutes. The coating technique is illustrated in Figure 15.1. After the coating process was completed, the coated fibers were end-capped in a 20% (v/v) methanolic solution of TMMS for 1 minute and then dried at room temperature for 1 day.

There are four major reactions that occur during the sol-gel formation [5], namely, hydrolysis, condensation and polymerization of monomers to form particles, growth of particles, and agglomeration of particles followed by the formation of network that extends throughout the liquid medium resulting in thickening, which forms a gel. Chong et al. [6] showed the reaction of coating process using PDMS as stationary phase, MTMOS as precursor, PMHS as

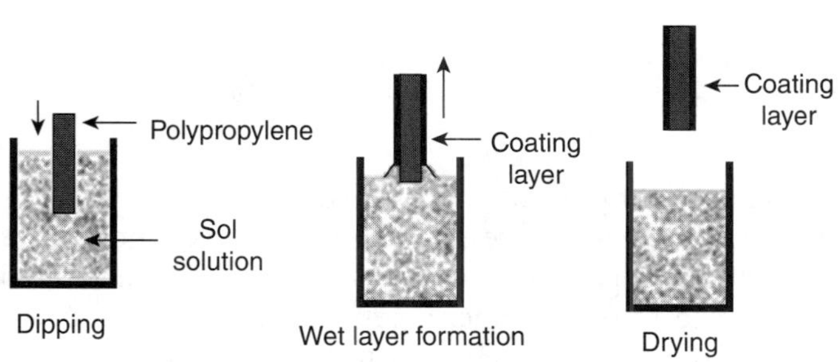

Figure 15.1. Sol-gel coating technique of PDMS fiber.

deactivation reagent, and TFA as an acid catalyst. The reactions can be represented by the following equations.

(a) Hydrolysis of the precursor, MTMOS catalyzed by TFA:

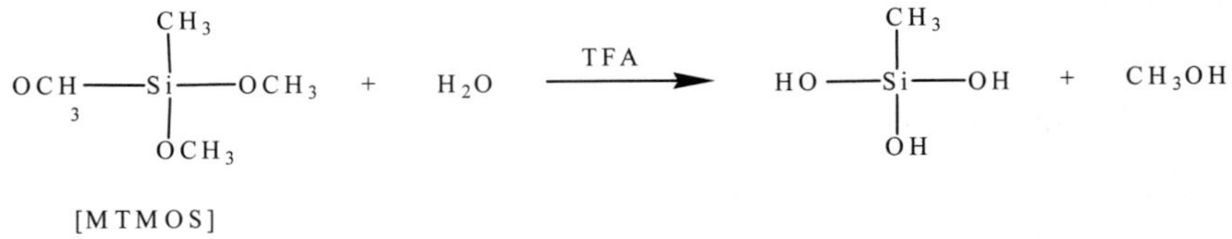

(15.1)

(b) Polycondensation of the hydrolyzed products and other sol-gel active moieties in the system. These reactions lead to the formation of a three-dimensional polymeric network:

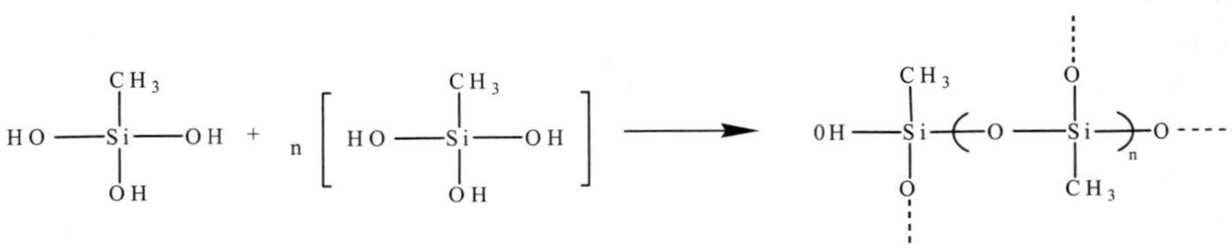

(15.2)

(c) Since OH-TPDMS is used as a coating material, this polymer is sol-gel active because it is hydroxyl-terminated. Selection of this polymer aimed at chemically binding the PDMS stationary phase to the growing silica network:

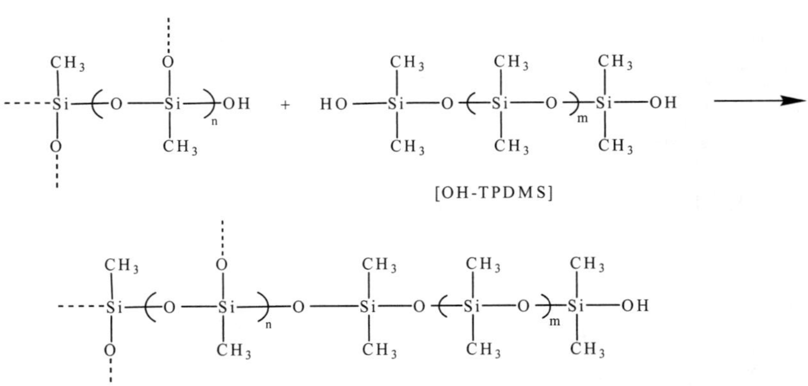

(15.3)

(d) Chemical anchoring of the evolving sol-gel polymer to the surface of the fiber to create a surface-bonded polymeric coating. The surface of fiber can also take part in the condensation reactions and provide chemical anchorage to the polymeric network in the immediate vicinity of the fiber surface. The PMHS was used as deactivation of the excess hydroxyl groups:

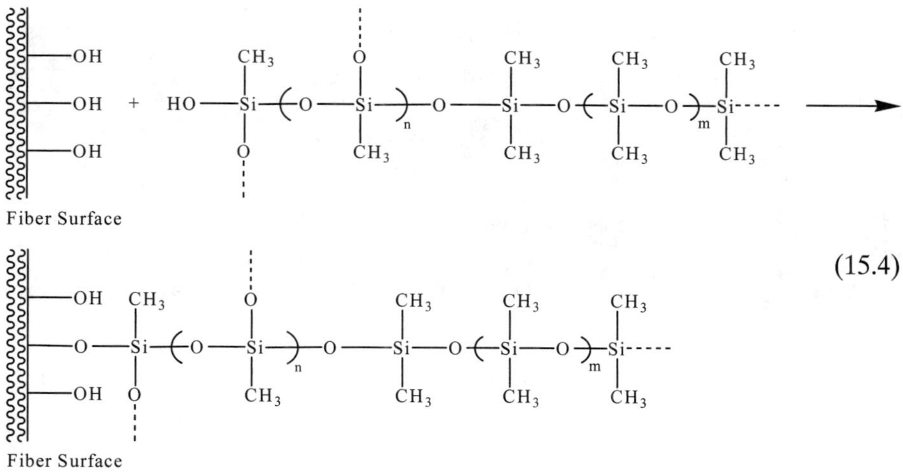

(15.4)

15.3 RESULTS AND DISCUSSIONS

15.3.1 Characterization of Physical and Chemical Properties

15.3.1.1 FTIR. Infrared spectrum of sol-gel solution used for the coating of the hollow fiber membrane is shown in Figure 15.2. The absorption band at 3350/cm shows the presence of O–H bond. It suggests that the hydroxyl group was not eliminated after the end-capping treatment. The hydroxyl group is used to allow reaction with activated polypropylene hollow fiber. The absorption band at 2963/cm shows the stretching of C–H bonds. The sharp absorption band at 1261/cm and 800/cm, and also small absorption band at 1412/cm, show the presence of Si–C bond from the sol-gel solution. The broad absorption bands around 1023/cm and 1094/cm correspond to stretching of Si–O bonds in the sol-gel solution.

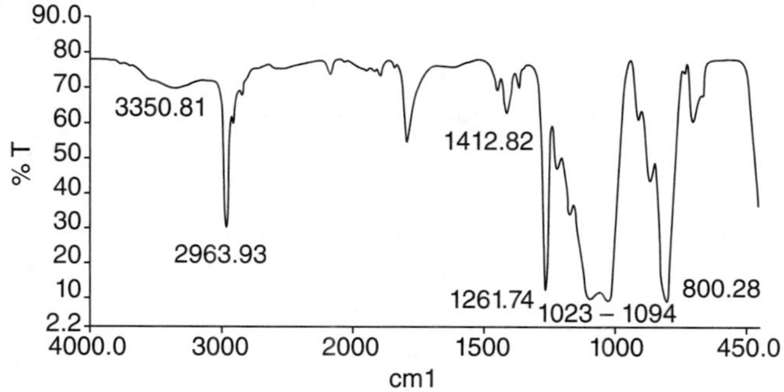

Figure 15.2. IR spectrum of sol-gel solution.

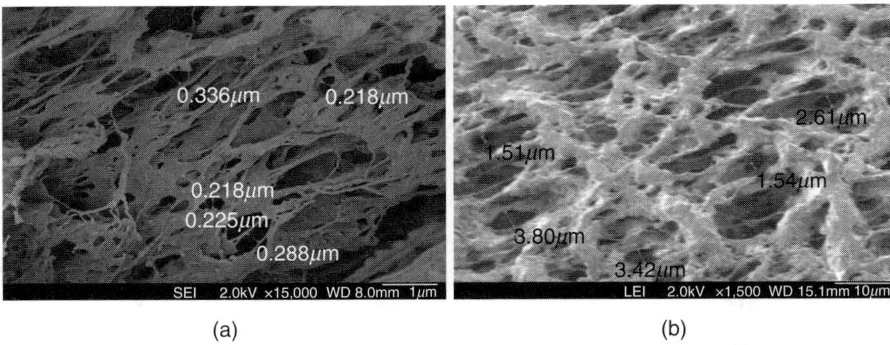

Figure 15.3. The SEM images: (a) original polypropylene fiber, (b) coated polypropylene fiber with dipping time of 2 minutes.

15.3.1.2 FESEM. In this technique, a fine electron beam (5–10 nm in diameter) was scanned across the sample surface in synchronization with a beam from a cathode ray. This instrument was used to identify the surface and the pore size of the coated fiber. The comparison between the original polypropylene fiber and coated fiber (under optimized conditions) was made. The surface morphology and the pore size of the original polypropylene fiber surface are shown in Figure 15.3a. It was found that the pore size values ranged from 0.2 to 0.3 μm (nominal value 0.2 μm).

Dipping time was optimized in order to obtain a coated hollow fiber with the best surface morphologies. It was found that when the membranes were dipped for 10 to 20 minutes, the pores were completely blocked. Therefore, short dipping time of 2 minutes was further studied, which resulted in coated membranes with good surface morphologies (Fig. 15.3b). The pore size was found to be essentially uniform by FESEM with values ranging from 1.5 to 3.8 μm. This dipping time (2 minutes) was therefore considered the optimum and was used for further experiments.

15.3.1.3 EDX. The analysis with EDX was carried out in order to determine the presence of hydroxyl groups which are crucial to allow the formation of covalent bonds with sol-gel solution. It was important to increase the ruggedness of the coated hollow fiber membranes during extraction process. In this work, EDX analysis was carried out on the original polypropylene hollow fiber and the results are presented in the form of plots of counts versus KeV and also percentage atom (Fig. 15.4a).

Based on the results obtained, it was found that only carbon was present (atom 100%) for the uncoated hollow fiber membranes (Fig. 15.4a). Pt (platinum) was also detected in the spectrogram because the polypropylene sample was coated with platinum before EDX analysis. On the other hand, as the membranes that have undergone Fenton's reaction (Fig. 15.4b) show, the EDX analysis contains 28.19% oxygen atom, 7.01% Fe atom that was produced from

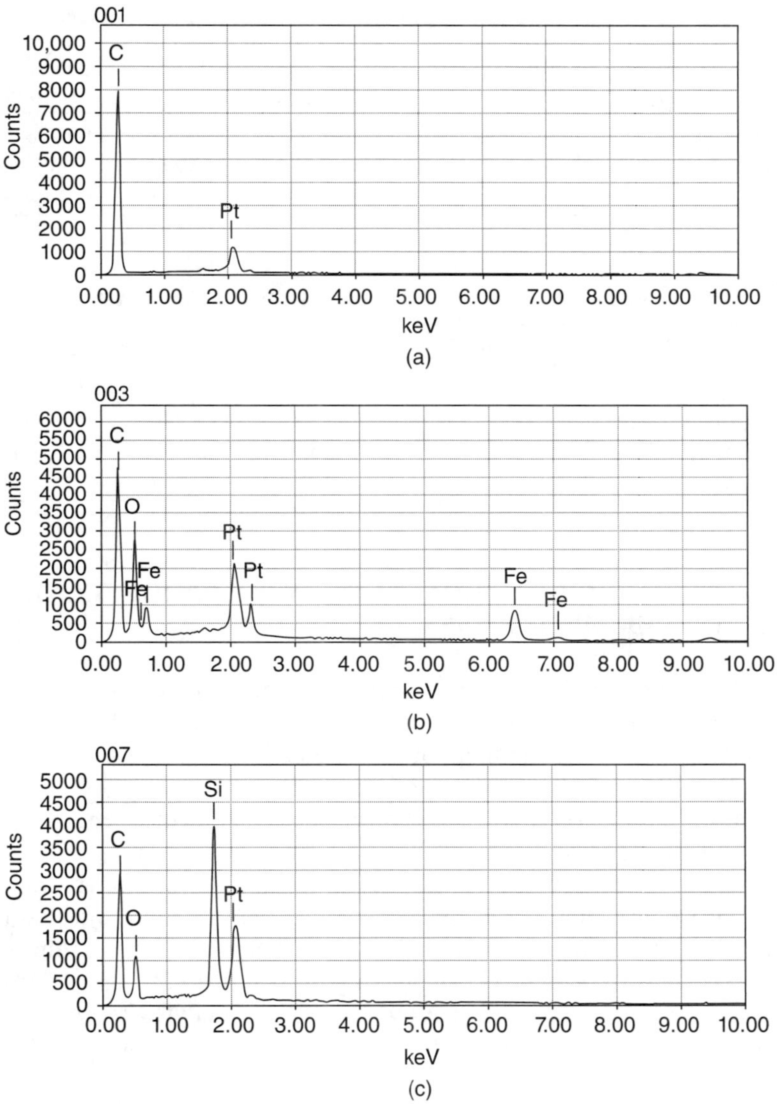

Figure 15.4. The EDX spectrogram of (a) uncoated polypropylene hollow fiber, (b) polypropylene hollow fiber after Fenton's reaction, (c) coated polypropylene hollow fiber.

the Fenton reaction residue, and 64.80% carbon atom. For the coated hollow fiber membranes (Fig. 15.4c), the atoms detected were carbon, oxygen, and silica at relative amounts of 72.84%, 18.24%, and 8.92%, respectively. Thus, these results confirmed that the polypropylene hollow fiber membranes were successfully coated using sol-gel technique.

TABLE 15.1. Analytical Performances of Uncoated Hollow Fiber LPME

Analytes	Recovery (%)	Enrichment Factor (EF)	Concentration Range (μg/L)	Correlation (r^2)	LOD (μg/L)	RSD (%)
Methidation	61.82	136	100–1250	0.9984	67.08	1.40
Quinalphos	62.99	138	25–750	0.9998	14.86	0.99
Profenofos	64.74	142	10–100	0.9982	5.42	0.80

TABLE 15.2. Analytical Performances of Coated Hollow Fiber LPME

Analytes	Recovery (%)	Enrichment Factor (EF)	Concentration Range (μg/L)	Correlation (r^2)	LOD (μg/L)	RSD (%)
Methidation	84.93	186	50–750	0.9997	16.90	1.73
Quinalphos	92.35	203	5–75	0.9996	1.96	1.41
Profenofos	97.85	215	1–50	0.9999	0.64	0.52

15.4 ANALYTICAL PERFORMANCES OF POLYPROPYLENE HOLLOW FIBER MEMBRANES

The PDMS-coated hollow fiber membranes were applied in liquid phase microextraction for the analysis of organophosphorus pesticides. A comparison of the analytical performances of uncoated (Table 15.1) and coated (Table 15.2) hollow fiber membranes were carried out.

Under optimum conditions (acceptor phase isooctane, length of fiber 1.5 cm, volume of acceptor phase 3 μL, donor phase 11 mL, stirring rate 1260 rpm, and extraction time 40 minutes), the liquid phase microextraction (LPME) procedure showed a satisfactory linear behavior in the tested range, with correlation coefficients ranging from 0.9982 to 0.9998 (Table 15.1). Acceptable recoveries were obtained ranging from 61.82% to 64.74%. The calculated limits of detection (LOD) of target analytes were less than 67.08 μg/L. The enrichment factors obtained were between 136- and 142-fold.

For the PDMS-coated hollow fiber membranes, analytical performances were carried out under the optimal conditions: acceptor phase *n*-nonane, length of fiber 1.5 cm, volume of acceptor phase 4 μL, donor phase 11 mL, stirring rate 1260 rpm, and extraction time 60 minutes.

The analysis of the analytes by coated fiber LPME exhibited good reproducibilities with relative standard deviation (RSD) ranging from 0.52% to 1.73%. The calibration curves studied were in the range of 50–750 ppb for methidation, 5–75 ppb for quinalphos, and 1–50 ppb for profenofos. Within the calibration range, the analytes showed good linearity with correlations coefficient (r^2) ranging from 0.9996 to 0.9999. The analyte recoveries obtained ranged from 84.93% to 97.85%. The results indicated that the analytes were not completely recovered from the acceptor solution probably due to slight acceptor phase leak-out during extraction. The calculated LOD of target analytes were less than 16.90 μg/L. Meanwhile, the enrichment factors achieved were between 186- and 215-folds.

The analyte recoveries for coated fiber were considerably higher (up to 97.85%) compared to those for uncoated fiber (up to 64.74%). The analyte enrichment factors for the coated fiber were between 186- and 215-fold whereas for uncoated fiber were only 136- to 142-fold. These results showed that the coated fiber offered higher extraction efficiency and higher enrichment factor of the target analytes compared to uncoated fiber. It is probably due to the fact that besides extraction via diffusion through pores of fiber, the analytes were also extracted via adsorption process on the coated fiber.

15.5 CONCLUSIONS

The sol-gel technology was successfully developed to prepare polypropylene hollow fiber membranes coated by PDMS. The coated hollow fiber membranes were applied in liquid phase microextraction for the analysis of selected organophosphorus pesticides. Based on the analytical performances for both uncoated and coated hollow fiber membranes, the PDMS-coated hollow fiber membranes provide better performances in term of analyte recoveries and enrichment compared to the uncoated hollow fiber membranes. The application of coated hollow fiber membrane was proven to be inexpensive, easy to operate, and robust.

REFERENCES

[1] Fried J. R. *Polymer Science & Technology*, 2nd ed. Prentice Hall, Upper Saddle River, NJ, 2003, pp. 80–105.

[2] Hiemenz P. C. *Polymer Chemistry (The Basic Concepts)*. Marcel Dekker, New York, 1984, pp. 49–63.

[3] Yu J., L. Dong, C. Wu, and J. Xing. Hydroxyfullerenceas a novel coating for solid phase microextraction fiber with sol-gel technology. *J. Chromatogr. A.* 978 (2002) 37–48.

[4] Grassie N. and G. Scott. *Polymer Degradation and Stabilization*. Cambridge University, New York, 1985, pp. 79–103.

[5] Wilson M., K. Kannangara, G. Smith, M. Simmins, and B. Raguse. *Nanotechnology, Basic Science and Emerging Technology*. Chapman and Hall/CRC, Australia, 2002, pp. 59–94.

[6] Chong S. W., D. Wang, J. D. Hayes, B. W. Wilhite, and A. Malik. Sol-gel coating technology for the preparation of solid phase microextraction fibers of enhanced thermal stability. *Anal. Chem.* 69 (1997) 3889–3898.

16

EFFECT OF DIFFERENT ADDITIVES ON THE PROPERTIES AND PERFORMANCE OF POROUS POLYSULFONE HOLLOW FIBER MEMBRANES FOR CO_2 ABSORPTION

Amir Mansourizadeh, Ahmad Fauzi Ismail, and Mohammad Ali Aroon

16.1 INTRODUCTION

Recently, porous membranes have attracted considerable attention in the area of membrane gas absorption [1–6]. The high hydrophobicity, low membrane transfer resistance, and excellent chemical resistance to various chemical-feed streams are the essential properties of these membranes. Polypropylene (PP), polyethylene (PE), polytetrafluourethylene (PTFE) and poly(vinylidene fluoride) (PVDF) are the most popular hydrophobic polymeric materials. However, since PE, PP, and PTFE cannot be dissolved in solvents, and the membranes are usually fabricated using stretching and thermal methods that give symmetric structure with relatively low porosity. On the other hand, because of critical surface tension of PVDF, the rate of exchange between solvent and nonsolvent is very low, and it is difficult to spin hollow fibers, particularly when using high percentage of solvent for the bore fluid. Asymmetric hollow fiber

Sustainable Membrane Technology for Energy, Water, and Environment, First Edition.
Edited by Ahmad Fauzi Ismail and Takeshi Matsuura.

membranes with low mass transfer resistance could be fabricated by increasing the outer surface porosity (high permeability) and removing the inner skin layer during spinning process. Moreover, for the membrane wetting pressure, despite depending on the property of the membrane material, it could also be adjusted by the membrane pore size, as suggested by the Laplace–Young equation.

Until now, only a few studies have been carried out to fabricate porous membranes with high permeability, using different additives [7–10]. Yeow et al. [11] have categorized additives used in the fabrication of PVDF membranes. Generally, low molecular weight water soluble additives that can be used to fabricate porous membranes include polymeric additives such as poly(vinyl pyrrolidone) and poly(ethylene glycol), organic and inorganic acids such as phosphoric acids, acetic acid and propionic acid, weak nonsolvents such as glycerol, ethylene glycol, and diethylene glycol, weak co-solvents such as methanol, ethanol, acetone, and inorganic salts such as lithium chloride (LiCl), lithium nitrate ($LiNO_3$), and lithium perchlorate ($LiClO_4$). Han [8] fabricated polysulfone (PSf) membranes using propionic acid in the polymer solution and found that the precipitation rate of the polymer solution was enhanced, resulting in the increase of membrane permeability. Addition of glycerol and phosphoric acid in the PVDF polymer dope showed a larger pore size and a higher value of MWCO compared to distilled water that improved the membrane permeability [12]. High molecular weight additive like polyvinylpyrrolidone (PVP) could enhance phase separation of spinning solution. Results of the experiments by Han and Nam [13] showed that adding 5 wt % of PVP in 15 wt % PSf solution enhanced exchange of solvent and nonsolvent during membrane fabrication. It also improved the membrane permeate flux by a factor of one. However, further increment of PVP in the solution resulted in delayed demixing and the permeate flux decreased, which reveals that the thermodynamic enhancement is overtaken by the rheological hindrance in demixing of the solution. Moreover, addition of LiCl into the PVDF polymer dope resulted in enhancement of porosity and permeation flux of the membrane. However, mechanical strength of the membranes decreased with increasing the amount of LiCl [14].

PSf has been widely used as a membrane material due to its mechanical strength, thermostability, stability against chemicals, and relatively high hydrophobicity. PSf is also an excellent material for spinning hollow fiber membranes. In addition, the above properties of this polymer provide a potential application in the membrane gas absorption processes. Thus, the main objective of this study is to prepare porous PSf hollow fiber membranes with high surface porosity (high permeability) and small pore size using different additives in the spinning solution, which are favorable for CO_2 absorption. The effects of additives (ethanol, PEG200, glycerol, and acetic acid) on the precipitation rate, measured by the cloud point method, on the membrane structure and on the CO_2 absorption performance are studied.

16.2 EXPERIMENTAL

16.2.1 Materials

PSf polymer pellets (Udel P-1700) supplied by Solvay Advance Polymers were predried at 60°C before use. *N*-methyl-1-pyrrolidone (NMP, >99.5%), was supplied by Merck and used as solvent without further purification. Glycerol (anhydrous extra pure), ethanol, PEG200, and acetic acid were purchased from Merck (Germany) and used as nonsolvent additive in the polymer solution. Tap water was used as coagulation medium in all cases.

16.2.2 Measurement of Cloud Point

Using constant weight ratio of the additive/solvent (1/6) in different polymer concentrations (range: 1–17% wt.) the polymer dopes were obtained. The cloud point data were measured using titration method. The polymer solutions were prepared in a sealed bottle using magnet stirrer. Nonsolvent (distilled water) was slowly added into the polymer dope under constant agitation at 25°C. In cases when local precipitation occurred, especially at higher polymer concentration, agitation was continued until the solution became homogeneous again. Then further addition of nonsolvent was performed until the titration end point was reached (the solution became permanently turbid). Compositions at the cloud point were calculated as weight percent of each component.

16.2.3 Fabrication of PSf Hollow Fiber Membranes

The PSf polymer (pellet form) was dried at 60 ± 2°C in a vacuum oven for 24 hours to remove moisture content. The spinning dope solutions of 17 wt% PSf, 6 wt% additive, and 77 wt% NMP were prepared (at room temperature 25–27°C) using stirring until the solution became homogeneous.

The hollow fiber spinning process by the dry-jet wet phase inversion was described elsewhere in detail [15]. The dope solutions were degassed before spinning. Aqueous solution of 95 wt% NMP was used as bore fluid in order to remove inner skin layer. Table 16.1 lists the detailed spinning parameters.

The as-spun fibers were immersed in water for 3 days to remove the residual NMP and the additives. Then they were posttreated by the solvent exchange method using methanol to minimize fiber deformation and pore collapse before drying at room temperature.

16.2.4 Gas Permeation Method

The determination of pore size and surface porosity over effective pore length is important in studying mass transfer in the porous asymmetric membranes for gas absorption. The total gas permeation mechanism through the

TABLE 16.1. Spinning Condition of Fabricating PSf Hollow Fiber Membranes

Dope extrusion rate (ml/min)	4.50
Bore fluid composition (wt %)	NMP/H_2O 95:5
Bore fluid flow rate (mL/min)	1.50
External coagulant	Tap water
Air gap distance (cm)	0.0
Spinneret o.d./i.d. (mm)	1.20/0.6
Coagulation temperature (°C)	25
Room relative humidity (%)	70–75

asymmetric porous membrane is the combination of Poiseuille flow and Knudsen flow [16]. Pure N_2 was used as the test gas. The test apparatus used was based on the volume displacement method [1]. The test module containing two hollow fibers with the length of about 10 cm was used to determine gas permeability. The upstream pressure was in the range from 0.5 to 4 bar (gauge). The N_2 permeation rate from the lumen side was measured at 25°C using soap-bubble flowmeter. The gas permeance (pressure normalized flux) was then calculated according to outer diameter of the hollow fiber.

16.2.5 Critical Water Entry Pressure (CEP_w), Maximum Pore Size, and Porosity

The same test module as above was used to determine the critical water entry pressure. Distilled water was fed into the lumen side of the hollow fibers using a diaphragm pump. The pressure was slowly increased at 0.5 bar interval. At each pressure, the membrane module was kept at the constant pressure for 30 minutes to check if any water droplet appeared in the outer surface of the fiber. Critical water entry pressure was considered as the pressure for the first water droplet in the outer surface of the hollow fiber.

Maximum pore size, d_{max}, of the hollow fibers was calculated using Laplace–Young equation:

$$\Delta P = \frac{4\gamma \cos\theta}{d_{\max}}, \tag{16.1}$$

where ΔP was replaced with CEP_w, θ and γ are contact angle and surface tension of water, respectively.

The membrane porosity, ε_m, was determined by gravimetric method. It is defined as the volume of the pores divided by the total volume of the membrane. The calculation method is available in the literature [17]:

$$\varepsilon_m = \frac{(w_1 - w_2)/\rho_w}{(w_1 - w_2)/\rho_w + w_2/\rho_p}, \tag{16.2}$$

where w_1 is the weight of the wet membrane, w_2 is the weight of the dry membrane, ρ_w is water density, and ρ_p is the polymer density.

16.2.6 CO_2 Absorption Experiment

The spun hollow fibers were used in a gas-liquid membrane contactor to measure CO_2 absorption flux. A total of 10 hollow fibers (i.d. = 0.55 mm, o.d. = 1 mm) were packed randomly in a stainless steel membrane module with 15 mm diameter and 150 mm effective length. Pure CO_2 was employed as the feed gas in the shell side and distilled water as liquid absorbent in the lumen side. In this study, countercurrent flow was used in which the gas and liquid flow rate were controlled using control valve and flowmeter. Using titration method, CO_2 concentration in the liquid outflow at various flow rates was measured to determine the CO_2 absorption flux.

16.3 RESULTS AND DISCUSSION

16.3.1 Effect of Additives on the Phase Inversion Behavior

Generally, the precipitation rate is an important parameter for fabricating membranes since rapid precipitation results in porous, anisotropic membranes [18]. The addition of additives to polymer dope was found to reduce miscibility area of the system and increase precipitation rate of the solution [19]. As it can be seen in Figure 16.1, with addition of the additives into the polymer solutions, the binodal lines shift toward the polymer-solvent axis, which means less nonsolvent is needed for precipitation of the polymer solution at the

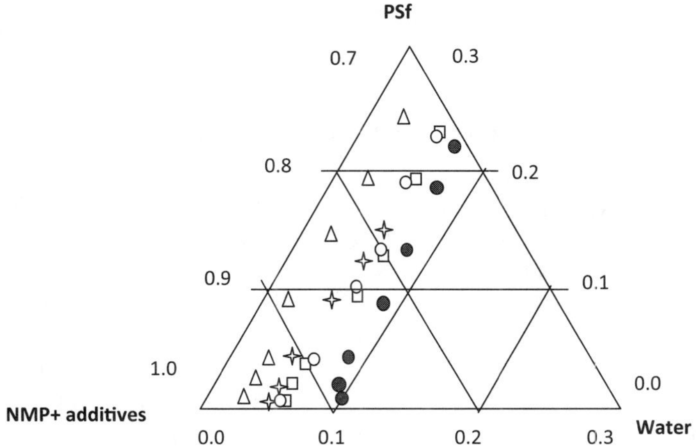

Figure 16.1. Effects of additives on the cloud point diagram of PSf/NMP/water/additive system at 25°C, •, without additive; □, PEG200; ○, ethanol; Δ, glycerol; ✧, acetic acid.

thermodynamic equilibrium. At 17 wt % PSf concentration, water content required for precipitation is 5.5 wt % for PSf/NMP solution compared with 4.05 wt % for the solution with PEG200, 3.9 wt % for ethanol, 3.4 wt % for acetic acid, and 1.15 wt % for glycerol. Therefore, the solution precipitation rates follow the trend of glycerol > acetic acid > ethanol > PEG200. Hence, these additives act as phase inversion promoter favoring for porous membrane fabrication.

16.3.2 Effect of Additives on the Structure and CEP_w

In this study, an attempt was made to prepare hollow fibers with improved permeability (high surface porosity) and small pore size (high hydrophobicity) favorable for gas absorption application. For this purpose, different kinds of small molecular weight additives were employed to fabricate porous membranes. The membranes were characterized in terms of critical water entry pressure, gas permeability, and porosity. The results are given in Table 16.2.

It can be seen that all the membranes possess a total porosity more than 70%. It can be attributed to low polymer concentration, since fabricated PSf membrane without additive also showed a high porosity. Among the additives, PEG200 made a structure more open with 75.3% porosity. Therefore, it seems that polymeric additives can form large microvoids and cavities near the outer skin layers, since inner skin layer was almost removed using 95 wt % NMP/H_2O in the bore side, a solution that has a composition very close to the binodal line on the ternary phase diagram. Same results were observed by adding PVP as polymeric pore-forming additive in PVDF dope solution by Kong and Li [20]. Results showed that effect of glycerol, ethanol, and acetic acid as additives on total porosity of the membranes was not significant.

However, the gas permeation method suggested by Yasuda and Tsai [21] was useful in determining the volume porosity of porous membranes. Li et al.

TABLE 16.2. Properties of Fabricated PSf Hollow Fiber Membranes

Membrane	Pressure Normalized Flux of N_2 at 1 Bar (cm^3/ $cm^2 \cdot s \cdot cmHg$) × 10^3	Average Pore Size (μm)	Effective Surface Porosity ε/L_p (10^{-2}/m)	CEP_w (bar)	Maximum Pore Size (μm)	Porosity %
PSf	2.90	0.012	285	3.0	0.486	71.05
PSf + PEG200	3.73	0.011	436	2.50	0.560	75.30
PSf + Ethanol	4.91	0.009	581	2.50	0.427	71.50
PSf + Glycerol	4.59	0.014	349	4.0	0.365	72.25
PSf + Acetic acid	7.90	0.013	775	2.50	0.40	71.35

[22] introduced a modified gas permeation method to determine the mean pore size and the effective surface porosity over the effective pore length of the asymmetric membrane. In general, the magnitude of mean pore size does not have any significant physical meaning, especially for the membranes prepared by phase inversion processes. It represents a parameter which can be compared quantitatively for the membranes prepared under different spinning conditions [23]. Average pore size and effective surface porosity of the fabricated membranes are shown in Table 16.2. The results revealed that addition of the additives increased surface porosity and gas permeability of the membranes. It can be attributed to enhancement of precipitation rate of the polymer solutions during spinning process. However, the additives did not affect the pore size significantly. Ethanol and acetic acid resulted in the membranes with high permeability and porosity. Glycerol increased the average pore size and gas permeability although it did not significantly affect the surface porosity. Based on the cloud point diagram, glycerol has the highest precipitation rate that can cause rapid phase inversion. As reported by Kesting [24], a large finger-like macrovoids with dense skin layer structure is generally formed when the phase inversion process is fast. It seems that adding glycerol to the spinning dope produces a membrane with porous dense skin layer. For PEG200, relatively low gas permeation flux can be attributed to the low porosity outer skin layer because of lower precipitation rate compared to the other additives.

As for the critical water entry pressure test, all the asymmetric membranes with an outer skin layer can withstand the pressure without water penetration into the pores during CO_2 absorption process in the membrane contactor, although as high critical entry pressure of water as possible is desirable for long-term gas absorption application. Generally, the liquid side operating pressure for gas-liquid membrane contactor should always be kept higher than the gas side by at least 0.2–0.5 bar to prevent bubble formation in the liquid phase, since it will result in loss of gas components and operating stability [25]. Among the membranes, ethanol, acetic acid, and PEG200 in the spinning dope gave the membrane with low CEP_w. It seems that these additives result in a thin porous skin layer due to higher precipitation rate. Moreover, polymeric additive may not be washed out completely in the water bath and remain in the membrane, decreasing the membrane hydrophobicity.

16.3.3 CO_2 Absorption Experiment

The PSf hollow fiber membranes were fabricated using a wet spinning method with four different additives in the spinning solution. For each membrane, same membrane contactor module was used to measure CO_2 absorption flux. The characteristics of the membrane contactor are listed in Table 16.3. Absorption of pure CO_2 in distilled water was carried out in continuous mode at 25°C. The gas flow rate on the shell side was controlled at 100 mL/min in counter-current with the liquid flow on the lumen side. The gas side pressure was set

TABLE 16.3. Characteristics of the Membrane Contactor Used

Module length (mm)	270
Fiber o.d. (mm)	1
Fiber i.d. (mm)	0.55
Effective fiber length (mm)	150
Number of fibers	10
Packing density (%)	8.8
Contact area (cm^2)	94.25

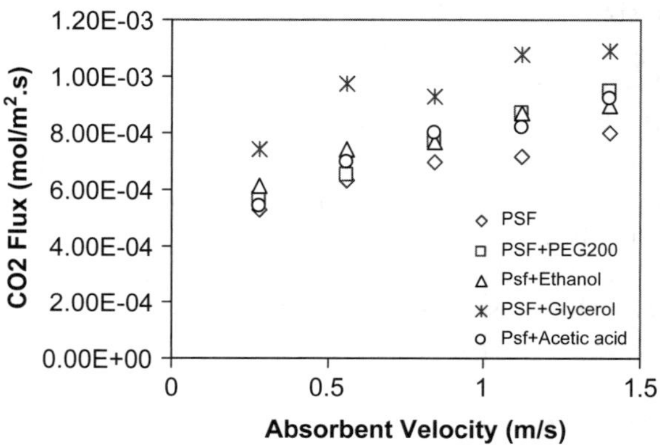

Figure 16.2. CO_2 absorption flux versus absorbent velocity in the gas-liquid membrane contactor.

at 1 bar while the liquid side pressure was maintained at a pressure 0.2 bar higher than the gas side to avoid bubble formation in the liquid. Before taking the samples, all the experiments were carried out for 30 minutes to achieve a steady-state condition. Effect of absorbent (water) velocity on CO_2 absorption flux is shown in Figure 16.2 for the different membranes.

From Figure 16.2, the CO_2 absorption flux increased with absorbent velocity for all the membranes in the same pattern. It can be attributed to disturb liquid boundary layer in the lumen side that prevents CO_2 saturation. In the case of physical absorption in membrane contactors, liquid phase resistance controls the mass transfer process [3, 26]. Therefore, increasing liquid velocity enhances the overall mass transfer coefficient. The membranes prepared without additive showed a low absorption flux. It can be attributed to lower porosity although the CEPw is more than the membranes with ethanol, PEG200, and acetic acid as additives. The results show that the additives contributed to the formation of more porous membranes with higher contact areas between gas and liquid. The membranes prepared with PEG200, acetic acid, and ethanol as additives showed sufficient permeabilities with small pore sizes (see Table 16.2). However, the relatively low CO_2 absorption flux can be attributed to the lower hydrophobicity of those membranes. It seems that these additives make the membrane less hydrophobic. CEP_w for these membranes are relatively low (2.5 bar), most likely due to the presence of residual additives in the membranes. Therefore, the membranes can become wet gradually during the CO_2 absorption experiment. The partial wetting makes the overall mass transfer resistance increase rapidly and significantly affects the absorption flux during operation [4]. Among the fabricated membranes, glycerol in the spinning solution resulted in higher CO_2 absorption flux compared to the other additives. A maximum CO_2 flux (1.09×10^{-3} mol/m^2 s) was achieved at 1.4 m/s absorbent velocity. As mentioned before, the precipitation rate of polymer solution with glycerol is so high that the cloud point curve is close to polymer-solvent axis (see Fig. 16.1). Therefore, rapid precipitation resulted in the membranes with a porous skin layer that can withstand wetting during CO_2 absorption operation. Comparing with the other additives, glycerol gave the membranes with $CEP_w = 4$ bar, which is higher than any other membranes.

16.4 CONCLUSION

Porous PSf hollow fiber membranes were fabricated with different additives such as PEG200, ethanol, glycerol, and acetic acid in the spinning dope. The effect of additives on the cloud point diagram, membrane structure, and CO_2 absorption performance were investigated. Cloud point results revealed that precipitation rate of the polymer solutions was followed the trend of glycerol > acetic acid > ethanol > PEG200. Results of gas permeation experiments showed that an increase in the precipitation rate by the presence of the additives in the polymer dope resulted in an increase in surface porosity (gas permeability). All the membranes had relatively small pore sizes, although the effect of precipitation rate of the polymer solution on the pore size was not significant. Results of CO_2 absorption experiments showed that CPE_w and surface porosity were the two factors that affected the absorption flux of the membranes. Among the additives, glycerol produced membrane with the highest CPE_w together with one of the highest surface porosity. Therefore, glycerol in the polymer dope resulted in the membrane with higher CO_2 absorption flux compare to the other additives.

REFERENCES

[1] Li K., D. Wang, C. C. Koe, and W. K. Teo. Use of asymmetric hollow fiber modules for elimination of H_2S from gas streams via membrane absorption method. *Chem. Eng. Sci.* 53 (1998) 1111–1119.

[2] Feron P. H. M. and A. E. Jensen. CO_2 separation with polyolefin membrane contactors and dedicated absorption liquids: performance and prospects. *Sep. Purif. Technol.* 27 (2002) 231–242.

[3] Dindore V. Y., D. W. F. Brilman, P. H. M. Feron, and G. F. Versteeg. CO_2 absorption at elevated pressures using a hollow fiber membrane contactor. *J. Memb. Sci.* 235 (2004) 99–109.

[4] Wang R., H. Y. Zhang, P. H. M. Feron, and D. T. Liang. Influence of membrane wetting on CO_2 capture in microporous hollow fiber membrane contactors. *Sep. Purif. Technol.* 46 (2005) 33–40.

[5] Bottino A., G. Capannelli, A. Comite, R. D. Felice, and R. Firpo. CO_2 removal from a gas stream by membrane contactor. *Sep. Purif. Technol.* 59 (2008) 85–90.

[6] Parka H. H., B. R. Deshwala, I. W. Kimb, and H. K. Lee. Absorption of SO2 from flue gas using PVDF hollow fiber membranes in a gas-liquid contactor. *J. Memb. Sci.* 319 (2008) 29–37.

[7] Wang D., K. Li, and W. K. Teo. Porous PVDF asymmetric hollow fiber membranes prepared with the use of small molecular additives. *J. Memb. Sci.* 178 (2000) 13–23.

[8] Han M. J. Effect of propionic acid in the casting solution on the characteristics of phase inversion polysulfone membranes. *Desalination* 121 (1999) 31–39.

[9] Xu Z. L. and F. A. Qusay. Polyethersulfone (PES) hollow fiber ultrafiltration membranes prepared by PES/non-solvent/NMP solution. *J. Memb. Sci.* 233 (2004) 101–111.

[10] Stropnik C., V. Kaiser, V. Musil, and M. Brumen. Wet-phase-separation membranes from the polysulfone/N,N-dimethylacetamide/water ternary system: the formation and elements of their structure and properties. *J. Appl. Polym. Sci.* 96 (2005) 1667–1674.

[11] Yeow M. L., Y. T. Liu, and K. Li. Morphological study of poly(vinylidene fluoride) asymmetric membranes: effects of the solvent, additive, and dope temperature. *J. Appl. Polym. Sci.* 92 (2004) 1782–1789.

[12] Atchariyawut S., C. Feng, R. Wang, R. Jiraratananon, and D. T. Liang. Effect of membrane structure on mass-transfer in the membrane gas–liquid contacting process using microporous PVDF hollow fibers. *J. Memb. Sci.* 285 (2006) 272–281.

[13] Han M. J. and S. T. Nam. Thermodynamic and rheological variation in polysulfone solution by PVP and its effect in the preparation of phase inversion membrane. *J. Memb. Sci.* 202 (2002) 55–61.

[14] Tomaszewska M. Preparation and properties of flat-sheet membranes from polyvinylidene fluoride for membrane distillation. *Desalination* 104 (1996) 1.

[15] Ismail A. F., I. R. Dunkinb, S. L. Gallivanb, and S. J. Shilton. Production of super selective polysulfone hollow fiber membranes for gas separation. *Polymer* 40 (1999) 6499–6506.

[16] Wang D. L., K. Li, and W. K. Teo. Effects of temperature and pressure on gas permselection properties in asymmetric membranes. *J. Memb. Sci.* 105(8) (1995) 9.

[17] Luo F., J. Zhang, X. L. Wang, J. F. Cheng, and Z. J. Xu. Formation of hydrophilic EAA copolymer microporous membranes via thermally induced phase separation. *Acta Polym. Sin.* 1 (5) (2002) 566–571.

[18] Baker R. W. *Membrane Technology and Application*, 2nd ed. John Wiley & Sons Ltd, West Sussex, England, 2004.

[19] Yewo M. L., Y. T. Liu, and K. Li. Isothermal phase diagrams and phase-inversion behavior of poly(vinylidene fluoride)/solvents/additives/water systems. *J. Appl. Polym. Sci.* 90 (2003) 2150–2155.

[20] Kong J. and K. Li. Preparation of PVDF hollow-fiber membranes via immersion precipitation. *J. Appl. Polym. Sci.* 81 (2001) 1643–1653.

[21] Yasuda H. and J. T. Tsai. Pore size of microporous polymer membranes. *J. Appl. Polym. Sci.* 18 (1974) 805.

[22] Li K., J. F. Kong, D. Wang, and W. K. Teo. Tailor-made asymmetric PVDF hollow fiber for soluble gas removal. *AIChE J.* 45 (1999) 1211–1219.

[23] Shih H. C., Y. S. Yeh, and H. Yasuda. Morphology of microporous poly(vinylidene fluoride) membranes studied by gas permeation and scanning microscopy. *J. Memb. Sci.* 50 (1990) 299–317.

[24] Kesting R. E. *Synthetic Polymeric Membranes*, 2nd ed. Wiley, New York, 1985.

[25] Malek A., K. Li, and W. K. Teo. Modeling of microporouse hollow fiber membrane modules operated under partially wetted condition. *Ind. Eng. Chem. Res.* 36 (1997) 784–793.

[26] Rangwala H. A. Absorption of carbon dioxide into aqueous solutions using hollow fiber membrane contactors. *J. Memb. Sci.* 112 (1996) 229–240.

17

ABSORPTION OF CARBON DIOXIDE THROUGH FLAT-SHEET MEMBRANES USING VARIOUS AQUEOUS LIQUID ABSORBENTS

Abdul Latif Ahmad, Sunarti Abd Rahman, and W. James Noel Fernando

17.1 INTRODUCTION

Microporous membrane gas absorption (MGA) became important in the late 1990s when global warming issues were identified, and it was realized that greenhouse gases needed to be treated as a waste and captured. As a way to address the greenhouse effect, MGA is considered a promising alternative to conventional and potential large-scale technology for the recovery and removal of carbon dioxide (CO_2) [1]. CO_2 is released into the atmosphere from the combustion of fossil fuels, and it is thought to contribute to undesired global warming. The operating principle of MGA is schematically shown in Figure 17.1.

Some of the more novel methods for CO_2 separation from flue and fuel gas streams include electrochemical pumps, chemical looping approaches, and gas separation membranes. Electrochemical membranes have the potential to economically separate CO_2 from flue gas. However, the material science issues such as reducing corrosion (molten carbonate pumps) and increasing conductivity (solid electrolyte pumps) need to be addressed. Meanwhile, the

Sustainable Membrane Technology for Energy, Water, and Environment, First Edition.
Edited by Ahmad Fauzi Ismail and Takeshi Matsuura.

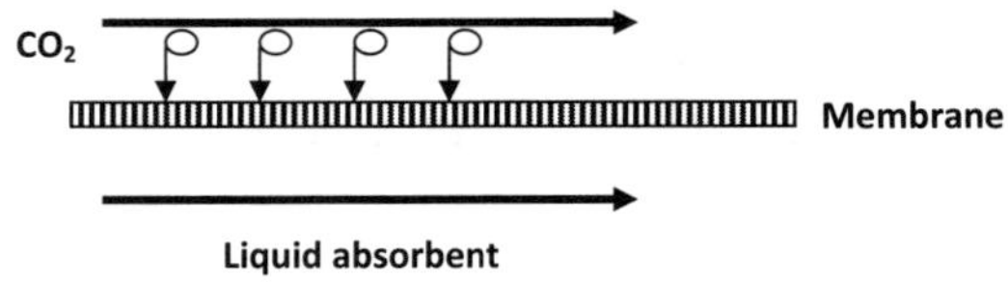

Figure 17.1. The operating principle of MGA.

limitations of the more active and robust oxide catalysts have caused research into the chemical looping combustion strategies to slow [2]. Membranes have been used with success in many gas separation processes. Many advances in membrane technology have been made recently, including the development of crosslinking techniques and composite membranes and the production of advanced materials [3–5]. In particular, polymeric CO_2/N_2 (nitrogen) gas separation membranes for use with power plant flue gases have been explained by Powell and Qiao [6]. The MGA process appears to be better than the membrane gas separation process, which involves no liquid absorbents. In gas separation processes, choosing a membrane with a higher permeability often means it will have a lower selectivity, or vice versa. Thus, a smaller flux will be produced. In contrast, the main advantage of MGA is that the membrane permeability can be chosen directly because the liquid absorbent provides the selectivity. As a result, there is a greater potential to achieve a higher flux with MGA.

MGA has been used for the removal of sulfur dioxide (SO_2), nitrogen dioxide (NO_2), nitrogen oxides (NOx), hydrogen sulfide (H_2S), ammonia (NH_3), tropospheric ozone (O_3), and many other compounds from gas streams, especially those that exist in flue gas, natural gas, indoor gas, and off gas [7–9]. In addition to being able to address the greenhouse problem, MGA can also be used in a large range of other applications, such as in biogas purification, food, submarine, space, medical, and air separation applications. In many instances, hollow-fiber and flat-sheet membrane modules are chosen in MGA applications. However, the significant advantages of hollow-fiber module are that the residence time is much shorter than that in conventionally packed towers and that the gas/liquid flow ratio can be increased easily by adjusting the gas or liquid stream independently [7]. Various factors, such as the porosity, the membrane dimensions, the liquid viscosity, and the chemical reaction that occurs during mass transfer in the membrane [10, 11], affect MGA performance.

The essential element in MGA is a porous, hydrophobic, and polymeric membrane. The gas phase remains separated from the liquid absorbent because of the hydrophobicity of the membrane. A gas stream is introduced to one side of the membrane. The components to be removed from the gas stream will diffuse through the gas-filled pores of the membrane. On the other side of the membrane, the components will be absorbed into the absorption liquid [8]. Various liquid absorbents have been studied experimentally. These include pure water and aqueous solutions of NaOH, KOH, amines, and amino acid salts. Normally, amines are classified as primary, secondary, and tertiary types. AMP (amino-2-methyl-propanol) a primary, sterically hindered amine in which the amino group is attached to a tertiary carbon atom, has been identified as a promising absorbent that might offer capacity and rate advantages over conventional amines [12]. For the secondary amines, diethanolamine (DEA) is a popular commercially used liquid absorbent with two ethanol groups attached to the nitrogen atom [13].

In the present study, mass transfer coefficients were obtained in flat-sheet membrane modules for feed streams with various CO_2 concentrations. The concentration of AMP and DEA solutions applied ranged from 1 M to 3 M. In addition, the separation efficiency of AMP and DEA were compared with that of deionized water.

17.2 METHODS/THEORY

17.2.1 Materials

The hydrophobic, microporous poly(vinylidene fluoride) (PVDF) flat-sheet membranes that were used in this experiment had an average pore diameter of 0.45 μm, a thickness of 100 μm, and a porosity of 70%, and were purchased from Millipore Asia Ltd. AMP and DEA (99% grades) were purchased from Merck (Malaysia) Sdn. Bhd. and were mixed with deionized water to prepare aqueous liquid absorbents with desired concentrations (1, 2, and 3 M). The experimental setup is schematically shown in Figure 17.2.

Pure CO_2 and pure N_2 were mixed in the following volume ratios: 10:90, 20:80, 30:70, 40:60, and 100:0. These ratios are in the composition range of flue gas. In the experiment, the flow rate of the feed gas supplied from a compressed gas cylinder was adjusted and controlled by Aalborg mass flow controllers, and the feed gas was then fed through the side of the membrane module at an atmospheric pressure.

The inlet and outlet gas volume flow rates were measured with a soap bubble flow meter. The pressure gauges were used to indicate the inlet and outlet pressure of the gas and the liquid. When a CO_2/N_2 mixture was used as the feed gas, gas chromatography (Thermal Conductivity Detector, PerkinElmer, Waltham, MA) was used to analyze the inlet and outlet gas compositions. All of the data were collected after the experiment had been operated for 10 minutes to ensure the system had reached steady state. The results of each run were averaged from five sampling instances. Either the pure or mixed gas flow rate was fixed at 1×10^{-4} m^3/s, while the liquid absorbent flow rate was fixed at 2.5×10^{-5} m^3/s. All of the experiments were performed at room temperature; therefore, the gas-phase temperature in this work was approximately 30°C. The volumes of the gas and liquid compartment were equal at 4.33×10^{-5} m^3.

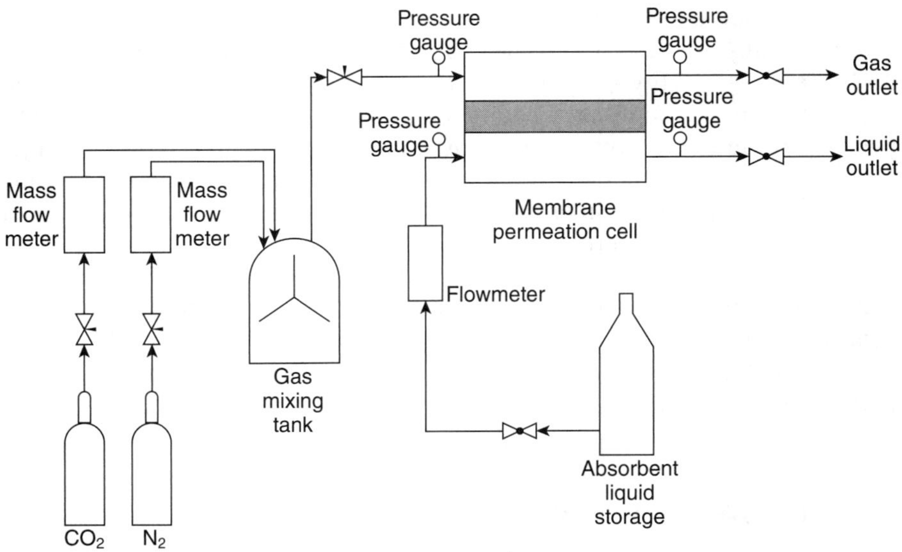

Figure 17.2. A schematic diagram of membrane contactor system.

17.3 RESULTS AND DISCUSSION

In the present study of CO_2 absorption using a nonwetted membrane, the overall mass transfer coefficient can be expressed using Equation 17.1 [5, 6]:

$$K_i = \frac{Q_G}{A}\ln\left(\frac{C_{in}}{C_{out}}\right), \tag{17.1}$$

where K_i is the overall mass transfer coefficient for CO_2 (m/s), A is the membrane area (m^2), Q_G is the feed gas flow rate (m^3/s), and C_{in} and C_{out} are the CO_2 concentrations in the gas phase at the inlet and outlet, respectively. It should be noted that Equation 17.1 is written with the assumption that the CO_2 concentration in the liquid phase is negligible. Meanwhile, the CO_2 flux, J_{CO2}, through the membranes can be expressed by Equation 17.2:

$$J_{CO_2} = K(c_g - c_l^*), \tag{17.2}$$

where J_{CO_2} is the CO_2 flux (mol/m^2 s), K is the overall mass transfer coefficient (m/s), c_g is the concentration in the gas phase (mol/m^3), and c_l* is the gas-phase concentration in equilibrium with the bulk liquid phase (mol/m^3).

The effects of the CO_2 concentration in the feed stream on the mass transfer coefficient were determined, as illustrated in Figure 17.3. As was expected, the CO_2 concentration in the feed stream influenced the mass transfer coefficients. For deionized water as a liquid absorbent, the results showed that the mass

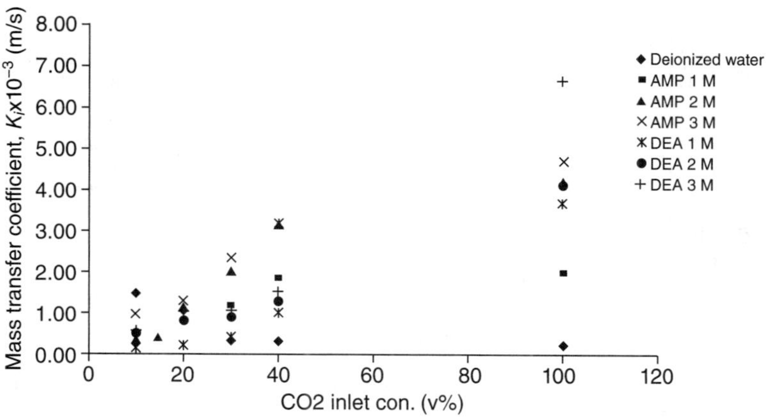

Figure 17.3. The effect of the CO_2 concentration in the feed stream on the mass transfer coefficient, K (gas flow rate: 1×10^{-4} m^3/s).

transfer coefficient decreased marginally with an increasing CO_2 concentration in the feed stream. This result suggests that deionized water is not a suitable solvent for this MGA system. The large driving force in the gas phase created a higher potential for the surface membrane to wet with deionized water when it was used as the liquid absorbent. Meanwhile, in the presence of alkanolamines (AMP and DEA) as liquid absorbents, absorption is facilitated by chemical reactions. AMP had a slightly faster rate than DEA because the absorption of CO_2 when using AMP as the liquid absorbent is much higher than that when using DEA. However, fluctuating reading might occur for pure CO_2 when using DEA as the liquid absorbent, which showed the highest mass transfer coefficient at 3 M. Because of the higher driving force for the absorption of CO_2 and the higher chemical reaction rate, the mass transfer coefficient for DEA should be lower than that for AMP at 3 M.

Figure 17.4 shows the CO_2 flux for different concentrations of CO_2 in the feed stream. For deionized water as the liquid absorbent, the CO_2 flux decreased with an increasing concentration of CO_2 in the feed stream because of the increase in the average CO_2 concentration in the liquid phase. Furthermore, the decreasing pattern of the CO_2 flux when deionized water was used as the liquid absorbent may also have resulted from the surface of the membrane becoming wetted from capillary condensation since the pores of the membrane have a small diameter of 0.45 μm. As expected, the CO_2 flux was greater for pure CO_2 than for 10, 20, 30, or 40% CO_2 for both amines. This result shows good agreement with the zwitterion mechanism. According to the zwitterion mechanism, the reaction of CO_2 with primary and secondary amines (AMP and DEA) occurs via a base to produce carbamate and a protonated base; the CO_2 flux would remain constant or go up with increasing concentrations of the amines in solution. In addition, a greater concentration of AMP or DEA produces a higher flux because of their higher reaction rate constants. The flux

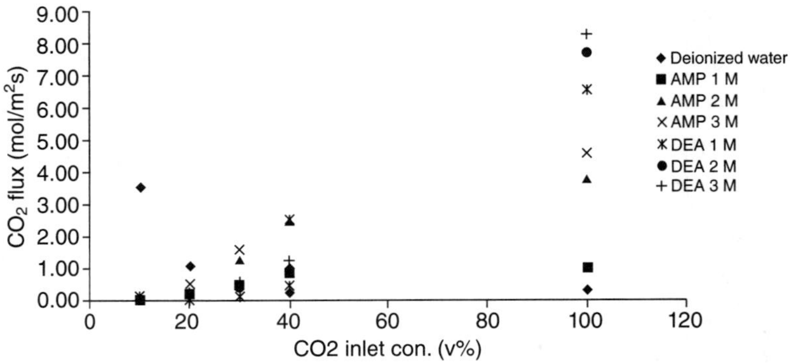

Figure 17.4. The effect of the CO_2 concentration in the feed stream on the CO_2 flux (gas flow rate: 1×10^{-4} m^3/s).

for the absorption of pure CO_2 is around 50% higher than that for the absorption of 40% CO_2. Figure 17.4 also shows that the aqueous solution of AMP had the best CO_2 flux capacity, followed by DEA, and then deionized water. This trend can be justified from the reaction kinetics of amines with CO_2.

17.4 CONCLUSIONS

The absorption of 10, 20, 30, 40, and pure CO_2 was theoretically studied in aqueous solutions of two single alkanolamine solvents (AMP and DEA) as well as in deionized water using flat-sheet membranes in an MGA application. For deionized water as the liquid absorbent, the CO_2 flux decreased with increasing feed concentrations of CO_2. In contrast, the CO_2 flux increased with increasing concentrations of both amines when either of them was used as the liquid absorbent. In addition, a greater concentration of AMP or DEA yielded a higher flux because of their higher reaction rate constant, according to the zwitterion mechanism.

ACKNOWLEDGMENTS

The authors would like to acknowledge that this work was supported by the Science Fund from the Ministry of Science and Technology, Malaysia (6013350).

REFERENCES

[1] Li J. L. and B. H. Chen. Review of CO_2 absorption using chemical solvents in hollow fiber membrane contactors. *Sep. Purif. Technol.* 41 (2005) 109–122.

[2] Evan J., T. Granite, and O. B. Thomas. Review of novel methods for carbon dioxide separation from flue and fuel gases. *Fuel Process. Technol.* 86 (2005) 1423–1434.

[3] Koros J. W. and R. Mahajan. Pushing the limits on possibilities for large scale gas separation: which strategies? *J. Memb. Sci.* 175 (2000) 181–196.

[4] Haiqing L. and D. F. Benny. Materials selection guidelines for membranes that remove CO_2 from gas mixtures. *J. Mol. Struct.* 739 (2005) 57–74.

[5] Pandey P. and R. S. Chauhan. Membrane for gas separation. *Prog. Polym. Sci.* 26 (2001) 853–893.

[6] Powell C. E. and G. G. Qiao. Polymeric CO_2/N_2 gas separation membranes for the capture of carbon dioxide from power plant flue gases. *J. Memb. Sci.* 279 (2006) 1–49.

[7] Wang D., W. K. Teo, and K. Li. Selective removal of trace H_2S from gas streams containing CO_2 using hollow fiber membrane modules/contractors. *Sep. Purif. Technol.* 35 (2004) 125–131.

[8] Feron P. H. M. and A. E. Jansen. CO_2 separation with polyolefin membrane contactors and dedicated absorption liquids: performances and prospects. *Sep. Purif. Technol.* 27 (2002) 231–242.

[9] Todorovića J., D. M. Krstića, G. N. Vataib, and M. N. Tekića. Gas absorption in a hollow-fiber membrane contactor with pseudo-plastic liquid as an absorbent. *Desalination* 193 (2006) 286–290.

[10] Matsumiya N., M. Teramoto, S. Kitada, and H. Matsuyama. Evaluation of energy consumption for separation of CO_2 in flue gas by hollow fiber facilitated transport membrane module with permeation of amine solution. *Sep. Purif. Technol.* 46 (2005) 33–40.

[11] Kreulen H., C. A. Smolders, G. F. Versteeg, and W. P. M. van Swaaij. Microporous hollow fiber membrane modules as gas-liquid contactors. Part 1. Physical mass transfer processes. A specific application: mass transfer in highly viscous liquids. *J. Memb. Sci.* 78 (1993) 197–216.

[12] Bosch H., G. F. Versteeg, and W. P. M. Van Swaaij. Kinetics of the reaction of CO_2 with the sterically hindered amine 2-amino-2-methylpropanol at 298 K. *Chem. Eng. Sci.* 45 (1990) 1167–1173.

[13] Xu S., Y. W. Wang, F. D. Otto, and A. E. Mather. Kinetics of the reaction of carbon dioxide with 2-amino-2-methyl-1-propanol solutions. *Chem. Eng. Sci.* 51 (1996) 841–850.

18

PREPARATION AND CHARACTERIZATION OF W/O EMULSION LIQUID MEMBRANE CONTAINING DIETHANOLAMINE (DEA) FOR CARBON DIOXIDE SEPARATION FROM GAS MIXTURES

Khairul Sozana Nor Kamarudin and Inamullah Bhatti

18.1 INTRODUCTION

Carbon dioxide is the major atmospheric contaminant leading to temperature increase caused by greenhouse effect. The increase of carbon dioxide (CO_2) levels in the atmosphere has spurred worldwide concerns of potential global climate change among international organizations, governments, and environmental scientists [1]. Carbon dioxide gas must be removed or separated from the gaseous stream in order to combat the greenhouse effect. It is also essential to remove it to avoid corrosion of pipeline and equipment as well as to meet fuel gas specifications [2]. For the last two decades, studies have been carried out to investigate the potential use of emulsion liquid membrane (ELM) in various separation systems including gas separation. In this system, ions in the external water phase are extracted into the organic membrane phase and then stripped into the internal water phase [3]. Thus, due to its simple operation

Sustainable Membrane Technology for Energy, Water, and Environment, First Edition.
Edited by Ahmad Fauzi Ismail and Takeshi Matsuura.

where it can accomplish both extraction and stripping in one step and high efficiency, it has attracted many researchers to develop the processes and study the characteristics of ELM [4]. In 1968, Li has attempted separation of hydrocarbons by liquid surfactant membrane [5]. It was followed by other researchers who had produced a number of articles on the separation of aromatic and aliphatic compounds by ELM [6, 7]. Bruining et al. has conducted some experiments for the absorption of oxygen into aqueous sodium sulfite solution in the presence of organic phase [8].

ELM offers a great potential to many industrial, environmental, and biomedical systems [9–13]. In the process, diffusion takes place from one liquid into another liquid with which it is incompletely miscible [14, 15]. ELM has also been referred to as surfactant liquid membranes and double emulsion membranes. They consist of an emulsion formed by organic solvent and water, which can be stabilized by the addition of surfactant [16]. This emulsion is then contacted with a continuous phase (sometimes referred to as the external phase) containing the desired solute [17]. The desired solute is present in the form of droplets and referred to as the internal (disperse) phase.

Nowadays, the application of ELM is found in food, cosmetics, pharmaceutical, and metallurgical processes [18–21]. However, the application of ELM in gas separation processes is still new and needs to be further investigated. Therefore, the objective of this study is to elucidate the optimal conditions of stable W/O ELM containing DEA as extractant by applying high speed homogenization to obtain fine water globules that are dispersed in the organic phase. In this W/O ELM, the oil phase functions as selective barrier or membrane phase and the aqueous solution is used as an internal stripping or receptor phase. An aqueous liquid phase dispersed in immiscible organic phase produces small droplets. In a system of W/O type emulsion, the dispersed aqueous phase contains the reactant, and the continuous phase is an organic solvent with higher solubility. In such case, the absorption rate will be increased due to the larger solubility and chemical reaction. Thin film is formed between the dispersed droplets and the bulk liquid phase. Gas enters in the bulk liquid phase and at the liquid–liquid interface, the chemical reaction of gas with the reactant starts. The bulk organic solvent will function as membrane phase where as the dispersed aqueous solution will work as stripping phase [22].

The particle size and the stability of the ELM were analyzed at various concentrations of extractor and surfactant, and different homogenization time. The effect of aqueous phase and organic phase was also assessed by changing the water oil ratio.

18.2 EXPERIMENTAL

18.2.1 Material

Kerosene (Sigma Aldrich) of low volatility is used to reduce the loss of evaporation at high speed homogenization. Span-80 (sorbitan monooleate), purchased from Merck, is a nonionic commercial surfactant. An analytical grade diethanolamine (DEA) used as an extractant was also purchased from Merck, and deionized water was used to prepare aqueous stripping phase solution of NaOH (0.1, 0.2, 0.3, 0.4, and 0.5 M). Analytical grade sodium hydroxide (NaOH) in pellet form was obtained from Ashland Chemical.

18.2.2 Experimental Method

Single ELM (W/O) is composed of diluent membrane phase and aqueous internal stripping phase. The membrane phase consists of organic solvent and the surfactant, whereas the aqueous internal phase comprises NaOH solution and the extractant. Kerosene was used as diluent to prepare a membrane phase in which Span-80 was dissolved by stirring at room temperature at 700 rpm for 15 minutes. An aqueous internal stripping phase was prepared by mixing extractant DEA in NaOH solution and by stirring at room temperature for 15 minutes at 700 rpm. The method was repeated for different concentrations of surfactant, extractant, and aqueous solution.

High performance disperser (Ultra Turrax® T25) with 18G mixing shaft was used for the preparation of the emulsion. Initially, the organic phase was homogenized at 10,000 rpm while the internal aqueous stripping phase was poured dropwise. After completely adding the aqueous phase, the speed of the disperser was increased to 20,000 rpm. Since emulsification time is one of the key factors affecting the particle size and stability of the emulsion, the time interval of 1, 2, and 3 hours was used for each sample preparation. Five sets of emulsion were prepared based on NaOH concentration (0.1, 0.2, 0.3, 0.4, and 0.5 M), and for each NaOH concentration, concentrations of extractant and surfactant were varied (2 wt %, 4 wt %, and 6 wt %). The effects of NaOH concentration, extractant and surfactant and the emulsification time on particle size and emulsion stability were investigated. The pH value of each sample was also measured. Table 18.1 summarizes the experimental study on the size and stability of the emulsion. Homogenization will be carried out at 60, 120, and 180 minutes.

Optical electronic microscope (Olympus® BX 50) with electronic image developer camera was used to analyze the size of the particles. Photographs of each sample were taken with the Five Digital Imaging System and the size of the particle was analyzed by using the Soft Imaging System Analysis Auto Five (SISauto) software. The stability of the emulsion was determined at room temperature (25 ± 2°C) where the emulsion was filled in the scaled test tubes. After 24 hours, the phase separation ratio of the water and oil was calculated using the following equation:

$$\text{Stability \%} = \frac{V_t - V_w}{V_t} \times 100 \tag{18.1}$$

TABLE 18.1. Parameters for Size and Stability Study of the Emulsion

Concentrations		Time (Min)		
		60	120	180
NaOH (M)	Span-80 and DEA%	Sample No.		
0.1	2	1a2	1b2	1c2
	4	1a4	1b4	1c4
	6	1a6	1b6	1c6
0.2	2	2a2	2b2	2c2
	4	2a4	2b4	2c4
	6	2a6	2b6	2c6
0.3	2	3a2	3b2	3c2
	4	3a4	3b4	3c4
	6	3a6	3b6	3c6
0.4	2	4a2	4b2	4c2
	4	4a4	4b4	4c4
	6	4a6	4b6	4c6
0.5	2	5a2	5b2	5c2
	4	5a4	5b4	5c4
	6	5a6	5b6	5c6

where V_t is the total volume of emulsion in the test tube and V_w is separated volume of aqueous internal phase in the test tube. The pH of the emulsion was also recorded.

18.3 RESULTS AND DISCUSSIONS

18.3.1 Effect of Concentrations

Figures 18.1–18.5 illustrate the effect of surfactant, extractant, and NaOH concentrations on the size of particle. It is observed that as the percentage of

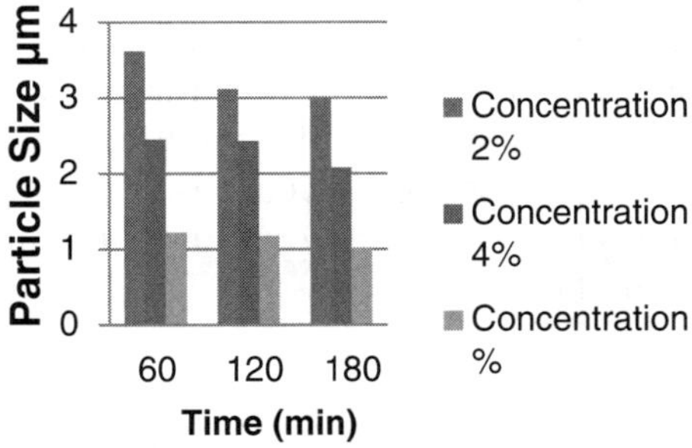

Figure 18.1. Effect of homogenization time (0.1 M NaOH).

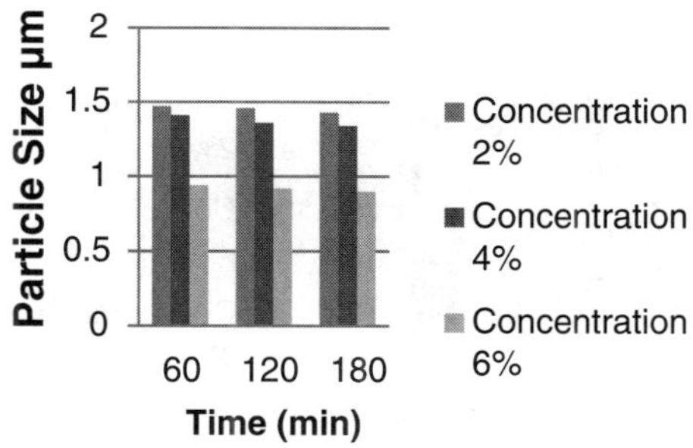

Figure 18.2. Effect of homogenization time (0.2 M NaOH).

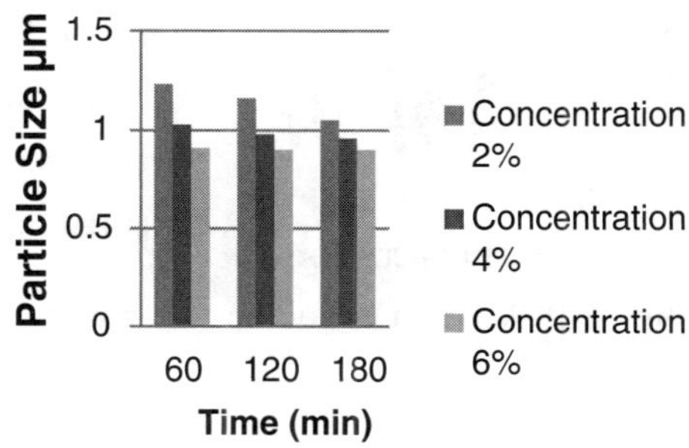

Figure 18.3. Effect of homogenization time (0.3 M NaOH).

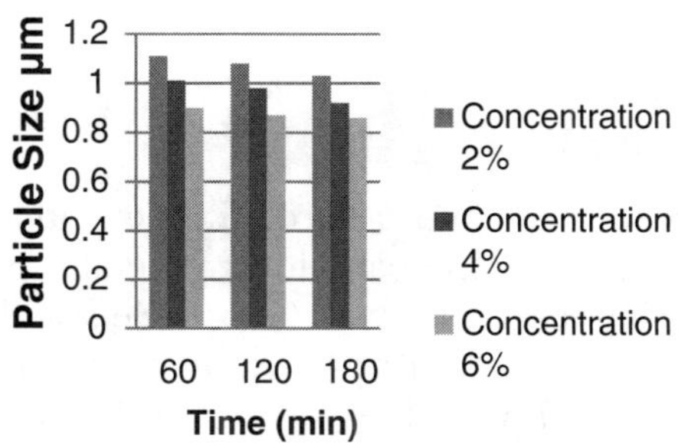

Figure 18.4. Effect of homogenization time (0.4 M NaOH).

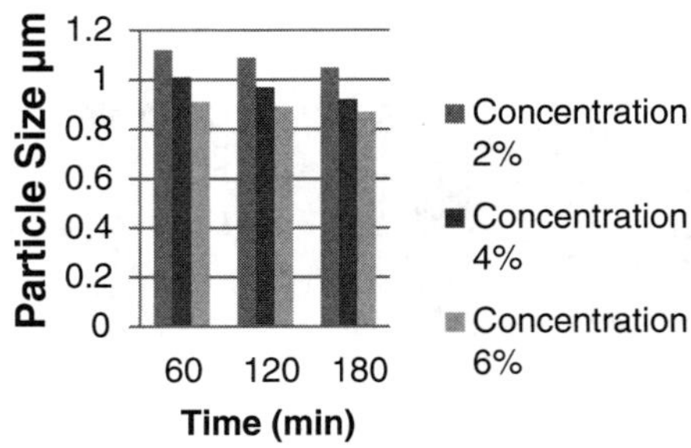

Figure 18.5. Effect of homogenization time (0.5 M NaOH).

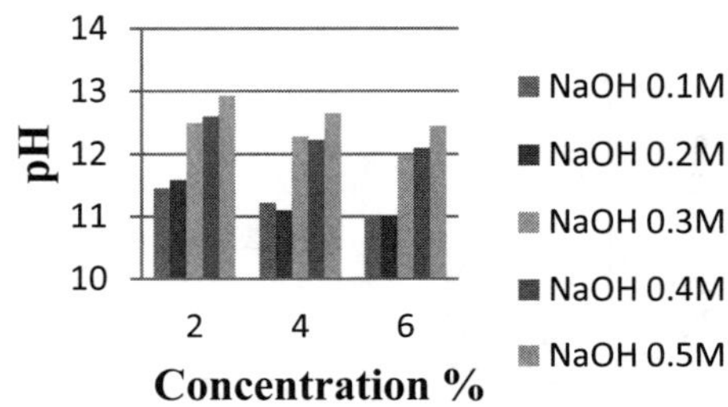

Figure 18.6. Effect of concentration on pH.

surfactant increases, the particles size decreases. Figure 18.6 shows the pH of the emulsion which is one of the important factors in the mass transport. High pH difference due to increasing the concentration of NaOH may increase the mass transfer rate of acidic gases.

18.3.2 Effect of Stirring Time

Stirring time and speed play important role in the stability of the emulsion. The effects of stirring time on the particle size are also presented in Figures 18.1–18.5. As the stirring time increases, the particle size decreases. However, Figure 18.7 shows the effect of stirring time and concentration of NaOH and DEA on emulsion stability. It was observed that for 120 minutes emulsification time, the prepared samples were more stable as compared to the 60 and 180 minutes. At 60 minutes emulsification, the quantity of emulsion droplet is very small, but the droplet size is relatively large. For this reason, the droplets are easier to coalesce with each other to form bigger drops within this short period, leading to separation of the dispersed from the continuous phase. On the other hand, when emulsification time is long (180 minutes), the excess amount of energy affects on the interfacial surface tension of the droplets. The smaller

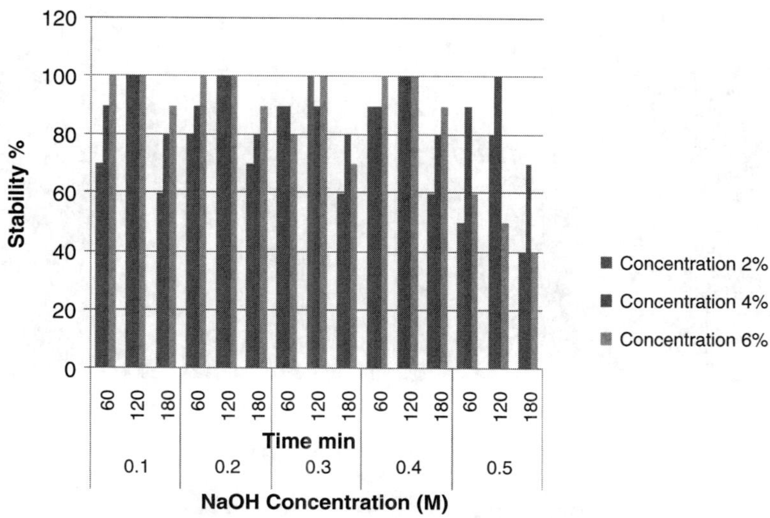

Figure 18.7. Effect of concentration and time in the emulsion stability.

droplets collide with each other to form bigger drops and hence, the phase separation takes place. The microscopic photographs (Fig. 18.8) show droplet formations at different homogenization times.

18.3.3 Effect of Strip Phase Concentration

In this study, NaOH is used in the stripping phase. It has been observed that the concentration of NaOH has a significant effect on the stability of emulsion It is obvious from Figure 18.7 that stability of emulsion increased as the concentration of Span-80 increased (from 2% to 6% for 0.1 M of NaOH). Similar effect is observed for 0.2 M of NaOH, but for 0.3 M NaOH, the stability decreased as the concentration of Span-80 increased from 4% to 6% when the homogenization time is 60 minutes. In the case of 0.4 M NaOH, stability behavior was the same as 0.1 M of NaOH. For 0.5 M of NaOH, the stability of the emulsion is relatively low. Therefore, considering the effect of pH and the stability of the emulsion, it was concluded that the optimal concentration of NaOH for the preparing of emulsion to absorb CO_2 was 0.4 M.

18.4 CONCLUSION

It is concluded that the concentration of surfactant affects the particle size and the stability of the emulsion. In addition, homogenization time also affects the size of the emulsion as well as the total number of globules formed. This study has found that the smallest particle size of 0.87 μm was obtained using 4% of

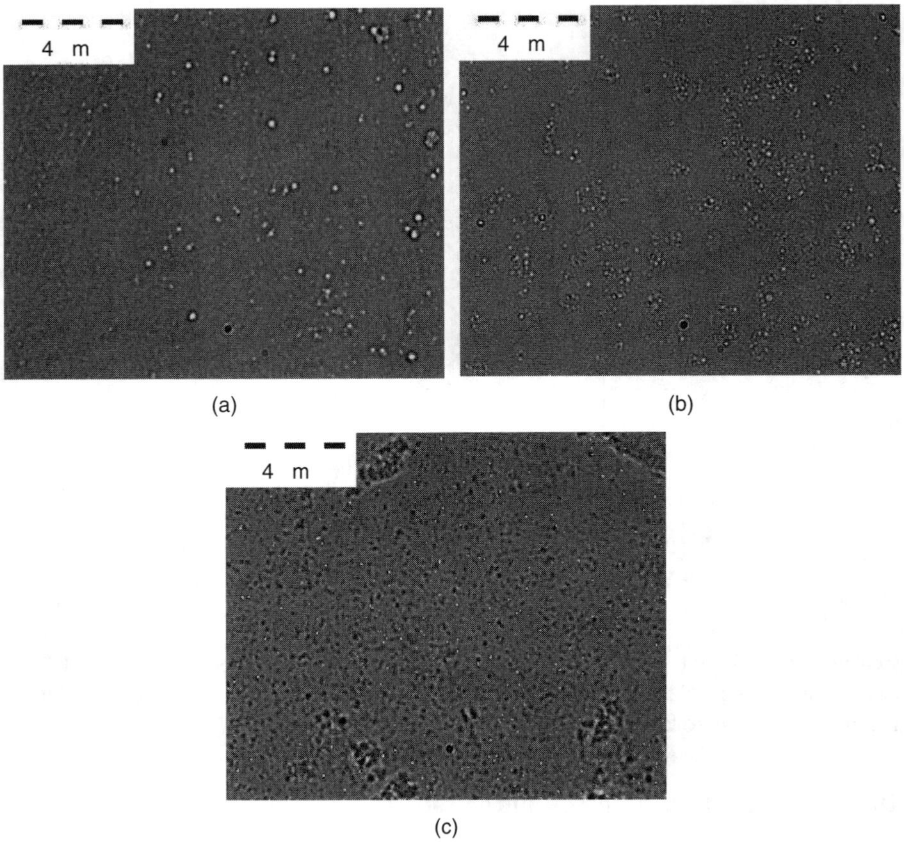

Figure 18.8. Optical microscope photograph (40X magnification) of particles at different homogenization times (a) 60 minutes, (b) 120 minutes, and (c) 180 minutes.

surfactant and extractant in a 0.4 M NaOH at 120 minutes homogenization time.

ACKNOWLEDGMENT

The authors would like to acknowledge the financial support from the Ministry of Science, Technology and Innovation, Malaysia and Universiti Teknologi Malaysia. Special thanks to Mrs. Siti Zalita of Bioprocess Engineering Laboratory for her help in carrying out this study.

REFERENCES

[1] Khoo H. H. and R. B. H. Tan. Life cycle investigation of CO_2 recovery and sequestration. *Environ. Sci. Technol* 40 (2006) 4016.

[2] Mandal B. and S. S. Bandyopadhyay. Simultaneous absorption of CO_2 and H_2S into aqueous blends of N-methyldiethanolamine and diethanolamine. *Environ. Sci. Technol* 40 (2006) 6076.

[3] Hirai T., N. Okamoto, and I. Komasawa. Preparation of spherical oxalate particles of rare earth in emulsion liquid membrane system. *AIChE J.* 44 (1998) 197.

[4] Chakrabort M. and H. J. Bart. Separation of toluene and n-heptane through emulsion liquid membranes containing Ag^+ As carrier. *Chem. Eng. Technol.* 28 (2005) 1518.

[5] Li N. N. Separation of hydrocarbons by liquid membrane permeation. *Ind. Eng. Chem. Process Des. Dev.* 10(2) (1971) 215.

[6] Ho S. V., P. W. Sheridan, and E. Krupetsky. Supported polymeric liquid membranes for removing organics from aqueous solutions: I. Transport characteristics of polyglycol liquid membranes. *J. Memb. Sci.* 112 (1996) 13.

[7] Zhendi W. and F. Merv. Separation and characterization of petroleum hydrocarbons and surfactant in orimulsion dispersion samples. *Environ. Sci. Technol.* 30 (1996) 3351.

[8] Bruining W. J., G. E. H. Joosten, A. A. C. M. Beenackers, and H. Hofman. Enhancement of gas-liquid mass transfer by dispersed second liquid phase. *Chem. Eng. Sci.* 41 (1986) 1873.

[9] Lee C. J. and C. C. Chan. Extraction of ammonia from a dilute aqueous solution by emulsion liquid membranes: 2. Theory and mass-transfer model. *Ind. Eng. Chem. Res.* 29(1) (1990) 101.

[10] Leal-Calderon F., O. Mondain-Monval, K. Pays, N. Royer, and J. Bibette. Water-in-oil emulsions: role of the solvent molecular size on droplet interactions. *Langmuir* 13 (1997) 7008.

[11] Kawakatsu T., G. Tragardh, C. H. Tragardh, M. Nakajima, N. Oda, and T. Yonemoto. The effect of the hydrophobicity of microchannels and components in water and oil phases on droplet formation in microchannel water in oil emulsification. *Colloids Surf. A Physicochem. Eng. Aspects* 179 (2001) 29.

[12] Skurtys O. and J. M. Aguilera. Formation of O/W macroemulsions with a circular microfluidic device using saponin and potato starch. *Food Hydrocolloids* 23 (2009) 1810.

[13] Kumbasar R. A. Separation and concentration of cobalt from aqueous thiocyanate solutions containing cobalt-nickel by emulsion liquid membranes using TBP as extractant. *J. Memb. Sci.* 338 (2009) 182.

[14] Ho W. S. and N. N. Li. *Membrane Handbook*, 1st ed. Van Nostrand Reinhold, New York, 1992.

[15] Ho W. S., T. A. Hatton, E. N. Lightfoot, and N. N. Li. Batch extraction with liquid surfactant membranes: a diffusion controlled model. *AIChE J* 28 (1982) 662.

[16] Kopp A. G., R. J. Marr, and F. E. Moser. A new concept for mass transfer in liquid surfactant membranes without carriers and with carriers that pump. *Int. Chem. Eng. Symp. Ser.* 54 (1978) 279.

[17] Teramoto M., N. Takeuchi, T. Maki, and H. Matsuyama. Gas separation by liquid membrane accompanied by permeation of membrane liquid through membrane physical transport. *Sep. Purif. Technol.* 24 (2001) 101.

[18] Ali K. R. and T. Osman. Separation of cobalt and nickel from acidic leach solutions by emulsion liquid membranes using almine 300 (TAO) as mobile carrier. *Desalination* 224 (2008) 201.

[19] Vasiljevic D., G. Vuleta, and M. Primorac. The characterization of the semi-solid W/O/W emulsions with low concentrations of the primary polymeric emulsifier. *Int. J. Cosmet. Sci.* 27 (2005) 81.

[20] Charcosset C. Preparation of emulsion and particles by membrane emulsification for the food processing industry. *J. Food Eng.* 92 (2009) 241.

[21] Velenzuela F., C. Araneda, F. Vargas, C. Basualto, and J. Sapag. Liquid membrane emulsion process for recovering the copper content of amine drainage. *Chem. Eng. Res. Des.* 87 (2009) 102.

[22] Park S. W., H. B. Cho, I. J. Sohn, and H. Kumzawa. CO_2 absorption into W/O emulsion with aqueous amine liquid droplets. *Sep. Sci. Technol.* 37 (2002) 639.

19

REMOVAL OF DYES FROM LIQUID WASTE SOLUTION: STUDY ON LIQUID MEMBRANE COMPONENT SELECTION AND STABILITY

NORASIKIN OTHMAN, NORLISA MILI, ANI IDRIS, AND SITI NAZRAH ZAILANI

19.1 INTRODUCTION

Large numbers of chemically different dyes are used for various industrial applications, and significant proportion appears in the form of wastewater and is spilled into the environment [1]. The release of synthetic dyes in the form of wastewater effluents poses severe hazards to the environment as these effluents contain large amount of surfactants, suspended solids, trace metals, color, and have high chemical oxygen demand (COD) [2]. Various efforts and attention had been paid on the treatment of colors which were caused by dye residues during the dyeing processes as most pollutants can be reduced by conventional treatment methods except color [3].

Synthetic dyestuffs are used extensively in textile, paper, printing industries, and dye houses. The effluents of these industries are highly colored and their disposal into receiving waters causes damage to the environment as they may significantly affect photosynthetic activity in aquatic life due to reduced light penetration and may also be toxic to some aquatic life in case they contain metals, chlorides, and so on. Even the presence of very low concentrations of

Sustainable Membrane Technology for Energy, Water, and Environment, First Edition.
Edited by Ahmad Fauzi Ismail and Takeshi Matsuura.

dyes (less than 1 mg/L) in the effluent is highly visible and is considered undesirable [4].

Nowadays, textile industry is the most important and rapidly developing industrial sectors. It has high importance in terms of its environmental impact, since it consumes considerably high amounts of processed water and produces highly polluted discharge water in large amounts. Textile dye and finishing processes are among the major industrial water users in many areas, and this industry has the wastes most difficult to treat satisfactorily. Textile dye wastewater is known to contain strong color, high pH, temperature, and COD and low biodegrability [5]. In Batik Industry, the dyeing process uses both organic and inorganic synthetic dyes that are available in variety of colors. Most of the batik producers use reactive dye as the coloring substance, and the plain fabrics are imported from China, India, and Japan [6].

The majority of synthetic dyes currently used in batik processing are the highly water-soluble azo-reactive dyes. Azo dyes are characterized by the existence of nitrogen–nitrogen double bonds (–N=N–) and the bright color of their aqueous solutions. The conventional treatments for these effluents include adsorption, biological oxidation, and chemical coagulation, each with its own advantages as well as limitations [2]. However, with the ever upgrading of dye compounds in terms of its stability and solubility, the conventional treatment methods tend to be rather ineffective, resulting in the discharge of highly intense colored effluents from treatment plants [7]. Therefore, an emulsion liquid membrane (ELM) process has enough ability to solve the above problems and is capable to separate and concentrate solutes very rapidly through its very thin liquid film having large interfacial area. An ELM process is reported as a more economic and also a superior process compared to conventional solvent extraction from the viewpoints of both capital and operating costs [8]. The ELM operation has already been applied in an industrial pilot plant in Austria and has reached the stage of practical application [9]. The advantages of this process are that extraction and stripping occur simultaneously in one step operation, and equilibrium limitation can be removed. It also can reduce the amount of expensive extractant, and high fluxes and high selectivity are possible. Unfortunately, it is a tailor-made system, meaning that there are differences in liquid membrane formulation and operating conditions for different systems. In addition, the major concern of using ELM in practical application is emulsion stability. The emulsion must be stable during the process and easy to break down for the recovery process. Despite much research on the extraction of various solutes using ELM, no one has attempted to use ELM for dyes recovery from batik industry wastewater.

This chapter presents the liquid membrane formulation and stability study for the extraction of dyes from batik industry wastewater. Various parameters influencing the transport of dye from liquid waste to membrane phase and then to receiving phase will be discussed in this chapter.

19.2 EXPERIMENTAL PROCEDURE

19.2.1 Reagents and Solutions

Extractants were obtained from various companies. These extractants were dissolved in kerosene or chloroform which was obtained from Acros. Hydrochloric acid (37% purity), sodium hydroxide and sulfuric acid (97% purity) were obtained from Merck (M) Sdn. Bhd. All chemicals used were of analytical grade and were used as received. The batik waste solution and the dyes such as turquoise blue (TB), red 3BS (R3BS), black B (BB), and methylene blue (MB) were obtained from Perusahaan Nozi Batik, Kuala Terengganu.

19.2.2 Carrier/Extractant and Stripping Screening

The extraction process was carried out by mixing an equal volume (10 mL) of aqueous simulated dye and organic solutions using a mechanical shaker with rotation at 250 rpm for a period of 18 hours. Then, the mixture was carefully poured into separation funnel. After the phase separation was completed (around 15–30 minutes), the aqueous phase was carefully separated from the organic phase. The procedures were repeated for different carrier and dye types. The absorbance of dye in the aqueous phase after extraction was measured by using a visible spectrophotometer. The concentration of dye extracted in the organic phase was calculated based on the standard curve of each dye.

For the stripping process, the organic phase (dye-loaded organic phase) was taken from the extraction process. Stripping process is the reverse extraction step. An equal volume of stripping agent at specified concentration and the dye-loaded organic phase were mixed by using mechanical shaker for 18 hours at 250 rpm and then carefully poured into a separation funnel where they were left for phase separation for about 15–30 minutes. After phase separation was completed, the aqueous phase was carefully separated to avoid any contamination by the organic phase, and the concentration of dye strip was measured by a visible spectrophotometer. The stripping procedures were repeated for each stripping agent. All experiments were carried out at room temperature (26°C).

19.2.3 Stability Test

The membrane phase was prepared by dissolving adequate amounts of Span 80 (1–9% w/v) in kerosene under a gentle mixing by a magnetic stirrer. Then, the prepared membrane phase was carefully poured into the internal aqueous phase (H_2SO_4) with a volume ratio of organic membrane solution to stripping solution (1:1). The experiment was carried out at various preset homogenizer speed (5000–12,000 rpm) for 5 minutes to obtain a white emulsion.

In 50 mL beaker, 10 mL of the prepared water in oil emulsion (organic phase and internal phase) is dispersed to 50 mL of external aqueous solution (real batik industry wastewater) to measure the emulsion stability during extraction process. The contents are stirred at speeds 200–350 rpm for 5 minutes. The double emulsion (W/O/W) is allowed to be spontaneously separated in separating funnel, and the feed phase (external aqueous phase) is filtered. These tests were conducted at room temperature (26°C). Each experiment was performed twice at least, and the mean values were presented.

19.3 RESULTS AND DISCUSSION

19.3.1 Carrier/Extractant Screening

The factor that needs to be considered in ELM process is the selection of carrier types and its concentration in order to give high selectivity and capacity of extraction. The high selectivity and capacity can only be achieved by selecting the right carrier for target ions as well as its concentration. The increase in carrier concentration will increase the extraction capacity, but to what concentration needs to be investigated in order to optimize the concentration to avoid excessive use of carrier, that is normally very costly.

From the above point of view, some experiments have been done to select the suitable carrier in liquid membrane formulation. There are a few types of acidic, chelating, basic, and solvating extractants go through the screening process to extract various types of dye such as TB, R3BS, BB, and MB. The results in Table 19.1, shows the potential of basic extractant to extract TB, R3BS, and BB, and only D2EHPA of acidic extractant provide high potential to extract MB. These indicate that basic extractant can extract reactive dyes (nonionic dyes) with high distribution coefficient, D, as shown in Table 19.2, and only acidic extractant can extract MB, a basic dye. The D value of

TABLE 19.1. Dye Extraction without pH Adjustment

				% Extraction			
Group	Extractant	Diluents	pH_{eq}	TB	R3BS	BB	MB
Acidic	Cyanex 272	Kerosene	6.92	35.88	0.00	19.01	0.00
	D2EHPA	Kerosene	6.51	22.55	0.00	18.21	98.84
Chelating	Tetramethylthiuram disulfide (TMTSD)	Chloroform	6.71	22.55	0.00	6.15	11.97
Basic	Trioctylamine (TOA)	Kerosene	7.52	11.75	6.20	5.56	0.00
	Tridodecylamine (TDA)	Kerosene	7.82	97.38	73.84	78.94	0.00
Solvating	Tri-n-octylphosphine oxide (TOPO)	Kerosene	6.77	40.20	0.97	16.92	0.00

Experimental condition: carrier concentration = 0.5 M, dye concentration = 100 ppm, agitation speed = 250 rpm, T = 26°C.

TABLE 19.2. Distribution Coefficient, D, of Each Type of Carrier to Dye Synthetic Dye

			Distribution Coefficient, D			
Group	Extractant	Diluents	TB	R3BS	BB	MB
Acidic	Cyanex 272	Kerosene	0.56	0.00	0.23	0.00
	D2EHPA	Kerosene	0.29	0.00	0.22	85.31
Chelating	Tetramethylthiuram disulfide(TMTSD)	Chloroform	0.29	0.00	0.07	0.14
Basic	Trioctylamine (TOA)	Kerosene	0.13	0.07	0.06	0.00
	Tridodecylamine (TDA)	Kerosene	31.28	2.82	2.24	0.00
Solvating	Tri-n-octylphosphine oxide (TOPO)	Kerosene	0.67	0.00	0.20	0.00

Experimental condition: carrier concentration = 0.5 M, dye concentration = 100 ppm, agitation speed = 250 rpm, T = 26°C.

TABLE 19.3. Type of Stripping Agent for Stripping Process of Each Dye

	% S			
Acid Types	MB	R3BS	TB	BB
H_2SO_4	76.59	0.00	0.13	0.30
HCl	64.27	7.66	10.24	5.16
NaOH	12.79	49.96	48.85	96.91

Experimental condition: H_2SO_4, HCl, NaOH concentrations = 1 M, initial organic phase concentration of MB loaded = 98.84 ppm, initial organic phase concentration of R3BS loaded = 75.47 ppm, agitation speed = 250 rpm, T = 26°C.

tridodecylamine (TDA) to TB, R3BS, and BB are 31.28, 2.82, and 2.24, respectively, while the D value of D2EHPA to MB is 85.3.

19.3.2 Stripping Selection and Screening

Several inorganic acids and solution at various concentrations were tested as a stripping agent for the dye recovery from the organic phase. As shown in Table 19.3, the results show that sulfuric acid has a good capability to strip methylene blue dye from dye loaded D2EHPA. On the other hand, NaOH could strip TB, R3BS, and BB from dye-loaded TDA.

19.3.3 Liquid Membrane Stability

The stability of emulsion is one of the important factors affecting the removal of pollutant in ELM process. When the W/O emulsions are dispersed by

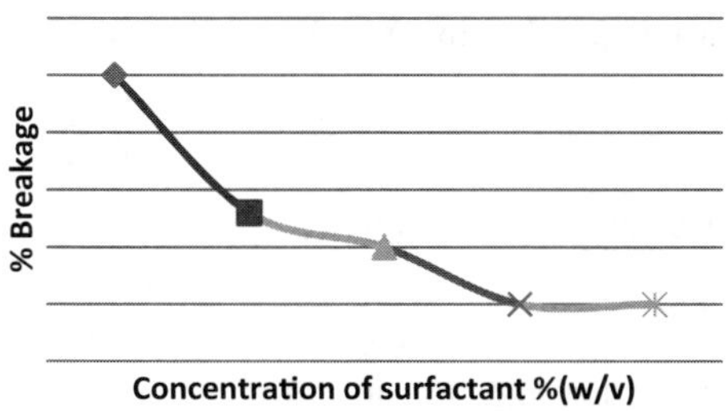

Figure 19.1. Effect of surfactant concentration on the W/O emulsions stability (experimental conditions: volume ratio of internal phase to organic phase, 1:1; emulsification time: 5 minutes; stirring speed: 240 rpm; volume ratio of W/O emulsions to external phase, 1:5; internal phase concentration [H_2SO_4]: 1 M; diluent: kerosene; homogenizer speed, 8000 rpm; contact time: 5 minutes).

stirring in the continuous aqueous solution, the emulsion must be sufficiently stable in order to extract dye ions into the internal aqueous droplets. The emulsion breakdown will result in a lowering of extraction efficiency and loss of the extracted species into external phase. In the present study, sulfuric acid was used as the internal phase, and the percentage of emulsion breakage (ε) was calculated using the following equation:

$$\varepsilon = \frac{V_i - V_S}{V_i} \times 100 \tag{19.1}$$

The emulsion breakage represents the ratio in percentage of the volume of internal phase leaked into the external phase by splitting (V_S) to the initial volume of the internal aqueous phase (V_i). The study investigates the effects surfactant, agitation time, and stirring speed on the stability of the prepared W/O emulsions.

19.3.3.1 Effect of Surfactant Concentration. Figure 19.1 shows the influence of surfactant concentration on the stability of emulsion. It was observed that the W/O emulsions stability proves to be more stable by increasing the surfactant concentration until 7% (w/v).For surfactant concentration lower than 3% (w/v), the rupture of the emulsion is higher up to 80%. This is because of insufficient layer of surfactant surrounding all the internal aqueous phase. Consequently, the amount of surfactant in the membrane phase must be minimal, but it must be enough to stabilize the emulsion. Additionally, it is known that the increase of surfactant concentration leads to a higher viscosity of the W/O emulsions and does not favor the extraction kinetics [10]. Hence, the surfactant concentration of 7% (w/v) is selected in this work.

19.3.3.2 Effect of Stirring Speed. Figure 19.2 shows the effect of stirring speed (160–400 rpm) on the stability of the W/O emulsions. It was observed

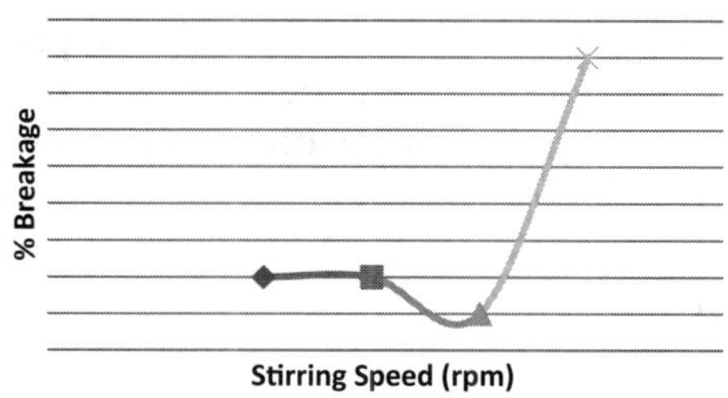

Figure 19.2. Effect of agitation/stirring speed on the W/O emulsions stability (experimental conditions: volume ratio of internal phase to organic phase, 1:1; emulsification time: 5 minutes; surfactant concentration, 7%; volume ratio of W/O emulsions to external phase, 1:5; internal phase concentration [H_2SO_4]: 1 M; diluent: kerosene; homogenizer speed, 8000 rpm; contact time: 5 minutes).

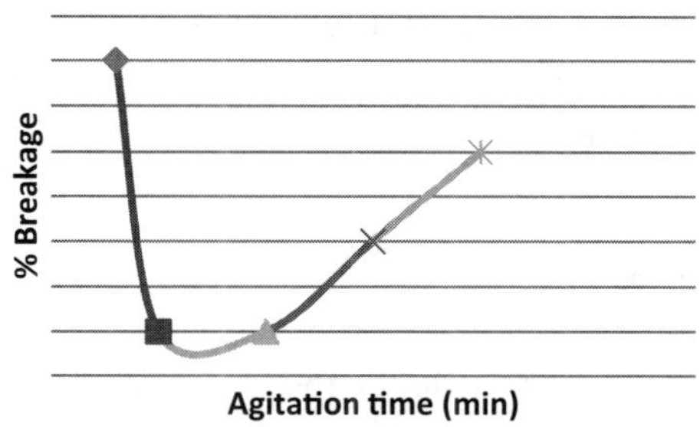

Figure 19.3. Effect of agitation/stirring time on the W/O emulsions stability (experimental conditions: volume ratio of internal phase to organic phase, 1:1; emulsification time: 5 minutes; surfactant concentration, 7%; volume ratio of W/O emulsions to external phase, 1:5; agitation speed, 320 rpm; internal phase concentration [H_2SO_4]: 1 M; diluent: kerosene; homogenizer speed, 8000 rpm).

that a good stability of the W/O emulsions is obtained for agitation speeds ranging from 160 to 320 rpm. When the level of agitation is increased (>160 rpm) up to 300 rpm, the percentage of emulsion breakage decreased indicate that the size of the emulsion globules become smaller, and the interfacial area available for mass transfer increased [11]. It is observed that by increasing agitation speed, the shear force which acts on the emulsion globules increases and this makes the globules smaller, but further increase to 400 rpm will affect the stability of the emulsion and the breakage of the membrane. For lower agitation speed, the size of the emulsion globules increases, and the interfacial area available for mass transfer decreases. Hence, the stirring speed was taken as 320 rpm to ensure a good stability of the W/O emulsion sand to enhance the interfacial area available for mass transfer.

19.3.3.3 Effect of Agitation Time. Figure 19.3 shows that an excellent stability of the W/O emulsions is obtained for an agitation time of 5–10 minutes, which implies that the steady state was reached. The prolonged exposure to homogenizer (15 minutes) results in a significant increase of the breakage, probably due to the coalescence (coagulation) of the internal phase droplets.

19.4 CONCLUSION

The feasibility of ELM extraction of dye from real liquid batik wastes was studied in this research. It was demonstrated that the ELM technique was very promising in extracting and recovering dye from batik processing wastes. The screening process obtained the suitable extractant for dye extraction for different types of dye. TDA from basic extractant is potential extractant to extract the reactive dyes. On the other hand, D2EHPA is an excellent extractant for methylene blue from basic dye. The results also showed that the production of W/O emulsions using homogenizer was a very attractive technique, and an excellent stability was obtained after 5–10 minutes contact time of the W/O emulsions with external phase.

ACKNOWLEDGMENTS

The authors would like to acknowledge the financial support from the Ministry of Science, Technology and Innovation, Malaysia (Science Fund grant, No. 79336) and the Universiti Teknologi Malaysia that made this study possible.

REFERENCES

[1] Meyer U. Biodegradation of synthetic organic colourant. *FEMS Symp.* 12 (1981) 371–385.

[2] Kim T. H., C. Park, J. Yang, and S. Kim. Comparison of disperse and reactive dye removals by chemical coagulation and Fenton oxidation. *J. Hazard. Mater.* 112 (2004) 95–103.

[3] Liu R., H. M. Chiu, C. S. Shiau, R. Y. L. Yeh, and Y. T. Hung. Degradation and sludge production of textile dyes by Fenton and photo-Fenton processes. *Dyes Pigm.* 73 (2007) 1–6.

[4] Clarke E. A. and R. Anliker. *Organic Dyes and Pigments, Handbook of Environmental Chemistry*, Anthropogenic Compounds. Part A, vol. 3. Springer, New York, 1980.

[5] Vlyssides A. G., D. Papaioannou, M. Loizidoy, P. K. Karlisand, and A. A. Zorpas. Testing an electrochemical method for treatment of textile dye wastewater. *Waste Manag.* 20 (2000) 569–574.

[6] Ahmad A. L., W. A. Harris, Syafiie, and B. S. Ooi. Removal of dye from wastewater of textile industry using membrane technology. *Jurnal Teknologi* 36 (2002) 31–44.

[7] Sulakova R., R. Herdina, and M. B. Soares. Oxidation of azo textile soluble dyes with hydrogen peroxide in the presence of Cu (II)-chitosan heterogeneous catalysts. *Dyes Pigm.* 73 (2007) 19–24.

[8] Hayworth H. C., W. S. Ho, W. A. Burns, and N. N. Li. Extraction of uranium from wet process phosphoric acid by liquid membranes. *Sep. Sci. Technol.* 18(6) (1983) 493.

[9] Ho W. S. and K. K. Sirkar. *Membrane Handbook Editions*. Van Nostrand Reinhold, New York, 1992, p. 701.

[10] Othman N. Selective emulsion liquid membrane extraction of silver from liquid photographic waste industries. PhD thesis, Universiti Teknologi Malaysia, 2006.

[11] Djenouhat M., O. Hamdaoui, M. Chiha, and M. H. Samar. Ultrasonication-assisted preparation of water-in-oil emulsions and application to the removal of cationic dyes from water by emulsion liquid membrane: Part 1. Membrane stability. *Sep. Purif. Technol.* 62 (2008) 636–641.

PART V

APPLICATIONS IN ENERGY

20

MODELING AND ANALYSIS OF SOLAR-POWERED MEMBRANE DISTILLATION UNIT FOR SEAWATER DESALINATION

Fawzi Banat and Mohammed Al-Jarrah

20.1 INTRODUCTION

Membrane distillation (MD) combines the use of distillation principles and membrane technology. Membranes used are hydrophobic porous membranes that allow vapors to migrate through the pores but not liquid water. The MD driving force is supplied by the vapor pressure gradient resulting from either by a temperature difference between the membrane sides or by applying vacuum in the permeate side. Most of the MD applications have been the concentration of nonvolatile solutes in aqueous solutions such as salts and sugars. In desalination, the passing vapor is condensed on a cooler surface to produce fresh water. Salt is not vaporized so it cannot pass through the pores.

To date, this process has not yet demonstrated any commercial success as the source energy costs involved to produce the water are still high. Hogan et al. [1] at the University of New South Wales in Australia was the first to introduce coupling of solar energy with MD process. The system was composed of a hollow-fiber MD module, and a solar collector of 3 m^2 area. The system was tested in Sydney. The production rate was 17 L/day/m^2 of collector area.

Sustainable Membrane Technology for Energy, Water, and Environment, First Edition.
Edited by Ahmad Fauzi Ismail and Takeshi Matsuura.

A photovoltaic (PV) and thermally driven small scale, stand-alone, compact desalination system with internal heat recovery was installed in Irbid, Jordan. The system was designed by Fraunhofer Institute for Solar Energy Systems ISE within the activities of "SMADES," EC-funded projects, and operated and tested by Banat et al. [2]. The desalination system is composed of a spiral-wound air-gap membrane desalination module, PV panels, and flat-plate solar collector. The plant produced as high as 120 L/d (20 L/d/m^2 ap. area) with an average distillate electrical conductivity of 5 μS/cm. The low-grade thermal energy required by the process was in the range of 200–300 kWh/m^3. The thermal recovery ratio was in the range of 1:2.

A larger autonomous desalination solar-driven MD plant was installed in the Aqaba city of Jordan [3] to produce potable water from untreated seawater. Four spiral-wound membrane modules with 40 m^2 effective area (10 m^2 for each membrane) coupled with 72 m^2 solar collector were used in the system. PV panels supplied the electrical energy required to drive the pumps. The unit found technically feasible with a production rate of 2–11 L/d/m^2 ap. area. The distillate conductivity was in the range of 20–250 μS/cm when the feed conductivity was about 55,000 μS/cm.

The main objective of this work was to develop a mathematical model capable of describing and predicting the performance of the solar powered spiral-wound MD unit used earlier by Banat et al. [2]. The model, after validation, was used to predict the unit performance under different operating conditions.

20.2 METHODS/THEORY

A schematic diagram of the solar powered membrane distillation (SPMD) pilot plant used by Banat et al. [2] is shown in Figure 20.1; the major components of the system are MD module, solar energy collector, PV panel, and feed tank.

The MD module is constructed of three channels: the condenser channel, the evaporator channel, and the distillate channel [2]. The condenser and the distillate channels are separated by an impermeable condensation foil. The evaporator and the distillate channels are separated by a hydrophobic porous membrane. The hot water coming from the solar collector enters the evaporation channel of the membrane where it cools down due to evaporation. The feed water is pumped from the feed tank (475 L) passing through the condenser channel in a counter flow to the hot water and thus warmed up before entering the solar collector. Water vapor that passes through the membrane pores is condensed on the condenser foil, making it possible to recover the latent heat of vaporization.

The hydrophobic nature of the membrane prevents water in its liquid state to pass but allows water vapor to pass through the air-filled membrane pores. The molar flux of water vapor (N_w) can be expressed as [4]:

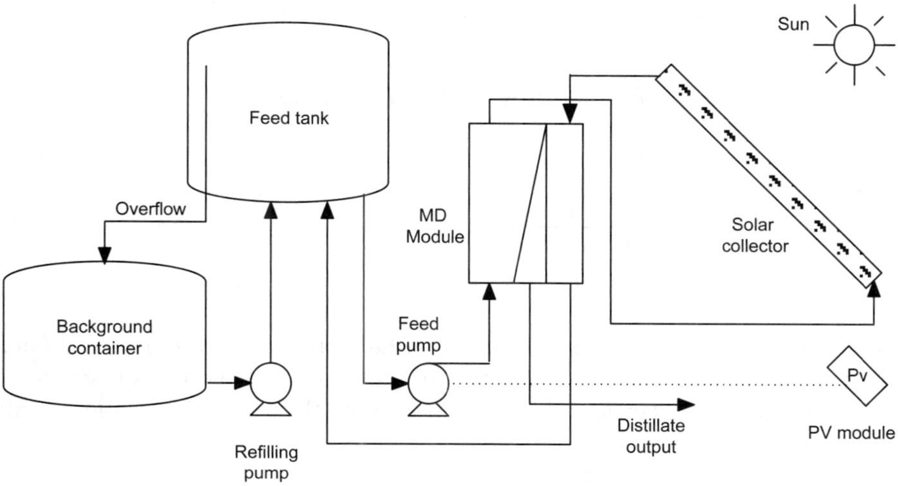

Figure 20.1. Flow sheet of the SPMD system.

$$N_w = \frac{\varepsilon D_{wA} P}{RT}\left\{\frac{1}{\delta\tau + b}\right\}\ln\frac{P - p_{wI}}{P - p_{wm}}, \tag{20.1}$$

where N_w is the molar flux of water vapor, P is the total pressure, R is the ideal gas constant, T is the average absolute temperature, ε is the membrane porosity, δ is the membrane thickness, τ is the membrane tortuosity, b is the air gap thickness, and, p_{wm} and p_{wI} are the partial pressure of water vapor at membrane hot side and condensation film interfaces, respectively. Heat transfer occurs in three consecutive steps: heat flux from the feed bulk to the membrane surface, from the membrane surface to the condensation interface, and from the condensation interface to the coolant bulk.

Heat flux from the feed bulk to the membrane surface

$$q_h = h_{hf}(T_b - T_m) + N_w Cp_{lw}(T_b - T_m) = h_h(T_b - T_m) \tag{20.2}$$

$$h_h = (h_{hf} + N_w Cp_{lw}), \tag{20.3}$$

where h_{hf} is the liquid film heat transfer coefficient in the hot channel, Cp_{lw} is the liquid phase heat capacity, T_b is the hot bulk temperature, and T_m is the hot side membrane interface temperature.

Heat transfer from the membrane surface to the condensation interface

$$q = q_s + N_w \lambda_w, \tag{20.4}$$

where q_s is the sensible heat flux, N_w is the molar flux of water vapor, and λ_w is the water latent heat of vaporization.

Heat transfer from the condensation interface to the coolant bulk

$$q_c = h_d(T_I - T_{c1}) = \frac{k_c}{l}(T_{c1} - T_{c2}) = h_{cf}(T_{c2} - T_c)$$
$$= h_c(T_I - T_c) \tag{20.5}$$

$$h_c = \frac{1}{\frac{1}{h_d} + \frac{k_c}{l} + \frac{1}{h_{cf}}}, \tag{20.6}$$

where h_d is the heat transfer coefficient in the condensation film, k_c and l are the thermal conductivity and thickness of the cooling plate, respectively, h_{cf} is the liquid film heat transfer coefficient in the cold channel, and h_c is the overall heat transfer coefficient.

An energy balance over the solar collector gives

$$\frac{d(mCp_{wl}T_b)}{dt} = A_c\alpha\beta G - A_A U_L(T_b - T_\infty) - F_b Cp_{wl}(T_b - T_c), \tag{20.7}$$

where A_c is the area of the collector, α is the absorptivity, β is the transmissivity, G is the solar radiation, A_A is the area of solar radiation to be absorbed, U_L is the collector heat losses coefficient, T_∞ is the ambient temperature, and F_b is the flow rate of the solution flowing in the collector.

Temperatures of the hot and cold streams vary along the membrane. The temperature of the hot solution decreases while it travels though the evaporation channel. On the other hand, the temperature of the feed solution increases while it travels through the condenser channel. Therefore, it is more accurate to divide the membrane area into subareas each of them called segment. To account for the temperature changes, the equations of mass and heat transfer are applied on each segment. Figure 20.2 shows the membrane segmentation.

Mass balance around segment s in the hot channel

$$F_{b,s} = F_{b,s+1} + \bar{N}_{w,s}, \tag{20.8}$$

where, $\bar{N}_{w,s}$ is the water flow rate from segment s, and is calculated by multiplying water flux (N_w) by the segmental membrane surface area (A_s). Rearranging Equation 20.8 gives

$$F_{b,s+1} = F_{b,s} - \bar{N}_{w,s}. \tag{20.9}$$

Energy balance around segment s in the hot chamber

$$q_{h,s} = F_{b,s}Cp_{wl}T_{b,s} - F_{b,s+1}Cp_{wl}T_{b,s+1}. \tag{20.10}$$

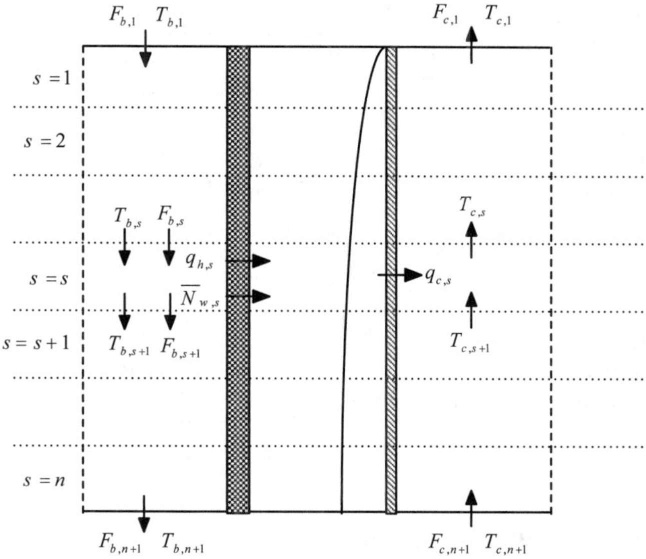

Figure 20.2. Membrane segmentation.

$$T_{b,s+1} = \frac{1}{F_{b,s+1}}\left(F_{b,s}T_{b,s} - \frac{q_{h,s}}{Cp_{wl}}\right). \tag{20.11}$$

Energy balance around segment s in the cold chamber

$$q_{c,s} = F_{c,s}Cp_{wl}T_{c,s} - F_{c,s+1}Cp_{wl}T_{c,s+1}. \tag{20.12}$$

$$T_{c,s} = \frac{1}{F_{c,s}}\left(F_{c,s+1}T_{c,s+1} + \frac{q_{c,s}}{Cp_{wl}}\right). \tag{20.13}$$

The flow rate of the distillate F_D, is the sum of the flow rates from all segments;

$$F_D = \sum_{s=1}^{n} \bar{N}_{w,s}, \tag{20.14}$$

where, $\bar{N}_{w,s}$ is the flow rate of the water vapor from each segment.

Computations were performed using the MATLAB package, version 6.1. The characteristics of the membrane and solar collector are shown in Tables 20.1 and 20.2, respectively.

TABLE 20.1. Membrane Characteristics

Characteristic	Value	Unit
Membrane material	PTFE	–
Membrane area (A_m)	10	m^2
Membrane thickness (δ)	35	μm
Membrane height (L)	0.6	m
Air gap thickness (b)	3	mm
Tortuosity (τ)	2	–
Pores diameter	0.2	μm
Porosity (ε)	0.8	–
Cooling plate material	PE	–
Cooling plate thickness(l)	0.001	m
Thermal conductivity of cooling plate (k_c)	0.12	W/m°C

PTFE, polytetrafluoroethylene.

TABLE 20.2. Solar Collector Characteristics

Characteristic	Value	Unit
Area of water to be heated (A_c)	5.07	m^2
Absorptivity (α)	0.92	–
Transmissivity (β)	0.84	–
Area of solar radiation to be absorbed (A_A)	6	m^2
Heat losses coefficient (U_l)	1.5	W/m^{2}°C

20.3 RESULTS AND DISCUSSION

20.3.1 System Performance

Figure 20.3 compares the model prediction and experimental data for the solar collector output temperature, T_b.

Figure 20.4 compares the model predictions and the experimental data of distillate flow rate for a full operation day.

Good agreement between the experimental data and the predictions of the model was observed with an error not exceeding 20%. The peaks of the experimental data were attributed to meteorological conditions that affect solar intensity, such as wind speed variation and cloud appearance in the sky [2] which were not taken into consideration in the model.

Figure 20.5 shows the effect of hot and cold solution temperatures on the permeate flux. The permeate flux is affected greatly by the hot solution inlet temperature, and to a smaller extent by the inlet temperature of the cold solution. As the difference between the two temperatures (T_b—T_c) increases, the value of the distillate also increases. Figure 20.5 clearly demonstrates that the distillate production increases exponentially with increasing the inlet hot temperature. This behavior can be explained by Antoine equation, which gives an

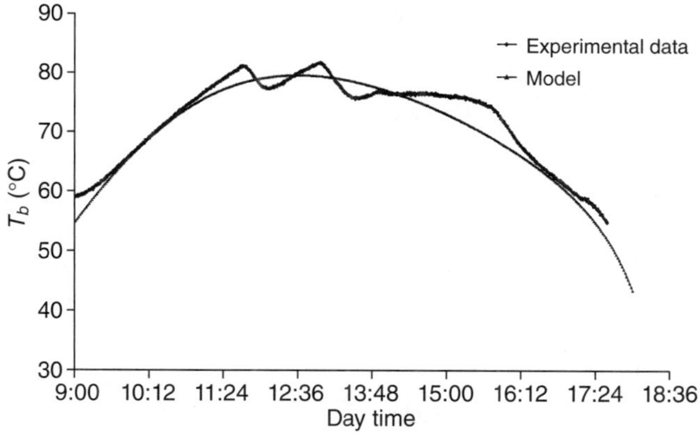

Figure 20.3. Solar collector output temperature.

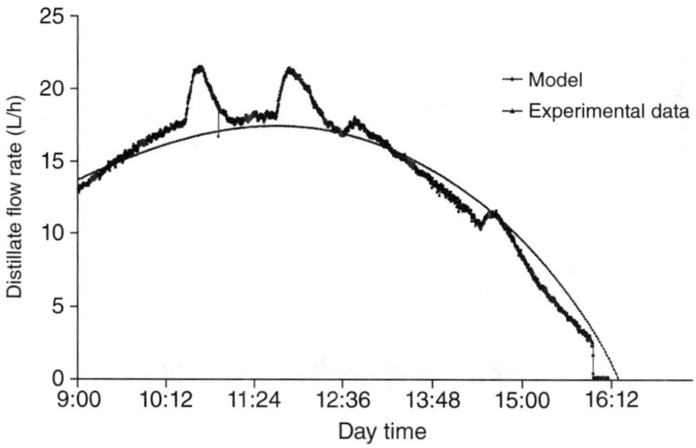

Figure 20.4. Hourly variation of distillate flow rate (August 16, 2006).

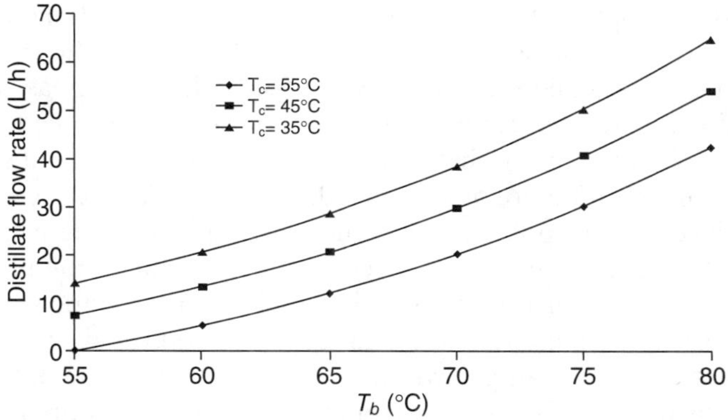

Figure 20.5. Effect of hot and cold solution inlet temperature on the distillate flow rate (salt fraction = 3.5 wt %, Re = 785).

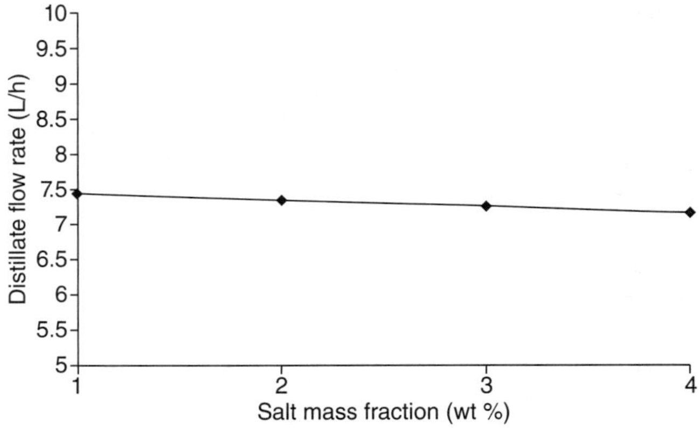

Figure 20.6. Effect of salt concentration on the distillate flow rate (Re = 785).

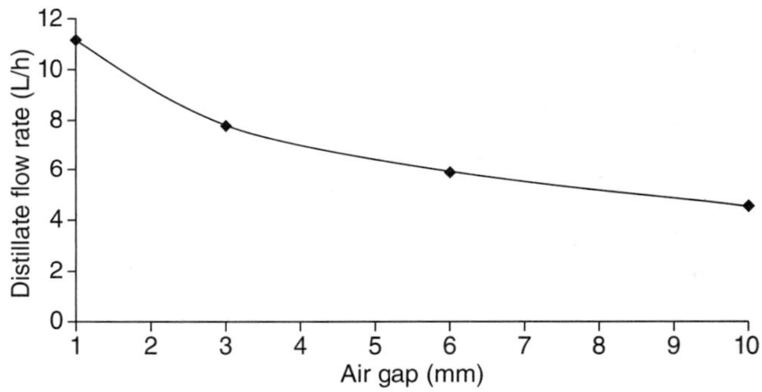

Figure 20.7. Effect of the air gap width on the distillate flow rate (salt fraction = 3.5 wt %, Re = 785).

exponential relation between the driving force (the vapor pressure difference) and the temperature. Banat and Simandl [4] and Alklaibi and Lior [5] found similar behavior.

Figure 20.6 illustrates the effect of salt concentration on the distillate flow rate. As shown, salt concentration has marginal effect on the distillate flow rate. In fact, as the salt concentration in the feed solution increases, the vapor pressure of the water decreases, resulting in lower driving force for evaporation. It is worthy to mention that when salt is present in the feed solution, an additional concentration boundary layer develops next to the feed membrane surface (concentration polarization). The effect of this layer increases with increasing the salt concentration and consequently reduces the mass transfer driving force. Banat et al. [6], and Alklaibi and Lior [5] reported similar trends.

Figure 20.7 shows the effect of the air gap width on the distillate flow rate. It can be seen that as the air gap width is reduced, the flow rate of the distillate increases; this behavior is expected from the diffusion equation. The total diffusion length is the sum of diffusion gap and membrane thickness. Increasing the air gap width increases the diffusion length and consequently, reduces the flux.

20.4 CONCLUSIONS

The model developed in this work provides useful basic information about the nature of the process and is helpful for process improvement. The main conclusions of this work can be summarized as follows:

- Good agreement was found between the model and the experimental data. The error of predictions was less than 20%.
- Distillate flow rate increased exponentially as a function of hot side temperature.
- Increasing the hot side flow rate decreased the temperature polarization effect and thus increased the distillate flow rate.
- Increasing the air gap width increased the mass transfer resistance and thus reduced the distillate flow rate.
- The salinity has slight effect on the distillate flow rate.

REFERENCES

[1] Hogan P., A. Sudjito, G. Fane, and G. Morrison. Desalination by solar heated membrane distillation. *Desalination* 81 (1991) 81–90.

[2] Banat F., N. Jwaied, M. Rommel, J. Koschikowski, and M. Wieghaus. Desalination by a "compact SMADES" autonomous solar-powered membrane distillation unit. *Desalination* 217 (2007) 29–37.

[3] Banat F., N. Jwaied, M. Rommel, J. Koschikowski, and M. Wieghaus. Performance evaluation of the "large SMADES" autonomous desalination solar-driven membrane distillation plant in Aqaba, Jordan. *Desalination* 217 (2007) 17–28.

[4] Banat F. and J. Simandl. Desalination by membrane distillation: a parametric study. *Separation Science and Technology* 33 (1998) 201–225.

[5] Alklaibi A. M. and N. Lior. Transport analysis of air-gap membrane distillation. *Journal of Membrane Science* 255 (2005) 239–253.

[6] Banat F., R. Jumah, and M. Garaibeh. Exploitation of solar energy collected by solar stills from desalination by membrane distillation. *Renewable Energy* 25 (2002) 293–305.

21

POLYSTYRENE IONOMERS FUNCTIONALIZED WITH PARTIALLY FLUORINATED SHORT SIDE-CHAIN SULFONIC ACID FOR FUEL CELL MEMBRANE APPLICATIONS

Ying Chang and Chulsung Bae

21.1 INTRODUCTION

Due to the strong electron-withdrawing effect of fluorine, fluorinated sulfonic acids are one of the most acidic moieties (i.e., superacids) and their ionic derivatives have found wide applications including acid catalysts, surfactants, ionic liquids, and electrolytes for batteries and fuel cells [1].

Proton exchange membrane (PEM), a key component of fuel cells, acts as an electrolyte to transport protons from anode to cathode and separates the fuel from oxidant. Currently, the state-of-the-art PEM technology is based on perfluorosulfonic acid polymer membranes, such as Nafion [2, 3]. It is generally accepted that the good performance of Nafion membrane in proton exchange membrane fuel cells (PEMFCs) is ascribed to (1) its strong acidity of the sulfonic acid group and/or (2) distinct nanophase-separated morphology that is derived from a highly hydrophobic perfluorinated backbone and the short flexible pendant perfluorinated hydrophilic sulfonic acid groups [4]. Recently,

Sustainable Membrane Technology for Energy, Water, and Environment, First Edition.
Edited by Ahmad Fauzi Ismail and Takeshi Matsuura.

researchers have synthesized sulfonated block copolymers composed of hydrophilic ionomeric blocks and hydrophobic polymer blocks to mimic the favorable phase separation of Nafion and reported the achievement of improved proton conductivity at low relative humidity conditions compared to their corresponding randomly sulfonated ionomers [5].

Although perfluorosulfonated cation exchange membranes such as Nafion have been dominantly used in PEMs and other applications in the past decades, they still suffer from the drawbacks, such as high cost and the rare availability of the fluorine-containing precursors, high fuel crossover, and poor mechanical stability at high temperature ($T > 100°C$). In addition, due to the lack of reactive sites, it is extremely difficult to modify the perfluorinated structure of Nafion further to enhance the desired properties. Compared with the perfluorinated polymers, hydrocarbon-based polymers are much cheaper, more easily available, and easier to modify to manipulate the properties [6]. Because hydrocarbon-based polymers containing pendant fluoroalkyl sulfonate groups are expected to keep the benefits of both hydrocarbon-based polymers (i.e., affordability and convenient accessibility) and perfluorosulfonated ionomers (i.e., superacidic property), they have been of significant interest of academia and industry for many years [7].

Yang and coworkers have reported alternative PEMs to Nafion which possess a partially fluorinated polymer backbone and a perfluorodiethylene ether sulfonic acid chain[7a]. The partially fluorinated PEMs, which were prepared by free radical copolymerization of tetrafluoroethylene, ethylene, and an olefin-containing fluorosulfonyl fluoride, exhibited comparable or slightly better performance than Nafion. However, the copolymerization has to be conducted in a pressure-withholding apparatus, and the resulting ionomers were insoluble in organic solvents. Most ionic polymers containing pendant fluoroalkyl sulfonic acid groups require similar synthetic procedures that involve high pressure equipment and/or highly toxic monomers [7]. In addition, because olefins containing fluorinated sulfonyl fluoride are less reactive when copolymerized with unfunctionalized olefins, only a low concentration (typically <10 mol %) of fluoroalkyl sulfonate group was incorporated. Thus, most of the reported fluoroalkyl sulfonated ionomers are still difficult to prepare in a practical way due to the high cost and inconvenient synthetic procedure, and new advanced membranes with lower cost and convenient accessibility are highly desired.

21.2 METHODS/THEORY

We have recently reported a mild and highly effective borylation reaction of polystyrene via activation of aromatic C–H bonds [8]. The content of incorporated pinacol boronic ester (Bpin) group can be well controlled up to ~50% by changing the loading of bis(pinacolato)diboron [$B_2(pin)_2$]. Using the borylated polymer as an intermediate, we herein introduce a convenient synthetic method of polystyrene ionomers containing partially fluorinated short side-chain sulfonic acid. The new ionomeric membranes exhibited high proton conductivity comparable to that of Nafion even at low relative humidity conditions. Syndiotactic polystyrene (sPS, **1**) gives much better resolved ^{1}H NMR spectrum and is easier to characterize than atactic polystyrene due to the stereoregularity of the phenyl group along the polymer main chain. Thus, we used sPS for postfunctionalization of this study.

Earlier study of the proton conductivity of membranes with similar perfluorinated polymer backbone and different side chain lengths (e.g., $-OCF_2CF(CF_3)OCF_2CF_2SO_3H$ and $-OCF_2CF_2SO_3H$) showed that the shorter side-chain membranes gave significantly higher proton conductivity and a current density at low to moderate water content [9]. Because of convenient availability of the fluorinated starting materials and the synthetic chemistry, we focused on the $-OCF_2CF_2SO_3H$ as our target fluorinated proton conducting moiety.

3,5-Dimethylphenyl 2-(4-bromophenoxy)tetrafluoroethanesulfonate ester (**6**) was synthesized via a multistep synthesis in 49% total yield using a modified procedure of literature method[7d]. Borylated polystyrene **2** (18, 26, and 41 mol % Bpin concentration) were prepared according to literature method [8]. Suzuki-Miyaura cross-coupling reactions of **2** with **6** allowed to introduce fluorinated sulfonate part to the sPS side chain and produced 1,3-dimethylphenyl-protected sulfonated polymer **3**. Hydrolysis of **3** in basic conditions afforded its sodium sulfonate ionomer **4**, which has good solubility in polar aprotic solvents such as dimethyl sulfoxide and dimethylacetamide. Membranes of **4** were readily prepared as dark transparent films by solution casting method, which were then acidified with 1 M H_2SO_4 solution to give their acid form polymer **5** (Scheme 21.1). Those ionic polymers were unambiguously characterized by NMR (^{1}H and ^{19}F) and IR spectroscopies, thermogravimetric analysis, viscosity, water uptake, ion exchange capacity (IEC), and proton conductivity.

21.3 RESULTS AND DISCUSSION

The mol % of borylated functional group in styrene repeating unit was calculated from the integral ratio of methine proton of the sPS main chain and the four methyl groups of the Bpin group in the ^{1}H NMR spectrum. After Suzuki–Miyaura coupling reaction, the resonance of Bpin at 1.16–1.19 ppm disappeared completely, and a new signal at 1.98 ppm that was assigned to two methyl groups of **3** appeared. Two clean peaks at −80.3 ppm ($-OCF_2-$) and −111.9 ppm ($-CF_2SO_3-$) were also observed in the ^{19}F NMR spectrum, confirming successful attachment of **6** to the polymer chain. Based on the relative integral ratio of the methine proton of the sPS main chain and the methyl groups of the side chain of **3**, the concentration of the fluoroalkyl sulfonated group was consistent with that of Bpin group in the borylated precursor polymer. Upon hydrolysis to **4**, the resonance of the methyl side groups

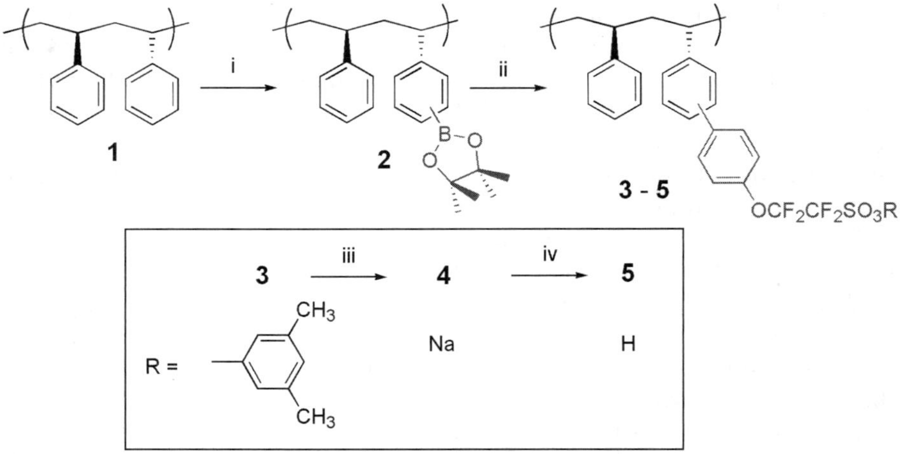

Scheme 21.1. Synthesis of syndiotactic polystyrene ionomer functionalized with fluorinated short side-chain sulfonic acid[a].

[a]Reagents and conditions: (1) $B_2(pin)_2$, $[IrCl(COD)]_2$ (1.5 mol %), d*t*bpy (3 mol %), cyclooctane, 150°C, 6 hours. (2) **6** (3 equiv), $Pd(PPh_3)_4$ (4 mol %), K_3PO_4 (4.5 equiv), THF/H_2O (10/1), 80°C, 12 hours. (3) NaOH (5 equiv), THF/H_2O, 80°C, 4 hours. (4) 1 M H_2SO_4.

disappeared in the ^1H NMR, and the fluorine signal of the difluoromethylene adjacent to sulfonate group shifted from −111.9 ppm to −116.2 ppm in the ^{19}F NMR spectrum. Two resonances of CH_2 and CH in the sPS main chain maintained a ratio of 2:1 in the ^1H NMR spectra of **2–4**, which suggests that main chain structure of sPS was intact during the postfunctionalization process.

IEC, water uptake, and intrinsic viscosity of sPS ionomers were measured and their results are summarized in Table 21.1. Experimental IEC values that were obtained by titration method ranged from 1.31 to 2.18 mequiv/g. As expected, the membrane with higher fluoroalkyl sulfonate content exhibited higher IEC value. The water uptake of **5** also increased as the mol % of sulfonated group increased from 18% to 41% in **5** (entries 1–3 in Table 21.1). All sPS ionomers prepared by this method have high molecular weight, as reflected in their intrinsic viscosity values (0.93–1.31 dL/g), and good film forming ability.

Figure 21.1 shows the humidity dependence of the proton conductivity of **5** and Nafion 112. Although many hydrocarbon-based sulfonated ionomers have shown good proton conductivity at high relative humidity conditions, they do not perform as well as Nafion at low relative humidity. One way to improve proton conductivity of hydrocarbon ionomers is to induce well-defined phase separations and create well-interconnected proton-conducting channels, such as in hydrophilic–hydrophobic block copolymer ionomers. Recently, there has been a great progress on this approach [5]. Another way to improve proton conductivity at low humidity conditions, which has been

TABLE 21.1. Ion Exchange Capacity, Water Uptake, and Intrinsic Viscosity of Syndiotactic Polystyrene Ionomers (4 and 5)

Entry	Sulfonate Content (mol %)	IEC (mequiv/g) exp[a]	IEC (mequiv/g) cal[b]	Water Uptake of **4** (wt %)[c]	Intrinsic Viscocity of **5** (dL/g)[d]
1	18	1.31	1.15	9	0.93
2	26	1.85	1.48	10	1.14
3	41	2.18	1.89	26	1.31
4	Nafion 112	0.90	–	18	–

[a] IEC measured from titration.

[b] IEC calculated from polymer structure based on the integral ratio of proton NMR spectra.

[c] Water uptake (%) = $(W_{wet} - W_{dry})/W_{dry}$ where, W_{dry} and W_{wet} are the weights of dried and wet membranes, respectively.

[d] Measured using an Ubbelohde viscometer at 30°C with **4** concentration 0.5 dL/g in dimethyl acetamide with 0.1 M NaI.

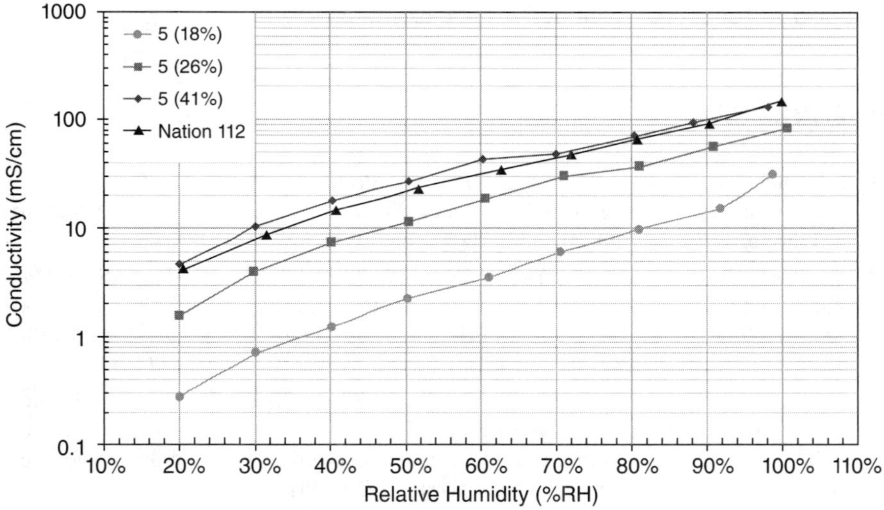

Figure 21.1. Proton conductivity of **5** under different relative humidity conditions at 80°C.

proposed but not investigated thoroughly, is to increase the acidity of ionomer by incorporating fluorine into the sulfonic acid moiety, such as in the case of Nafion. As can be seen in Figure 21.1, the proton conductivity of **5** did not decrease sharply and its conductivity slope with respect to relative humidity is almost identical to that of Nafion which is unprecedented in randomly sulfonated hydrocarbon-based ionomer.

21.4 CONCLUSIONS

In conclusion, we have introduced a new efficient synthetic methodology to incorporate partially fluorinated short side-chain sulfonic acids into the aromatic ring of polystyrene for fuel cell applications. This new class of membranes could be conveniently prepared from a low-cost commercial polymer and tetrafluoroethane precursors in high yields, and their performance in proton conductivity was comparable or slightly better than that of Nafion even at low relative humidity. We believe these results reported herein would be a significant step toward the research of Nafion alternatives, and further research on the variations of the polymer backbones and the fluorinated pendant side chains will provide comprehensive understanding of the relationship of sulfonic acid structure–proton conductivity at molecular level.

ACKNOWLEDGEMENT

We thank the NSF (CAREER DMR-0747667) and Nevada Renewable Energy Consortium for generous support, Frontier Scientific Co. for a gift of B_2pin_2, Tae Soo Jo for proton conductivity measurement.

REFERENCES

[1] (a) Olah G. A., P. S. Iyer, and G. K. S. Prakash. *Synthesis* (1986) 513–531.; (b) Paillard E., F. Toulgoat, J. Y. Sanchez, M. Médebielle, C. Iojoiu, F. Alloin, and B. Langlois. *Electrochim. Acta* 53 (2007) 1439–1443.

[2] (a) Steele B. C. H. and A. Heinzel. *Nature* 414 (2001) 345–352.; (b) Haile S. M. *Mater. Today* 6 (2003) 24–29.; (c) Winter M. and R. J. Brodd. *Chem. Rev.* 104 (2004) 4245–4269.; (d) Carrette L., K. A. Friedrich, and U. Stimming. *Fuel Cells* 1 (2001) 5–39.; (e) Dresselhaus M. S. and I. L. Thomas. *Nature* 414 (2001) 332–337.

[3] (a) Heitner-Wirguin C. *J. Memb. Sci.* 120 (1996) 1–33.; (b) Costamagna P. and S. Srinivasan. *J. Power Sources* 102 (2001) 242–252.; (c) Doyle M. and G. Rajendran, in W. Vielstichm, A. Lamm, and H. A. Gasteiger (eds.), Handbook of Fuel Cells, vol. 3. John Wiley & Sons Ltd, New York, 2003, p. 351.

[4] Mauritz K. A. and R. B. Moore. *Chem. Rev.* 104 (2004) 4535–4585.

[5] (a) Einsla M. L., Y. S. Kim, M. Hawley, H. S. Lee, J. E. McGrath, B. Liu, M. D. Guiver, and B. S. Pivovar. *Chem. Mater.* 20 (2008) 5636–5642.; (b) Roy A., M. A. Hickner, X. Yu, Y. Li, T. E. Glass, and J. E. McGrath. *J. Polym. Sci. [B]* 44 (2006) 2226–2239.; (c) Nakabayashi K., K. Matsumoto, T. Higashihara, and M. Ueda. *J. Polym. Sci. A* 46 (2008) 7332–7341.

[6] (a) Rikukawa M. and K. Sanui. *Prog. Polym. Sci.* 25 (2000) 1463–1502.; (b) Hickner M. A., H. Ghassemi, Y. S. Kim, B. R. Einsla, and J. E. McGrath. *Chem. Rev.* 104 (2004) 4587–4612.

[7] (a) Yang Z. Y. and R. G. Rajendran. *Angew. Chem. Int. Ed. Engl.* 44 (2005) 564–567.; (b) Yang Z. Y., L. Wang, N. Drysdale, M. Doyle, Q. Sun, and S. K. Choi. *Angew.*

Chem. Int. Ed. Engl. 42 (2003) 5462–5464.; (c) Yang Z. Y., L. Wang, N. E. Drysdale, and M. Doyle. *Macromolecules* 36 (2003) 8205–8207.; (d) Feiring A. E. and E. R. Wonchoba. *J. Fluor. Chem.* 105 (2000) 129–135.; (e) Feiring A. E., S. K. Choi, M. Doyle, and E. R. Wonchoba. *Macromolecules* 33 (2000) 9262–9271.; (f) Sauguet L., B. Ameduri, and B. Boutevin. *J. Polym. Sci. A* 44 (2006) 4566–4578.; (g) Ni H. B., Z. H. Li, H. Y. Dou, H. Li, and C. X. Zhao. *J. Fluor. Chem.* 127 (2006) 1036–1041.

[8] Shin J., S. M. Jensen, J. Ju, S. Lee, Z. Xue, S. K. Noh, and C. Bae. *Macromolecules* 40 (2007) 8600–8608.

[9] Gebel G. and R. B. Moore. *Macromolecules* 33 (2000) 4850–4855.

22

CONTRIBUTION OF NANOCLAYS TO THE BARRIER PROPERTIES OF SPEEK/CLOISITE15A® NANOCOMPOSITE MEMBRANE FOR DMFC APPLICATION

Juhana Jaafar and Ahmad Fauzi Ismail

22.1 INTRODUCTION

There are a number of studies carried out in order to enhance the proton exchange membrane (PEM) performance for direct methanol fuel cell (DMFC) application, which consist of modifications of Nafion® membrane, development of membranes from nonfluorinated polymers, and composite membranes.

Currently, polymer-inorganic nanocomposite materials are extensively investigated for DMFC, and have the potential to provide a solution to the trade-off problem of polymeric membranes. For example, many polymer-inorganic nanocomposite membranes showed much lower methanol permeability but similar or even improved proton conductivities compared to the corresponding pure polymer membranes and even commercial Nafion membranes. The nanocomposite materials may combine the advantages of each material: for instance, the flexibility and processability of polymers, and the selectivity and thermal stability of the inorganic fillers [1].

Sustainable Membrane Technology for Energy, Water, and Environment, First Edition.
Edited by Ahmad Fauzi Ismail and Takeshi Matsuura.

Attempts have been made to improve the potential of polymer inorganic membrane to be an alternative PEM for DMFC by modifying the external surface of inorganic fillers and by modifying solution nanocomposite preparation techniques. Since silicate clays are hydrophilic and have little affinity for hydrophobic polymers, an organic modification is desirable to enhance their compatibility with polymeric materials. For instance, the introduction of organic functional groups on an inorganic filler surface sometimes contributes to not only better dispersion of the inorganic material in the polymer membrane, but also better transportation of penetrants, which contributes to favorable conductivity and permeability. Additionally, the compatibility between polymers and inorganic fillers can be improved by evaporating the solvent used to prepare the polymer solution above the glass transition temperature (T_g) of the polymers during membrane fabrication [2].

Besides modification of external surface of organic fillers, the composite membrane preparation technique also affected the performance of the nanocomposite membranes. Kim et al. [3] produced Nafion organoclay (S-MMT) composite membrane by using film coating process and by varying the ratios of cosolvents: DMAc and *N*-methylpyrrolidone (NMP), the S-MMT loading amounts, and the concentrations of composite dispersions in the lab scale. They reported that the small amount of high boiling point solvent (NMP) made the membranes dense through azotropic vaporization of two solvents and reorganization of the sulfonic acids to form ion clusters, but, with large amounts of NMP, vaporization of the solvents was retarded, leaving NMP droplets in the drying membranes and forming coarse membranes. Table 22.1 summarizes the performance of polymer-inorganic materials as PEM for fuel cell application.

Although there were improvement in performance by external surface modification of inorganic filler and by producing homogeneous solution, there are still problems that appear to be failure in producing exfoliated nanocomposite membranes [4]. The idea of producing homogeneous nanocomposite membranes was developed recognizing the poor compatibility of the inorganic surface and the polymer where the polymer chains could not tightly contact the inorganic nanoparticles, thus forming a narrow gap surrounding the inorganic nanoparticles, as shown in Figure 22.1. The methanol diffusion path was shortened and thus, the apparent methanol diffusivity and permeability were increased.

Once the nanoparticle surface became compatible with the polymer, the nanogaps could not form any more due to the tight contact between the polymer and the filler particles. Therefore, in the present study, a compatibilizer is used in order to enhance the compatibility between polymer matrices (SPEEK) and the inorganic filler. 2,4,6-Triaminopyrimidine (TAP) is used as the compatibilizer due to its ability to induce strong bonding such as hydrogen bonding between available functional groups of fillers and polymers. Cloisite 15A® clay is used as inorganic filler in order to produce nanocomposite membranes for DMFC. As an organoclay, cloisite is the material prepared from

TABLE 22.1. Summary of the Characteristics of Polymer-Inorganic Membranes

Reference	Materials	Proton Conductivity (S/cm)	Methanol Permeability (cm^2/s)
[2]	Nafion + S-MMT + POP		
	Nafion	100×10^{-3}	2.6×10^{-6}
	Nafion/5 wt % MMT	40×10^{-3}	2.7×10^{-6}
	Nafion/5 wt % MMT/POP	92×10^{-3}	1.6×10^{-6}
[3]	Nafion + S-MMT		
	Nafion	0.09	1.95×10^{-6}
	Nafion/5 wt % S-MMT	0.083	1.26×10^{-6}
[5]	Nafion + MMT		
	Nafion	$29–62 \times 10^{-3}$	8.7×10^{-6}
	Nafion/5 wt % MMT	$23–32 \times 10^{-3}$ at 40–100°C, 100% RH	8.7×10^{-7} at RT
[6]	PBI + MMT		
	PBI	–	3.28×10^{-8}
	PBI/5 wt % MMT	–	6.2×10^{-9}
[7]	Nafion + MMT		
	Nafion	0.08	2.3×10^{-6}
	Nafion/5 wt % MMT	0.05	2.5×10^{-8}

POP, poly(oxypropylene); PBI, polybenzimidazole; RH, relative humidity; RT, room temperature.

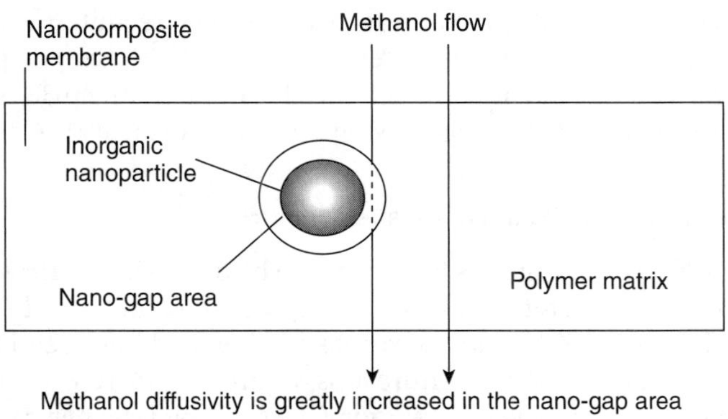

Figure 22.1. Illustration of nanogap formation in nanocomposite membrane.

montmorillonite and cation, di-tallow. Tallow is the mixture of octadecyl, hexadecyl, and tetradecyl. Octadecyl is the major component (>60%) of the tallow. Cloisite is a very promising additive in nanocomposite membranes since it is already being organically modified from the natural MMT, which is expected to enhance the compatibility with organic polymers.

The developed composite membranes were characterized in terms of thermal stability, physical properties, and membrane performance.

22.2 METHODS/THEORY

22.2.1 Materials

Poly (ether ether ketone) (PEEK) in powder form was obtained from Vitrex Inc. (West Conshohocken, PA). Concentrated sulfuric acid (95–97%) purchased from QRex was used as a sulfonating agent in sulfonating process. Dimethylsulfoxide (DMSO) was obtained from Sigma-Aldrich and was used as supplied. DMSO was used as a solvent to dissolve SPEEK, Cloisite15A, and TAP. Cloisite15A was obtained from Southern Clay Products, Inc. and was used as received. TAP in powder form was obtained from Sigma-Aldrich.

22.2.2 Sulfonation Reaction Process

Sulfonation reactions were conducted using sulfuric acid as the sulfonating agent. A mixture of 50 g PEEK and 1000 mL sulfuric acid was magnetically stirred at room temperature for 1 hour. Then the solution was continually stirred at 70°C for 9 hours to obtain SPEEK of 91% DS. In order to obtain SPEEK with 48% of DS, the reaction time of sulfonation reaction was reduced to 3 hours. The sulfonated polymer was recovered by precipitating the product solution into a large excess of ice water. The resulted SPEEK polymer was filtrated and washed thoroughly with deionized water until the pH became ~6–7. Finally, the sulfonated PEEK was dried in the drying oven at 80°C for 24 hours.

22.2.3 Preparation of Nanocomposite Membrane

10 wt % of SPEEK solution was first prepared by dissolving SPEEK in DMSO. Another solution was prepared by dissolving desired amounts of Cloisite15A and TAP in DMSO at 60°C under vigorous stirring for 2 hours, and then added to the SPEEK solution. The mixture was vigorously stirred for 24 hours at 60°C to produce a homogeneous solution. The solution was then cast on a glass plate with a casting knife to form a flat sheet nanocomposite membrane. The resultant membrane was then dried in a vacuum oven for 24 hours at 80°C. The membrane was further dried for 6 hours at 100°C to completely remove the residual solvent. After the membrane was detached from the glass plate, by immersing the membrane into water, it was dried for 3 days in a vacuum oven at 80°C. Finally, the membrane was treated with 1 M sulfuric acid solu-

tion for 1 day at room temperature and subsequently rinsed with water several times to remove the remaining acid to ensure the H form of sulfonate.

22.2.4 X-Ray Diffraction Analysis (XRD)

XRD analysis was monitored using Philip PW1710 XRD with nickel-filtered Cu-Kα source ($\lambda = 0.154056$ nm) at 30 kV and 30 mA. The diffractograms were obtained with a scanning rate of 2°/min in a 2θ range of 1.5° to 10° at room temperature. The *d*-spacing of Cloisite 15A in nanocomposites was calculated from Bragg's equation using XRD results:

$$d = \frac{n\lambda}{2\sin\theta}, \qquad (22.1)$$

where d is the spacing between layers of the clay, λ is the wave length of X-ray equal to 0.154056 nm, θ is the angle at the maximum point of the first peak (lowest θ) in the spectra, and n is a whole number, representing the order of diffraction, taken 1 in our calculation.

22.2.5 NMR Spectroscopy

Nuclear magnetic resonance (H^1NMR) spectroscopy was used to determine the DS of SPEEK. In the present study, H^1NMR spectra were recorded on a Varian Unity Inova spectrometer at a resonance frequency of 399.961 MHz at room temperature. For each analysis, 3 wt % polymer solution was prepared in DMSO-d_6. The DS was determined by comparative integration of distinct aromatic signals according to the following equation:

$$\frac{n}{12-2n} = \frac{SH_{13}}{\Sigma SH_{(1,2,3,4,5,6,7,8,9,10,11,12,14,15)}} \quad (0 \le n \le 1) \qquad (22.2)$$

where n is the number of H_{13} per repeat unit. SH_{13} is the area of under the graph for H_E region, and $\Sigma SH_{(1,2,3,4,5,6,7,8,9,10,11,12,14,15)}$ is the total areas of under the graph for all the other aromatic hydrogen (H_1, H_2, H_3, H_4, H_5, H_6, H_7, H_8, H_9, H_{10}, H_{11}, H_{12}, H_{14}, and H_{15}) regions. An estimation of the degree of sulfonation, DS, is obtained as DS = $n \times 100\%$.

The IEC of the SPEEK in H form can be calculated by

$$DS\% = \frac{288(IEC)}{1000-80(IEC)} \times 100. \qquad (22.3)$$

It should be mentioned that the unit molecular weight of SPEEK in hydrogen form and PEEK are 368 and 288, respectively. The number (80) arises from the difference between these two unit molecular weights.

22.2.6 Thermogravimetric Analysis (TGA)

Thermogravimetric analysis (TGA) was performed on a Mettler Toledo TGA/SDTA851e Thermogravimetric Analyzer. TGA is used to measure weight loss of membranes in order to determine the thermal stability in the range of the operating temperature of the DMFC system, that is 60–120°. Approximately 5–10 mg samples were first dried at 210°C for 30 minutes to remove any moisture/solvent, and then heated from 90 to 900°C at a heating rate of 10°C/min under nitrogen atmosphere.

22.2.7 Water Uptake Measurement

The swelling behavior of the resulting nanocomposite membranes was studied in terms of water uptake and methanol uptake. Water uptake is determined by measuring the weight difference of the membranes in wet and dry conditions. The membranes were dried in an oven at 60°C for 48 hours. The weighed films with diameter ~5.0 cm were then soaked in deionized water overnight at room temperature and then blotted dry with absorbent paper to remove any surface moisture, and then reweighed. Methanol uptake was determined in a similar fashion. Later, the water/methanol uptake was calculated by

$$\text{Water/methanol uptake} = \frac{G_w - G_d}{G_d} \times 100\%, \tag{22.4}$$

where, G_w and G_d are the weight of the dry and the wet membrane, respectively.

22.2.8 Proton Conductivity Measurement

Transverse proton conductivities of the developed membranes were measured by AC impedance spectroscopy over a frequency range of 1–10^7 Hz with 50–500 mV oscillating voltage, using a Solatron 1260 gain phase analyzer. Films having 18 mm diameter sandwiched between two stainless steel block electrodes with ~3 kg/cm^2 pressure were placed in an open temperature-controlled cell. The most crucial step prior to proton conductivity measurements is that SPEEK membrane samples and the Nafion sample must be soaked in water at room temperature for hydration. The thickness of the hydrated membranes was then measured by using micrometer several times to obtain the average membrane thickness. The conductivity, σ, of the samples in the transverse direction was calculated as follows:

$$\sigma = \frac{d}{RS}, \tag{22.5}$$

where, *d* and *S* are the thickness and face area of the membrane sample, respectively, and *R* is derived from the low intersection of the high frequency semicircle on a complex impedance plane with the Re (Z) axis.

22.2.9 Methanol Permeability Measurement

The methanol permeability of SPEEK and its nanocomposite membranes were determined according to Equation 22.6 by carrying out each experiment for 3 hours at room temperature:

$$P = \alpha \times \frac{V_B}{A} \times \frac{L}{C_A}, \tag{22.6}$$

where, P is membrane permeability, $\alpha = C_B(t)/(t\text{-}to)$ is the slope of linear interpolation of C_B (methanol concentration in permeate compartment at time t) versus t, V_B is the volume of the water compartment, A is the membrane cross-sectional area (effective area), L is the membrane thickness, and C_A is the concentration of methanol in the feed compartment.

22.3 RESULTS AND DISCUSSION

22.3.1 Preparation of SPEEK/Cloisite 15A/TAP Nanocomposite Membranes

Degree of sulfonation of SPEEK was determined using 1H NMR analysis and the value of DS was 47.83 and 91% corresponding to low and high DS, respectively. The IEC value calculated from Equation 22.3 was 1.47 meq/g and 2.52 meq/g, respectively.

In order to produce homogeneous SPEEK/Cloisite 15A/TAP nanocomposite membranes, the intercalation method was employed by preparing the SPEEK and Cloisite 15A/TAP solutions separately before mixing them up together. Table 22.2 depicts the compositions of the casting solutions to prepare SPEEK and SPEEK/Cloisite 15A/TAP nanocomposite membranes. It is important to produce homogeneous nanocomposite membranes in order to fulfill the needs of high membrane performance, particularly in improving the membrane barrier properties.

TABLE 22.2. Composition of Nanocomposite Membrane Preparation Solution

Sample Designation	Polymer (g)	Cloisite 15A (g)	TAP (g)
SP48	10	0	0
SP91	10	0	0
SP/Cl	10	0.1	0
SP/Cl/TAP	10	0.1	0.1
SP/TAP	10	0	0.1

22.3.2 Dispersion of Cloisite 15A in SPEEK Matrix

Although the Cloisite 15A has already been organically modified from the natural clay (montmorillonite), Chen [8] claimed that among the commercially available Cloisite clays, Cloisite 15A exhibited the highest hydrophobicity when it was incorporated into polyurethane matrix [8]. Therefore, in the present study, a compatibilizer was introduced into the SPEEK/Cloisite 15A nanocomposites in order to enhance the compatibility between the materials. There are required characteristics for a compatibilizer to be incorporated with polymer and filler. A compatibilizer should be from a low molecular weight material which can interact simultaneously with fillers and polymers. It also should have the ability to induce strong bonding such as hydrogen bonding with functional groups available in fillers and polymers. Besides making effective bonding, they must be soluble in the solvent used to make the polymer dope solution. In order to prevent the compatibilizer from evaporating during dope formulation, they also must be in solid at room temperature. TAP is a kind of compatibilizer that possesses all the characteristics that a compatibilizer should have [9].

It is expected that TAP will form hydrogen bonding with both the polymer (SPEEK) and the inorganic filler (Cloisite 15A) to improve the compatibility between them. Therefore, the role of TAP was interpreted in relation to the number of oxide and hydroxyl groups on the external surface of the inorganic particles.

We also proposed that the porous layered Cloisite 15A exfoliated in SPEEK matrix can act as selectively permeable barriers that block methanol permeation but do not reduce or may potentially even increase the proton conductivity. Cloisite 15A will behave in such a way because of the combination of small pore size of a few angstroms that can permeate protons selectively rather than methanol as well as a high aspect ratio that creates high tortuous pathway toward methanol molecules [10].

XRD measures the degree of dispersion by estimating the distance between individual platelets after mixing with polymer. If the clay interlayers diffraction peak in the composite is equal to or lower than the value observed for the pure clay alone, an ordinary composite is obtained. A shift to smaller angles of the characteristic diffraction peak for the composite suggests an increase in the interlayer spacing or gallery of the clay, which is referred to as intercalation. When the peak corresponding to the clay interlayer diffraction peak is not observed in the polymer/clay diffractograms, it indicates that enough polymer chains are inserted into the silicate galleries giving rise to the formation of a nanocomposite with an exfoliated structure in which the polymer is nanoscopically confined [11].

Figure 22.2 clearly shows that the primary diffraction peaks for the basal spacing (d_{001}), that is 3.30 nm of Cloisite 15A layers disappeared completely in the XRD patterns of SPEEK/Cloisite 15A/TAP (abbreviated as SP/Cl/TAP) nanocomposite membranes. This indicates an exfoliated dispersion of

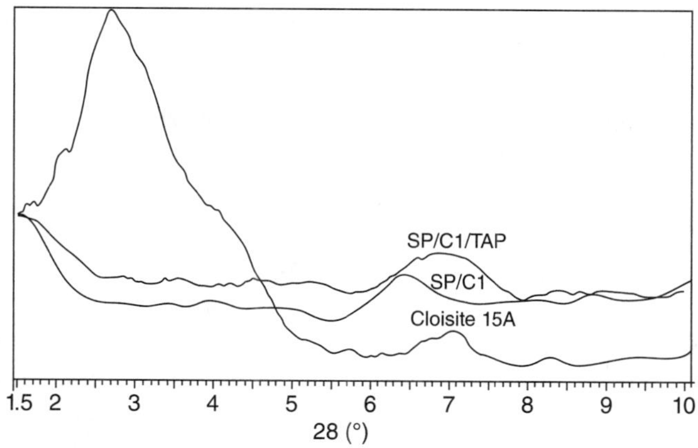

Figure 22.2. XRD patterns of Cloisite 15A, SP/Cl, and SPCl/TAP.

Cloisite 15A in the SPEEK matrix. It was also found that the nanocomposite membrane was almost transparent, indicating the good dispersion of Cloisite 15A [12]. However, SPEEK/Cloisite 15A (abbreviated as SP/Cl) still shows a small peak that corresponds to the Cloisite interlayer peak. This suggests that a typical composite membrane has been fabricated.

22.3.3 Thermal Stability Study

Figure 22.3 shows the TGA profiles for SPEEK, SP/Cl, SP/Cl/TAP, and SP/TAP membranes. The SPEEK samples started to lose weight in the range of temperature 100–300°C. The SPEEK membrane is said to undergo thermal degradation when the membrane loses water that interacts with sulfonic acid group during the sulfonation process [13]. This phenomenon is delayed by the addition of Cloisite 15A and TAP in the SPEEK polymer matrix, which indicates that the out-diffusion of water is slowed down by the good interaction of nanoclay in the polymer matrix. It seems that all SPEEK-nanocomposite membranes have three weight losses regime; that is, they showed first weight loss at 350°C, which was due to the decomposition of sulfonic acid group in the SPEEK. The second and third weight loss regimes were observed at 550 and 750°C, respectively. The midpoint temperature of the degradation at 550°C for SP/Cl and SP/Cl/TAP may be attributed to the release of olefin and amine of Cloisite 15A nanoclay [14]. TAP started to degrade at 750°C. It was hence concluded that the addition of the inorganic filler into the SPEEK matrix significantly increased the degradation temperature of the native SPEEK. It was also found that such thermal decomposition was well above the normal operating temperature of DMFC (i.e., 60–120°C). It is proven that the thermal stability of SPEEK membranes is sufficient for direct methanol fuel cell application.

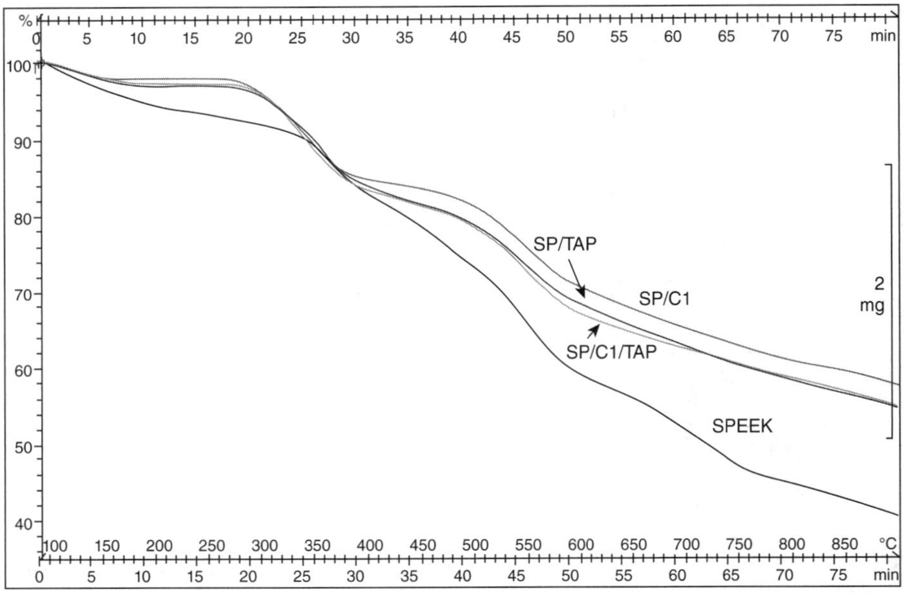

Figure 22.3. TGA curves for SPEEK, SP/Cl, SP/Cl/TAP, and SP/TAP.

TABLE 22.3. Physical Properties of SPEEK Nanocomposite Membranes

Sample	Water Uptake, $\times 10^3$ (wt %)	Methanol Uptake, $\times 10^3$ (wt %)
SP91	0.095	0.522
SP91/Cl	2.260	NA
SP91/Cl/TAP	0.884	6.150
SP91/TAP	0.511	3.270
SP48	0.010	0.014
SP48/Cl	0.034	0.043
SP48/Cl/TAP	0.020	0.030
SP48/TAP	0.029	0.031
Nafion	0.021	0.041

22.3.4 Liquid Uptake of the SPEEK/Cloisite 15A/TAP Nanocomposite Membranes in Water and in Methanol Aqueous Solution

The membrane stability in water and aqueous methanol solution at room temperature was studied with respect to SPEEK and its nanocomposite membranes. Table 22.3 shows that among the nanocomposite membranes with 91% DS, the SPEEK/Cloisite 15A membrane (SP91/Cl) showed the highest liquid uptakes. Moreover, the membrane was dissolved in methanol during the uptake experiment. This might be due to the apparent hydrophilic domain of the SPEEK membrane that was further increased by the hydrophilic domain of Cloisite 15A clay itself [15].

The methanol uptake of the nanocomposite membranes for 91% DS showed the SPEEK membrane with Cloisite 15A alone is an undesirable membrane since it was dissolved in methanol during the experiment. Looking into the data for the composite membranes, the water uptake of SP48/Cl/TAP was as low as 1% of the SP 91/Cl membrane. This indicates that the nanocomposite membranes with 48% DS were more stable and gave an appropriate uptake value after 1 day immersion in water. Methanol uptake of SP48/Cl/TAP membrane was also the lowest among all the composite membranes. This might be due to the polymer-clay nanocomposite formation. When SPEEK is intercalated or exfoliated in Cloisite 15A layers, the polymer chain movement is hindered, which results in chain packing. From Table 22.3, it was also found that many of the nanocomposite membranes tested showed methanol uptake lower than the commercial Nafion 112 membrane. It was also found that most of the nanocomposite membranes showed higher liquid uptakes, especially the water uptakes, as compared to the parent SPEEK membrane. This might be due to the stronger hydrogen bonding between water or methanol molecules and the hydrophilic groups, that is, OH and NH_2, on the TAP and Cloisite 15A layers [5].

22.3.5 Proton Conductivity

Table 22.4 clearly shows that SPEEK membrane with higher DS exhibited higher proton conductivity than SPEEK membrane of lower DS. This was due to the larger amount of sulfonic acid (SO_3H) group in the SPEEK of higher DS. The water uptake also rose with increasing DS (Table 22.3). These results suggest that the water-swollen ionic domains in the membrane pores were interconnected to form a network structure. The lower proton conductivity observed for membranes with lower DS was due to the diffusion limitation caused by segregation in the ionic domains [16]. However, when the SPEEK membrane was composited with the inorganic fillers, this effect was no longer valid. It seems that the effect of inorganic fillers loading is more pronounced

TABLE 22.4. PEM Performance of SPEEK Nanocomposite Membranes

	Proton Conductivity, σ ($\times 10^{-3}$ S/cm)		Methanol Permeability ($\times 10^{-7}$ cm^2/s)		Overall Membrane Characteristics, Φ ($\times 10^3$)	
DS	48	91	48	91	48	91
SPEEK	1.91	2.00	3.06	5.59	6.241	3.558
SP/Cl	1.23	1.10	3.41	13.9	3.607	0.791
SP/Cl/AP	3.87	3.03	1.29	1.81	30.000	16.740
SP/TAP	1.97	2.07	2.65	9.78	7.434	2.117
Nafion	12.00		15.72		7.643	

than the effect of the ionic group attached to the polymer ring since the addition of fillers into SPEEK of 48% DS showed higher proton conductivity as compared to the composite SPEEK membranes of 91% DS (with an exception of SP/TAP). This may be due to the better dispersion of inorganic fillers in the SPEEK 48% DS.

From Table 22.4, SP/Cl/TAP exhibited the highest proton conductivity value, that is 3.87 and 3.03 mS/cm for 48% DS and 91% DS, respectively. It seems that the TAP has successfully enhanced the dispersion of Cloisite 15A and consequently improved the compatibility of the Cloisite 15A and SPEEK polymer. This behavior facilitates the migration of protons through the membrane. It is also suggested that the porous Cloisite 15A when dispersed homogenously in nanocomposite membranes increased the mobility of protons through the Cloisite 15A particle itself via free diffusion mechanism. In addition, the proton conductivity of the native SPEEK has been further increased by the presence of Cloisite 15A that possesses good proton conductivity, that is, 10^{-4} S/cm at room temperature [2].

When we compare the proton conductivity result for SPEEK nanocomposite membranes, it was found that the conductivity dramatically increased for SPEEK/Cloisite 15A/TAP nanocomposite membranes. Although the DS of SPEEK used was at low degree, the membranes still exhibited high proton conductivity as compared to other membrane samples. It seems that the ability of TAP to improve the compatibility between SPEEK and Cloisite 15A is more pronounced as compared to the amount of sulfonic acid (SO_3H) group attached to the SPEEK polymer ring. This can be understood by comparing the conductivity value of the representative membrane with the nanocomposite membrane without TAP that showed the lowest conductivity value. A similar behavior was observed by Yong et al. [11] who studied the influence of TAP on the membrane properties for gas separation application [10]. Unfortunately, none of the studied SPEEK nanocomposite membranes could achieve proton conductivity as high as or higher than that of commercial Nafion membrane. This might be due to the inappropriate amount of Cloisite 15A and TAP loadings to the SPEEK polymer for producing a desirable proton conductivity value.

22.3.6 Methanol Permeability

According to Table 22.4, it can be seen that all SPEEK composite membranes showed lower methanol permeability than that of Nafion 112 membrane. The corresponding methanol permeability for Nafion 112 was 15.72×10^{-7} cm²/s. A comparable value was reported by Ismail et al. [17].

The high methanol permeability value for Nafion 112 membrane obtained was possibly due to difference in microstructure between SPEEK and Nafion 112 membrane. The high hydrophobic/hydrophilic domains of Nafion membrane not only allowed water and protons to migrate through the membrane but also small polar molecule such as methanol to pass through [18].

The result also indicates that the introduction of Cloisite 15A has reduced the methanol permeability of native SPEEK. It indicates that the addition of Cloisite nanofiller with good compatibility has improved the barrier properties of parent SPEEK membrane toward methanol. This is because the well dispersed Cloisite 15A nanofiller increases the tortuosity path for methanol across the membranes due to its high length-to-width ratio [19]. It was reported that Cloisite 15A possesses a higher aspect ratio than other nanoclays, that is, 70–150 [20]. However, the methanol permeability value for the SP/Cl membrane was higher than the native SPEEK and the other SPEEK composite membranes tested. This might be due to the poor compatibility between the SPEEK and Cloisite 15A nanofiller.

In order to perform as an excellent PEM for DMFC, PEM of high proton conductivity and low methanol permeability is desirable. Table 22.4 summarizes the overall membrane characteristics of SPEEK and SPEEK composite membranes together with Nafion 112 membrane. Although Nafion 112 possesses the highest proton conductivity, its high methanol permeability restricts its ability to perform as a good PEM for DMFC application. Interestingly, SPEEK membrane with a very low amount of Cloisite 15A and TAP loaded into SPEEK of low DS has performed as a very good candidate for DMFC PEM owing to the acceptable proton conductivity value and very low methanol permeability, that is, the overall performance of the SP/Cl/TAP membrane was 75% higher than the Nafion 112 membrane.

22.4 CONCLUSIONS

TAP was used as a compatibilizer to improve the compatibility between SPEEK and Cloisite 15A by enabling the preparation of an exfoliated SPEEK/Cloisite 15A/TAP nanocomposite membranes. The prepared homogenous SPEEK/Cloisite 15A/TAP nanocomposite membranes have successfully improved the membrane barrier properties due to the unique feature of Cloisite 15A that contributed longer pathways toward methanol through the membrane. The appearance of inorganic aggregation in the polymer matrix was observed by XRD. The stability of the prepared membranes in water and in methanol aqueous solution was enhanced by the introduction of TAP into the SPEEK/Cloisite 15A nanocomposite membrane. The SPEEK nanocomposite membrane with a lower DS, that is, 48% was a favorable PEM owing to considerable proton conductivity value and significantly improved barrier properties that contributed to higher overall membrane characteristics than that of conventional SPEEK and also commercial Nafion membrane.

REFERENCES

[1] Thomassin J. M., C. Pagnoulle, G. Caldarella, A. Germain, and R. Jerome. Contribution of nanoclays to the barrier properties of a model proton exchange

membrane for fuel cell application. *Journal of Membrane Science* 270 (2006) 50–56.

[2] Lin Y. F., C. Y. Yen, C. C. M. Ma, S. H. Liao, C. H. Hung, and Y. H. Hsiao. Preparation and properties of high performance nanocomposite proton exchange membrane for fuel cell. *Journal of Power Sources* 165 (2007) 692–700.

[3] Kim T. K., M. Kang, Y. S. Choi, H. K. Kim, W. Lee, H. Chang, and D. Seung. Preparation of nafion-sulfonated clay nanocomposite membrane for direct methanol fuel cells via a film coating process. *Journal of Power Sources* 165 (2007) 1–8.

[4] Duval D., A. Jean, and H. A. Gutierrez. Ras signaling and transcriptional synergy at a flexible ets-1/Pit-1 composite DNA element is defined by the assembly of selective activation domains. *Journal of Biological Chemistry* 278 (2003) 39684–39696.

[5] Gosalawit R., S. Chirachanchai, S. Shishatskiy, and S. P. Nunes. Krytox-montmorillonite-nafion nanocomposite membrane for effective methanol crossover reduction in DMFCs. *Solid State Ionics* 178 (2007) 1627–1635.

[6] Cong H., M. Radosz, B. F. Towler, and Y. Shen. Polymer-inorganic nanocomposite membranes for gas separation. *Separation and Purification Technology* 55 (2007) 281–291.

[7] Chuang S. W., S. L. Chung, and H. C. L. Hsu. Synthesis and properties of fluorine-containing polybenzimidazole/montmorillonite nanocomposite membranes for direct methanol fuel cell applications. *Journal Power Sources* 168 (2007) 172–177.

[8] Chen S. C. Synthesis and characterization of polyurethane/clay nanocomposite: melt compounding. Individual Enquiry, The University of Queensland, St. Lucia, Australia, 2002.

[9] Song M. K., S. B. Park, Y. T. Kim, K. H. Kim, S. K. Min, and H. W. Rhee. Characterization of polymer-layered silicate nanocomposite membranes for direct methanol fuel cells. *Electrochimica Acta* 50 (2004) 639–643.

[10] Yong H. H., H. C. Park, Y. S. Kang, J. Won, and W. N. Kim. Zeolite-filled polyimide membrane containing 2, 4, 6-triaminopyrimide. *Journal of Membrane Science* 188 (2001) 151–163.

[11] Hudiono Y., S. Choi, S. Shu, W. J. Korros, M. Tsapatsis, and S. Nair. Porous layered oxide/Nafion nanocomposite membranes for direct methanol fuel cell applications. *Microporous and Mesoporous Materials* 118 (2009) 427–434.

[12] Frounchi M., S. Dadbin, Z. Salehpour, and M. Noferesti. Gas barrier properties of PP/EPDM blend nanocomposite. *Journal of Membrane Science* 282 (2006) 143–248.

[13] Gaowen Z. and Z. Zhentao. Organic/inorganic composite membranes for application in DMFC. *Journal of Membrane Science* 261 (2005) 107–113.

[14] Jaafar J., A. F. Ismail, and A. Mustafa. Physicochemical study of poly(ether ether ketone) electrolyte membranes sulfonated with mixtures of fuming sulfuric acid and sulfuric acid for direct methanol fuel cell application. *Materials Science and Engineering: A* 460–461 (2007) 475–484.

[15] Sangrungsri P. Development of polypropylene nanocomposites for flexible packaging in order to replace a conventional barrier material. MSc. Thesis, University of Massachusetts Lowell, 2006.

[16] People B. C. In-situ production of polyolefin-clay nanocomposites. University of California: Dissertation Thesis, 2008.

[17] Ismail A. F., N. H. Othman, and A. Mustafa. Sulfonated polyether ether ketone composite membrane using tungstosilicic acid supported on silica-aluminium oxide for direct methanol fuel cell (DMFC). *Journal of Membrane Science* 329 (2009) 18–29.

[18] Nasef M. M., N. A. Zubir, A. F. Ismail, M. Khayet, K. Z. M. Dahlan, H. Saidi, R. Rohani, T. I. S. Ngah, and N. A. Sulaiman. PSSA pore-filled PVDF membranes by simultaneous electron beam irradiation: Preparation and transport characteristics of protons and methanol. *Journal of Membrane Science* 268 (2006) 96–108.

[19] Kreuer K. D. On the development of proton conducting polymer membranes for hydrogen and methanol fuel cells. *Journal of Membrane Science* 185 (2001) 29–39.

[20] Chang J. H., J. H. Park, C. S. Kim, and O. O. Park. Proton-conducting composite membranes derived from sulfonated hydrocarbon and inorganic materials. *Journal of Power Sources* 124 (2003) 18–25.

23

PURIFICATION OF BIOGAS USING CARBON NANOTUBES MIXED MATRIX MEMBRANE: EFFECT OF FUNCTIONALIZATION OF CARBON NANOTUBES USING SILANE AGENT

Tutuk Joko Kusworo, Abdullah Busairi,
Ahmad Fauzi Ismail, Azeman Mustafa, and Budiyono

23.1 INTRODUCTION

Off-gas from anaerobic digestion and landfills has significant potential as an alternative energy source. At present it can be used only at the place where it is produced, most often for steam generation. However, the use of biogas still has limitation due to low quality of biogas containing substantial amounts of CO_2 (20–45 vol%) and H_2S (0–1 vol%) in addition to main component CH_4 (55–80 vol%). In addition, biogases are also frequently saturated by water vapor. The presence of CO_2 as an incombustible gas reduces its calorific value and makes it uneconomical to compress and transport to longer distances [1]. Biogas purified and enriched in methane can be used for household applications, automobile fuel (liquefied), or electricity generation. There is a great need to make biogas transportable. This can be done by compressing the gas in cylinders, which is possible only after removing CO_2, H_2S, and water vapor. There is a lot of potential if biogas could be made viable as a transport vehicle

Sustainable Membrane Technology for Energy, Water, and Environment, First Edition.
Edited by Ahmad Fauzi Ismail and Takeshi Matsuura.

fuel like compressed natural gas (CNG), by compressing and filling the cylinders in it after scrubbing and drying. Enrichment of methane in biogas to have fuel of higher calorific value can be achieved by removal of carbon dioxide. Elimination of carbon dioxide from the flue gas helps to increase its calorific value as well as to eliminate the greenhouse gas. The carbon dioxide thus generated can be utilized as an effective refrigerant.

Currently, a variety of processes are being used for removing CO_2 and H_2S from natural gas (NG) in petrochemical industries. Several basic mechanisms are involved to achieve selective separation of gas constituents. These may include physical or chemical absorption, adsorption on a solid surface, membrane separation, cryogenic separation, and chemical conversion. Chemical methods are based on absorption under elevated pressure (in water, 30% solution of potassium carbonate, solution of monoethanolamine, etc.). Other methods are based on adsorption in which the choice of suitable adsorbent material plays a vital role. A very important stage of the biogas upgrading is its purification from H_2S traces. The biogas desulfurization is based on a chemical reaction of H_2S with a suitable substance, such as quicklime, slaked lime in solid form, or slaked lime in liquid form. High concentration of CO_2 which is present in biogas makes the satisfactory removal of H_2S difficult since the CO_2 also reacts with the quick and slaked lime and uses it up quickly. At present, membrane separation processes have emerged during the last two decades. This is due to the fact that membrane separation processes may offer more capital and energy efficiency when compared to the conventional separation processes [2]. In addition, the advantage of membrane technology is its simplicity, that is, no absorbent, which has to be regenerated; it can be easily retrofitted, modularized, and scaled up for several applications [3].

At present, polymeric membranes dominate the membrane separation field due to the fact that they are well developed and are quite competitive in separation performance and economics. Since the separation of gas mixtures depends on the relative solubility and diffusivities of individual components within a particular membrane, no single membrane material is capable of successfully separating all gas mixtures. The choice of membrane material will be dictated by the nature of the gas mixture to be separated. Glassy polymers such as polysulfone, polyethersulfone (PES), polyimide, and ethyl cellulose which are known to have high mechanical stability and desirable inherent transport properties at high temperature are more commonly used [4].

Despite concentrated efforts to tailor polymer structure to affect separation properties, current polymeric membrane materials have seemingly reached a limit in the trade-off between productivity and selectivity. Pure polymer membranes oftentimes show several limitations such as low selectivity, high temperature instability and swelling, and decomposition in organic solvents. Hybrid membranes composed of organic–inorganic has attracted attention as the future membrane material. The concept of hybrid membranes or mixed matrix membranes combines the advantages of high separation capabilities of the molecular sieves and the desirable mechanical properties and economical processing capabilities of polymers [4–6]. Many studies have reported that the separation performance of mixed matrix membrane could be improved by integrating porous or nonporous inorganic filler such as zeolite, silica, carbon molecular sieve, and activated carbon. Currently, significant efforts have been devoted in fabrication of mixed matrix membranes by using carbon nanotubes (CNTs) as great potential fillers. The properties of polymer nanocomposites containing CNT depend on several factors in addition to the polymer: synthetic process used to produce nanotubes; nanotube purification process (if any); amount and type of impurities in the nanotubes; diameter, length, and aspect ratio of the nanotube objects in the composite (isolated, ropes, and/or bundles); nanotube orientation in the polymer matrix. Kim et al. [7] studied the effect of incorporating CNTs on polyimidesiloxane matrix on gas separation performance. They observed that the addition of small CNTs to the copolymer matrix reduced the permeability of helium and hindered the diffusion of nitrogen due to the impermeable properties of CNTs.

Currently, different approaches are reported to improve the dispersion of CNTs in solvents or polymers in order to fabricate the composite mixed matrix nanotubes membrane. Recent reports on the chemical compatibility and dissolution properties of CNTs have promoted a great deal of interests in developing modification and functionalization of their surface. The chemical functionalization of multiwall CNT can be done by using novel silane coupling agents, which helps to covalently link polymers. Besides that, treatment by carboxylic acid groups also can be used in this modification, but for the strong acid used, it can form defects on CNT sidewalls and decrease the aspect ratio of CNTs. Other possible surface modification techniques included plasma, thermal, and laser ablation [8]. By using silane agents, the coupling process can be accomplished via the chemical reaction between the trialkoxy groups and the hydroxyl groups of silane molecules and the hydroxyl groups on the glass substrates, whereas other functional group of silane molecules, which are generally ethylene, amine, epoxy, and thiohydroxy can remain [9].

Therefore, in the present study, the feasibility study of fabrication and characterization of mixed matrix membranes using particles of CNTs as selective inorganic fillers was investigated. The CNTs were functionalized using Dynasylan Ameo silane coupling agent. The aim is to get CNTs linked with the coupling agent having a functional group such as a double bond which can be utilized further for copolymerization. Hence, this reaction product (CNTs silanization) can be chemically explored for improving their compatibility with polymer matrix.

23.2 METHODS

Polyethersulfone was supplied by Solvay Advanced Material (Shanghai, China). The polymers were dried in a vacuum oven at 120°C overnight before dope preparation; *N*-methyl pyrrolidone (NMP) from Merck was used as the solvent due to its low toxicity. The Single-wall nanotubes (SWNTs) were produced using thermal catalytic chemical vapor deposition (CCVD). The particle size of CNTs was about 25.76 nm. In this study, the polymer solution consists of 25 wt % polymer, 75 wt % NMP, and 1 wt % nanotubes in the total solid. Flat sheet membrane was prepared according to the dry/wet phase inversion technique. The solution was poured onto a clear, flat, and smooth glass plate that was placed on the trolley. Stainless steel support casting knife was used to spread the solution to a uniform thickness by pneumatic force. CNTs were additionally modified by a silane coupling agent. First, the CNTs were again dried and pretreated in H_2SO_4. Double bonds present in the aromatic nanotube structure were oxidized with a sulfuric acid. This pretreatment is necessary to create carboxylic acid moieties on nanotube sidewalls that can react with the silane and also to remove impurities from the supported catalyst. Finally, the modified CNTs were dehydrated at 110°C for 5 hours in a vacuum oven to remove the adsorbed water vapor or other organic vapors before it was ready to be used in preparation of the casting dope.

The membrane sheets were coated with highly permeable elastomeric silicone polymer (Sylgard 184 Dow Corning). The membrane coating was done after the uncoated membranes were tested. The intention of coating is to fill any surface pinholes or defects on membrane surface. Membranes were submerged in the 3%w/w solution of silicone in n-hexane for 24 hours and subsequently placed in an oven for 3 days at 120°C to allow curing before permeation testing. A Supra 35 VP field emission scanning electron microscope was used to ensure the asymmetric structure and to determine the dimension of the fibers. The glass transition temperature of each cast film was determined using differential scanning calorimetry (Mettler Toledo DSC 822e). A small piece of membrane or pure polymer sample was first stored under vacuum at 100°C for 24 hours to remove adsorbed water, then weighed and placed into aluminum differential scanning calorimetry (DSC) pans. The scanning range was 50–320°C with scanning rate of 10°C/min in the first DSC cycle to remove thermal history and then cooled from 320 to 25°C at the rate of 10°C/min, Finally, the second cycle was carried out with the same procedure.

The permeation test involved the use of gas permeation cell in which the membrane was placed on a sintered metal plate and pressurized at the feed side. Pure gas permeation rates were measured by a constant pressure system using a soap bubble flowmeter. Pressure normalized gas permeation flux or permeance for gas i, $(P/l)_i$, in gas permeation units (GPU), can be calculated as follows:

$$\left(\frac{P}{l}\right)_i = \frac{Q_i}{A\Delta p}, \tag{23.1}$$

where Q_i is the volumetric flow rate of gas i, Δp is pressure difference across membrane (cmHg), A is membrane-affective surface area (cm^2), and l is membrane skin thickness (cm). The ideal separation factor $\alpha_{i/j}$ can be calculated by using the equation below:

$$\alpha_{i/j} = \frac{(P/l)_i}{(P/l)_j}. \tag{23.2}$$

The gas permeation experiments were carried out at the transmembrane pressure difference of 10 bar and at room temperature.

23.3 RESULTS AND DISCUSSIONS

In order to further investigate the effect of silanization of CNTs on the dispersion of CNTs in mixed matrix membrane, careful field emission scanning electron microscopy (FESEM) inspections were carried out. The FESEM micrographs of the cross sectional and the surface of the neat PES, PES-unmodified nanotubes, and PES-modified nanotubes mixed matrix membranes are shown in Figures 23.1–23.3. As shown in Figures 23.1, the nanotube composite membranes show finger-like structures which consist of a dense skin layer supported by a spongy porous substructure. The particles of unmodified CNTs seem to adhere well to polymer matrix as shown in Figures 23.2. The smooth surface of the unmodified nanotubes-PES mixed matrix membrane might be due to the shape of the particles of CNTs which were oblong; therefore, no sharp edges could easily perforate the active surface skin. However, in the case of unmodified CNTs, the unselective voids appeared, indicating that the CNTs did not completely adhere onto the surface of polymer matrix. The small surface ruptures and aggregation also occurred on the mixed matrix membrane with unmodified CNTs as shown in Figure 23.2b.

The cross section and surface of PES-modified CNTs using 15 wt % of silane are shown in Figure 23.3a,b. The porous substrate of the fabricated membrane and the active skin layer can be seen in Figure 23.3. Cross section of porous substrate region for PES-modified CNTs clearly shows the compatibility between modified CNT particles with PES. Moreover, from the cross

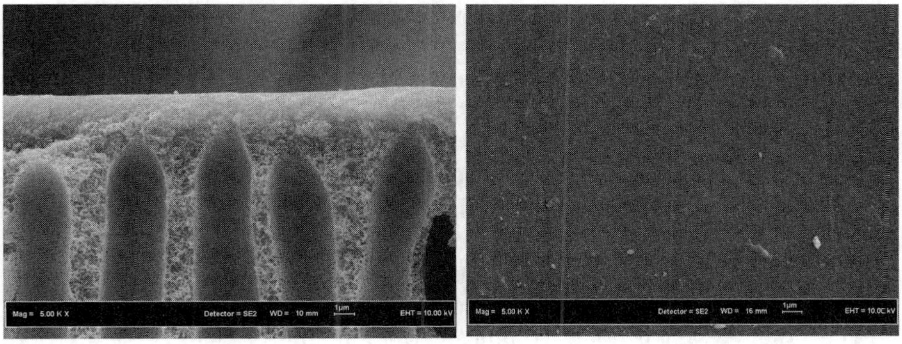

(a) (b)

Figure 23.1. SEM picture of asymmetric polyethersulfone membrane at the (a) cross section and (b) top surface image.

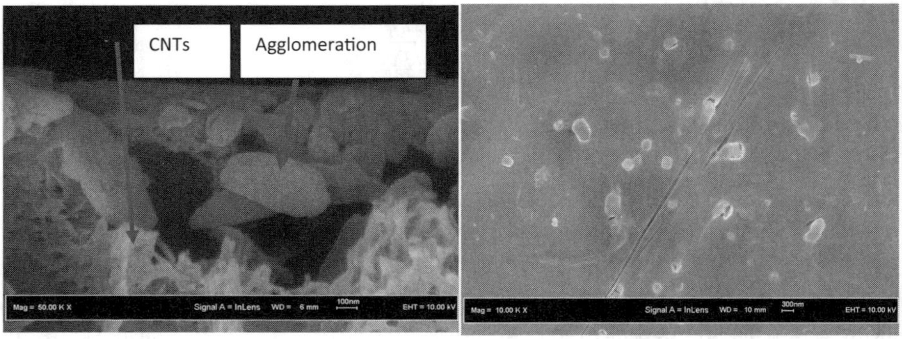

(a) (b)

Figure 23.2. SEM picture of asymmetric polyethersulfone-unmodified CNTs membrane at the (a) cross section and (b) top surface image.

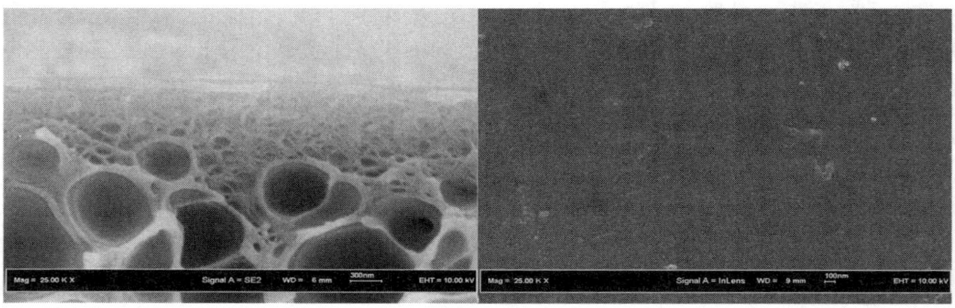

(a) (b)

Figure 23.3. SEM picture of asymmetric polyethersulfone-modified CNTs membrane at the (a) cross-section and (b) top surface image.

section can be seen that CNTs clearly act as the filler into the PES polymer host. The chemical and physical properties of amino functional groups in silane could enhance the bond strength which provided a stable and strong bond between the modified CNTs with PES matrix and finally reduced the presence of voids surrounding the CNT surface. Therefore, the good compatibility between polymer matrix and CNTs would finally diminish the transport of gas penetrants via unselective voids formed between CNTs and polymer matrix, and simultaneously high gas separation performance of membrane is able to be achieved. The smooth surface of CNTs might also enhance the adhesion between the nanotubes and the host of polymer.

The DSC has been done to indicate whether the CNT-PES blends were indeed miscible by observing the glass transition temperatures of the blend composition. The T_g for each membranes was determined from the heat flow versus temperature curve using an onset method. The influence of silane

TABLE 23.1. Effect of Functionalization of CNTs on the Glass Transition Temperature of PES-CNTs Mixed Matrix Membrane

Membrane	T_g (°C)
Neat PES	219.05
Untreated CNTs	221.07
15 wt.% DA silane	224.65

modification on the glass transition temperature of PES-CNTs mixed matrix membranes is tabulated in Table 23.1. As shown in Table 23.1, the amino-functionalized CNTs have increased the glass transition temperature T_g. This can be seen in the effect of T_g for CNT modification with 15 wt % silane. The increasing of T_g in amino-functionalized CNTs MMM is because of the strong interaction between the PES matrix and amino-functionalized CNTs, which reduces the matrix mobility. The strong interaction expresses itself in a shift of T_g [10]and beside that the diffusion of small molecules can be retarded under high temperature. It can also be seen that compared with neat PES polymer, the T_g of the mixed matrix membranes increased about 5°C with incorporation of only 1 wt % modified CNTs using 15 wt % silane treatment into the matrix. This phenomenon indicates that the mobility of polymer chains is reduced due to the restriction effect of CNTs. This result is in agreement with the previous study [11]. Based on Table 23.1, it can be concluded that the CNT surface modification using Dynasylan Ameo silane coupling agent could induce the adherence between polymer matrix and the particles of CNTs. Therefore, it can be concluded that chemically modified CNTs, when incorporated into PES matrix, will make the composite film stiffer and stronger due to strong interfacial interaction between the nanotubes and PES host matrix. Moreover, with addition of nanotubes, the movement of polymer chain in polymer host matrix will be restricted. This phenomenon looks like physical cross-linking.

In the gas separation performance, the criteria which is important in determining whether the membrane is considered as good or not is the ability of the membrane to give the high gas separation performance than other membranes. Thus, this testing was carried out in order to study the effectiveness of chemical modification on CNTs surface. The membrane effectiveness in the gas separation performance was determined by the membrane permeability and selectivity for major biogas components. In this study, we used pure CO_2 and CH_4 to represent the biogas product. Table 23.2 summarizes the permeance and selectivity data of neat PES, PES-unmodified, and PES-modified CNT membranes at a silane concentration of 15 wt %.

Generally, the idea to insert inorganic fillers into an organic polymer is to enhance gas permeability of the polymer due to the disturbed polymer chain packing by the nanofillers. Therefore, the well-dispersed and well-adhered

TABLE 23.2. Results of Gas Permeation Experiments for Neat PES, PES Mixed Matrix Membrane with Untreated CNTs, and PES Mixed Matrix Membrane with 15 wt % DA Saline Treated CNTs

	Single Gas Permeance (GPU)		Selectivity
Membrane	CO_2	CH_4	CO_2/CH_4
Neat PES	20.36	0.70	28.75
Untreated CNTs	25.18	1.96	12.84
15 wt.% DA silane	22.18	0.73	30.42

Experiments carried out at room temperature and 10 bars.

CNTs will effectively increase the gas permeability due to the insertion of the CNT particles between polymer chains of the matrix. The addition of 1 wt % of modified CNTs loading to PES resulted in slight increase in permeability of CH_4. However, for the small gas molecules such as CO_2, permeability also increased slightly with the addition of CNTs in the PES host matrix. The main pathways of gas transport through the mixed matrix membranes are through dense layer of PES matrix, CNTs, and nonselective gaps or voids between the matrix and sieve particles. From FESEM data in Figures 23.1 and 23.3, the CNTs are well dispersed in polymer matrix and serve as channels to transport gas molecules. As a result, as presented in Table 23.2, the ideal selectivity for CO_2/CH_4 of the PES-1 wt % CNTs increased. On the other hand, the permeability for the PES-unmodified CNT membrane increases for both gases from the neat PES membrane, while the CO_2/CH_4 selectivity decreased. Therefore, the increase in permeability in the PES-unmodified CNTs membrane was not due to the gas transportation inside the nanotubes. As shown in Figure 23.2, the unmodified CNTs were not well distributed in the polymer matrix, and unselective voids between carbon nanoparticles and polymer matrix appeared. The selectivity decreased due to the presence of these void spaces.

23.4 CONCLUSION

In this study, PES membranes with and without incorporation of CNTs were fabricated, and their gas permeation performance was tested. In one case, CNTs were modified by silanization using Dynasylan Ameo silane coupling agent. Based on the experimental results and analysis, the following conclusions can be drawn:

- The FESEM cross-sectional and surface images of mixed matrix membranes showed that CNTs were well dispersed in the polymer matrix when modified by salinization. The surface ruptures did not occur when the CNTs without modification were incorporated due to the shape of CNTs which appeared to be oblong. The smooth surface of CNTs might have also helped to enhance adhesion between the nanotubes and the host polymer.

- Both permeance and selectivity of the PES membranes increased slightly by incorporating modified CNTs.
- The CNTs have potential as an organic filler for fabrication of mixed matrix membranes in the future biogas purification.

REFERENCES

[1] Kapdi S. S., V. K. Vijay, S. K. Rajesh, and R. Prasad. Biogas scrubbing, compression and storage: perspective and prospectus in Indian context. *Renewable Energy* 30 (2005) 1195–1202.

[2] Lin W. H. and T. S. Chung. Gas permeability, diffusivity, solubility, and aging characteristics of 6FDA-durene polyimide membranes. *Journal of Membrane Science* 186 (2001) 183–193.

[3] Baker R. W. Membrane Technology and Applications. McGraw-Hill, New York, 1999.

[4] Hacarlioglu P., L. Toppare, and L. Yilmaz. Polycarbonate–polypyrrole mixed matrix gas separation membranes. *Journal of Membrane Science* 225 (2003) 51–62.

[5] Zimmerman C. M., A. Singh, and W. J. Koros. Tailoring mixed matrix composite membranes for gas separations. *Journal of Membrane Science* 137 (1997) 145–154.

[6] Kusworo T. D., A. F. Ismail, A. Mustafa, and T. Matsuura. Dependence of membrane morphology and performance on preparation conditions: the shear rate effect in membrane casting. *Separation and Purification Technology* 61 (2007) 249–257.

[7] Kim S., T. W. Pechar, and E. Marand. Poly(imide siloxane) and carbon nanotube mixed matrix membranes for gas separation. *Desalination* 192 (2006) 330–339.

[8] Ajayan P. M., O. Stephen, C. Colliex, and D. Trauth. Aligned carbon nanotube arrays formed by cutting a polymer resin–nanotube composite. *Science* 265 (1994) 1212–1214.

[9] Lau K. T. and D. Hui. The revolutionary creation of new advanced materials-carbon nanotubes composites. *Composites Part B* 33 (2002) 263–277.

[10] Ma P. C., J.-K. Kim, and B. Z. Tang. Effects of silane functionalization on the properties of carbon nanotube/epoxy nanocomposites. *Composites Science and Technology* 67 (2007) 2965–2972.

[11] Zhu B.-K., S.-H. Xie, Z.-K. Xu, and Y.-Y. Xu. Preparation and properties of the polyimide/multi-walled carbon nanotubes (MWNTs) nanocomposites. *Composites Science and Technology* 66 (2006) 548–554.

24

SELECTIVITY OF POLYMERIC SOLVENT RESISTANT NANOFILTRATION MEMBRANES FOR BIODIESEL SEPARATION

RAHIMAH OTHMAN, ABDUL WAHAB MOHAMMAD, MANAL ISMAIL, AND JUMAT SALIMON

24.1 INTRODUCTION

Polymeric solvent resistant nanofiltration (SRNF) membranes provide a potentially viable alternative where separation and purification process was one of the major concerns in the biodiesel production. Not only do they operate at a lower energy consumption, they also have been widely used for aqueous system at greater separation and purification efficiencies as compared to the commercial biodiesel separation [1]. In addition, there are two main hindrances of the commercial biodiesel production process: (1) the removal of residual triglyceride (TG), glycerol, and unreacted alcohol from the biodiesel product due to the reversible nature of transesterification process in biodiesel production; (2) the catalyst removal by washing steps that is applied in commercial biodiesel production will tend to produce wastewater containing small amount of soap and free glycerol which should be carefully handled before discharge [2–5]. Both these problems can potentially be overcome through the use of an SRNF membranes application.

Recently, Dube et al. alternatively developed a membrane reactor for biodiesel production. This process has many advantages, such as improved product yield, elimination of feed loss, reduced effluent quantity, and thus, low energy consumption [2]. The reactor is supposed to remove unreacted oil from

Sustainable Membrane Technology for Energy, Water, and Environment, First Edition.
Edited by Ahmad Fauzi Ismail and Takeshi Matsuura.

the fatty acid methyl ester (FAME) product, yielding biodiesel of high purity and shifting the reaction equilibrium to the product side. The membrane acts a phase barrier which limits the presence of TG and nonreacting lipids in the product. Four different carbon membrane pore sizes have been tested: 0.05 μm, 0.2 μm, 0.5 μm, and 1.4 μm [2, 3]. However, inconsistent results have been obtained for the existence of those bonded glycerides in the permeate. For example, diglyceride (DG) was observed in the permeate when the FAME concentration was larger than 25–35 wt %, but neither TG nor monoglyceride (MG) were detected by Cao et al. in the permeate stream [3]. On the other hand, Wang et al. used ceramic membrane to refine the crude biodiesel instead of using hot water washing step [5]. Most of the reported works used inorganic membranes compared to organic membranes in the biodiesel separation and purification system where inorganic membranes have the advantages in the industrial operations in terms of high-temperature durability, sufficient mechanical strength, chemical inertness, organic solvent resistance, unique surface characteristics, [6] and less likelihood of bacteria contamination [7]. While there are so many advantages for inorganic membranes, some disadvantages are very critical and serious enough to hamper the progress of industrial application of inorganic membranes such as brittleness (easy to crack), difficult sealing problems at high temperature applications, low surface-to-volume ratio, few membranes with high selectivity, and expensive capital and repair costs [8]. Therefore, Bournay and Baudot recommended the use of polymeric nanofiltration and reverse osmosis (RO) membranes of different types of polymers such as polyethersulphone, silicone, polyvinyl alcohol (PVA), polyamide (PA), polyimide (PI) to separate an alcohol-rich effluent, and alkyl esters-rich effluent from the transesterification product [9].

In this work, the selection and experimental verification of the polymeric SRNF membrane performance on the permeation properties of the modification transesterification product was investigated as well as the selectivity of SRNF polymeric membrane for biodiesel separation and purification process. Biodiesel or also known as methyl ester was produced by transesterification process in which refined, bleached, and deodorized (RBD) palm olein was reacted with anhydrous methanol in the presence of strong base catalyst of sodium hydroxide. The reversible reaction of the transesterification process was described in Figure 24.1. Based on the stoichiometry, the reaction needs

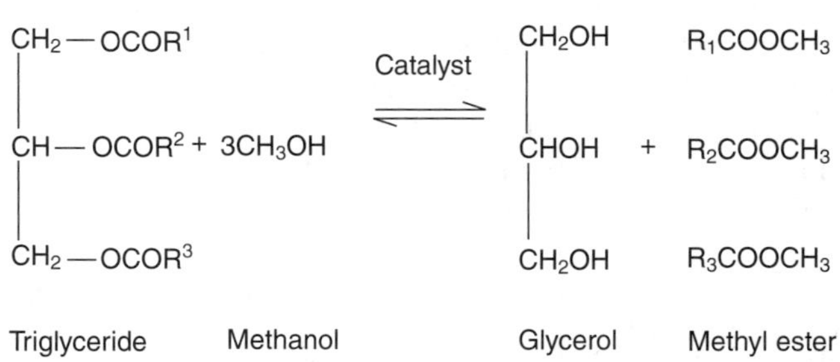

Figure 24.1. General equation of transesterification process.

a 3:1 molar alcohol-to-oil ratio, but the excess methanol is usually added to drive the equilibrium toward the products side [10]. Ideally, a membrane should be able to remove the methyl ester as a product, allowing recycling of methanol back to the reactant.

24.2 MATERIALS AND METHODS

24.2.1 Chemicals

The following chemicals were used: anhydrous methanol (high performance liquid chromatography [HPLC]-grade) and sodium hydroxide (pellets; R&M Chemicals). RBD olein was kindly supplied by Intercontinential Speciality Fats Berhad (ISF), Malaysia.

24.2.2 Membranes

Eight SRNF membranes from different manufacturers were examined, and their properties are shown in Table 24.1. MPF-34 and MPF-44 were supplied in a "wet" form in a preserving solution of 0.7% Roccal, while the other membranes were supplied in a "dry" form.

24.2.3 Transesterification Procedure

The synthesis of biodiesel was carried out in a 1-L three-neck round-bottomed flask fitted with thermometer, reflux condenser, and glass stopper on a magnetic stirrer [11]. Filtered RBD palm olein, methanol (10:1 methanol:oil mole ratio) with dissolving of base catalyst, sodium hydroxide (1% wt of oil) were then transesterified in the round bottom flask. The reaction was performed at the agitation speed of 600 rpm and temperature of 333.15K for 2 hours. At the normal pressure, the methanol vapor was condensed by the cooling water. After 2 hours reaction, the transesterification product from the reaction was

TABLE 24.1. SRNF Membranes and Their Properties

Manufacturer	Membrane Type	MW Cutoff	Membrane Property	Polymer Type
Osmonics	Desal-DL	150–300	Hydrophilic	Polyamide
	Desal-DK	150–300	Hydrophilic	
Koch membrane	MPF-34	200	Hydrophilic	PDMS
	MPF-44	250	Hydrophilic	
MET	STARMEM 120	200	Hydrophobic	Polyimide
	STARMEM 240	400	Hydrophobic	
Solsep BV	Solsep 030306F	[a]	Hydrophobic	Silicone type
	Solsep 030705	[a]	Hydrophobic	

[a] Not specified; PDMS, Polydimethylsiloxane.

transferred to a SEPA ST Stirred Cell (Osmonic, Minnetonka, MN) for permeation experiment.

24.2.4 Modification of Transesterification Product Alkalinity

The alkalinity modification of the transesterification product was done by using phosphoric acid, H_3PO_4 (100% purity), where 0.1 mL of H_3PO_4 was added to 200 mL of transesterification product. Potassium phosphate, Na_3PO_4 was produced after the modification and can be used as valuable by-product, for instance, as a fertilizer [12, 13]. The pH value of the transesterification product was modified from 12.43 (after reaction) to 8.68 whereby this final pH value was suitable for the polymeric membrane separation process.

24.2.5 Permeation Experiment

A SEPA ST Stirred Cell (Osmonic) was used in the permeation experiments. It has a volume capacity of 300 mL and can hold a membrane disc of 49 mm in diameter. The effective area of the membrane A is 15.2 cm^2. The maximum operating pressure of the cell is 3100 kPa (SI unit). The stirring speed was fixed at 400 rpm and range of operating pressures of 600 kPa to 3000 kPa. The volume flux of permeate was expressed as (m^3/m^2 s) [14]:

$$J_v = \frac{v}{A\Delta t} \tag{24.1}$$

where J_v (m^3/m^2 s) is the solvent flux, v (m^3) is the permeate volume, and t (s) is the separation time. The experiments were carried out at fixed separation temperature of 313.15K and the range of operating pressures of 600 kPa to 3000 kPa. The rejection of the membranes for the methanol content was calculated according to the following equation [14]:

$$R = 1 - \frac{C_{i,p}}{C_{i,R}} \times 100\%, \tag{24.2}$$

where $C_{i,p}$ is the methanol permeate concentration and $C_{i,R}$ is the methanol retentate concentration of each separation process. The rejection of other component is also calculated by the equation similar to Equation 24.2.

24.2.6 Analyses

24.2.6.1 Methanol Content. To evaluate the percentage rejection of methanol content in feed and separation permeate samples, the analysis was carried out using Agilent 7890 A GC with the flame ionization detector (FID) employing a DB-WAX capillary column of 30 m length, 0.25 mm ID, and 0.25 μm film capillary column thickness. Acetone was used as the internal standard and solvent for this analysis. One microliter of the analyzed sample was injected into the column. The carrier gas was nitrogen with a flow rate of 0.3 mL/min.

The temperature of the GC injector was fixed at 473.15K with the column detector of 573.15K, and the oven temperature was set isothermally at 308.15K. One methanol peak with the retention time of 3.3 minutes would be used as a method indicator in the chromatogram. The concentration of methanol in the samples was determined by means of the calibration curve. The resulting mass percentage of methanol in the samples were calculated through the following equation:

$$W_M\% = (a_M \times A_M/A_{is} + b_M) \times W_{is} \times 100/W, \tag{24.3}$$

where W_M was the weight percentage of methanol in sample, a_M was the slope calibration function, b_M was the intercept of the calibration function, A_M was the peak area of methanol, A_{is} was the peak area of the internal standard, W_{is} was the weight of internal standard in milligram unitm and W was the sample weight in mg unit.

24.2.6.2 TGs, DGs, MGs, Free Glycerin, and Ester Content. The composition of TG, DG, MG, free glycerin, and ester in the permeate and feed samples were analyzed by using Agilent 6890N GC with a FID employing a DB-5HT capillary column (Agilent) of 30 m length, 0.32 mm ID, 0.1 μm film thickness. The method used was based on ASTM-D6584 [15]. Data collection and analysis were performed using Agilent Chemstation software. 1,2,4-Butanetriol and tricaprin were used as the internal standard 1 and 2, respectively. A stock solution of *n*-heptane was added to the vial after the sample was silylated with *N*-methyl-*N*-trimethylsilyltrifluoracetamide (MSTFA). Samples were prepared for analysis by adding 1 mg of oil to 10 mL septa vials. One microliter of the sample was injected into the column. The oven temperature program starts at 50°C (hold 1 minute) and ramps at 15°C/min to 180°C, 7°C/min to 230°C, and 3°C/min to 38°C (hold 30 minutes). Methyl ester of oleic acid, 1-monoolein, 1, 3-diolein, and triolein were used as standard samples to determine the retention time of FAME, MG, DG, and TG.

24.2.6.3 Fourier Transform Infrared Spectroscopy (FTIR). All tested membrane samples for each separation process were examined of their functional group changes by using FTIR spectroscopy Model Perkin-Elmer Spectrum One at wave numbers between 4000/cm and 400/cm. FTIR spectra in the transmission mode were recorded using a Nicolet Nexus connected to a PC, in which the number of scan was 32 and the resolution was 4/cm.

24.3 RESULTS AND DISCUSSION

24.3.1 Permeability Experiment with the Modified Alkalinity of Transesterification Product

The permeation experiments were carried out to examine the permeate flux of the modification transesterification product through the polymeric SRNF

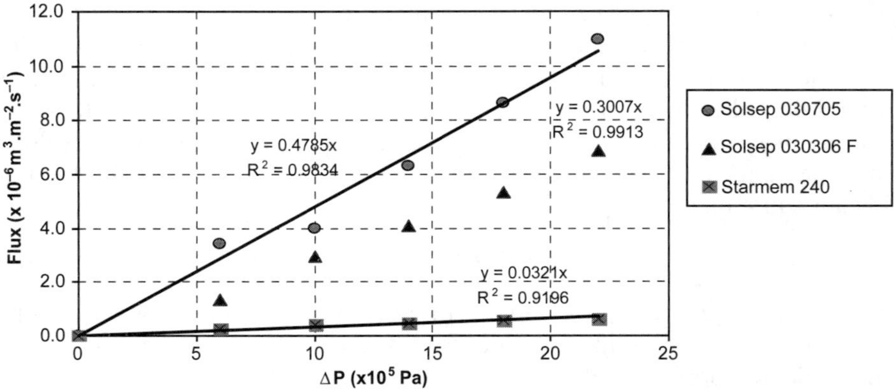

Figure 24.2. Modified transesterification product flux versus applied pressure.

membranes listed in Table 24.1 at a constant separation temperature of 313.15K and in a range of operating pressures of 600 kPa to 2200 kPa. The feed weight percentage of transesterification product after neutralization was TGs, 1.0011 wt%; MGs, 0.0036 wt%; DGs, 1.4179 wt%; free glycerin, 5.0440 wt%; methyl esters, 92.3963 wt%; unreacted excess methanol was 0.1370 wt%. The pH value of the transesterification product after neutralization was recorded at 8.68. Figure 24.2 shows that at the moderate levels of pressure (≤2200 kPa), the flux behavior for the modified transesterification product was increased linearly with respect to the applied pressure through Solsep 030705, Solsep 030306 F, and Starmem 240 membranes. However, no permeate flux was obtained through the other five listed membranes. Out of all tested membranes, it was found that Solsep 030705 membrane gave the highest permeate flux compared to the other membranes, and it was followed by Solsep 030306 F and Starmem 240 membrane. The data obtained show a linear profile which conforms to the Hagen Poiseuille equation. Commonly, this behavior has been correlated by the linear expression given by [16]:

$$J = L_p \Delta P, \tag{24.4}$$

where J is the volume flux, L_p is the membrane permeability constant, and ΔP is the applied pressure.

24.3.2 FTIR Analysis

The ester frequency range of the functional group (C=O) that would point to the presence of methyl ester is between 1830 and 1800/cm and 1780 and 1640/cm [17]. Figure 24.3 depicts the FTIR spectra for crude methyl esters and pure

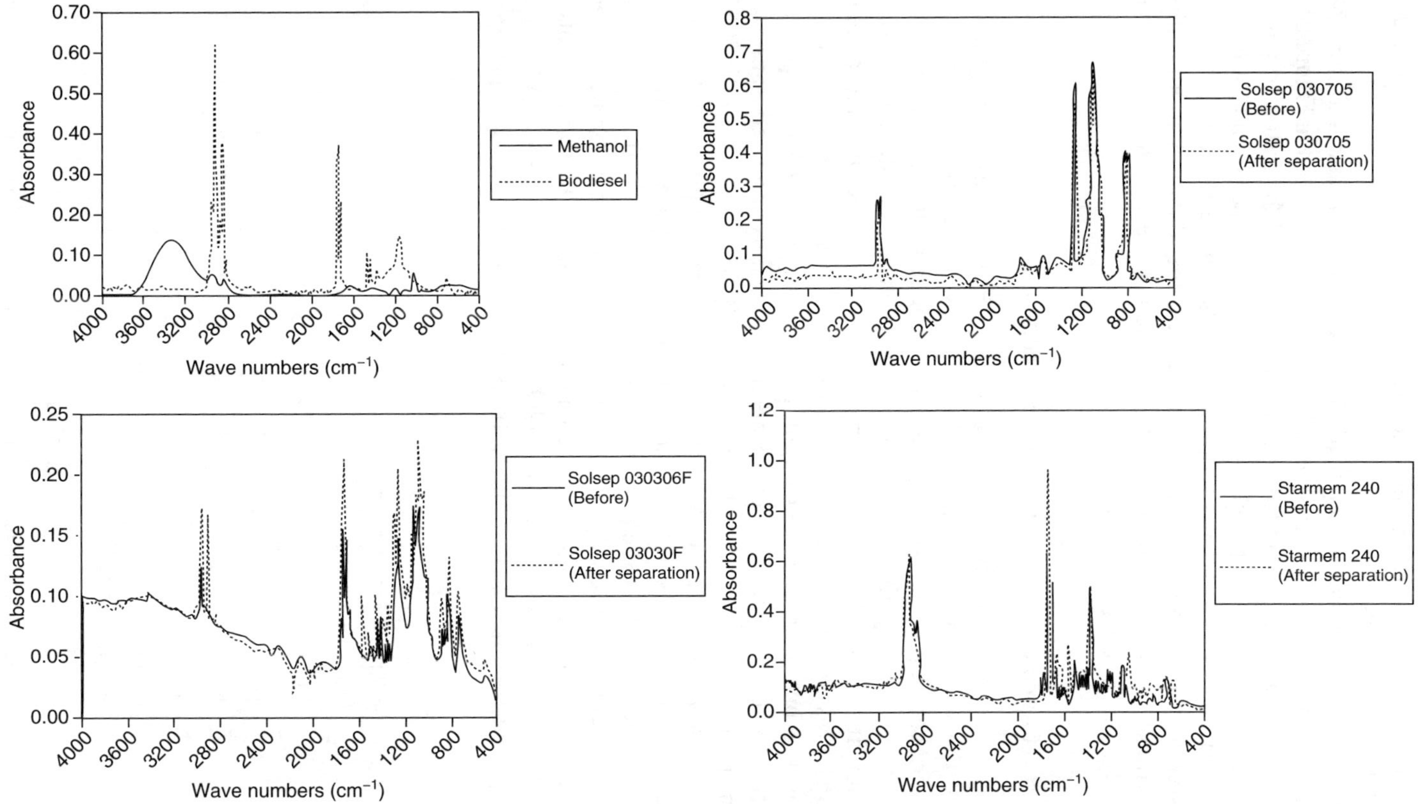

Figure 24.3. FTIR spectrum of pure methanol, crude biodiesel, and tested SRNF membranes.

methanol as a reference to examine the differences on the FTIR spectrums for all tested membranes.

The FTIR spectrums of Solsep 030705, Solsep 030306 F, and Starmem 240 membranes for before and after separation were also presented in Figure 24.3. As can be seen, with respect to the functional group of all the tested polymeric SRNF membranes, minor changes were found between before and after the transesterification product separation process. These results indicate that the transesterification product can be separated after the alkalinity modification on the transesterification product without affecting the tested SRNF membranes after each separation process.

24.3.3 Performance of Polymeric SRNF Membrane

The performance of Solsep 030705 and Solsep 030306 F was investigated to evaluate the efficiency of the membrane to separate the alkalinity modified transesterification product. The experiment was done by using pure methanol solvent before testing the same membrane for transesterification product separation at a constant separation temperature of 313.15K and in a range of operating pressures of 600–3000 kPa. This result was used as a control to examine the behavior of tested membranes for transesterification product separation. After the separation experiment for alkalinity-modified transesterification product was completed, both Solsep 030705 and Solsep 030306 F membrane were used again for pure methanol permeation test to check whether the membranes were fouled during the separation experiment. Figures 24.4 and 24.5 show the pure methanol flux before and after the separation experiment for Solsep 030705 and Solsep 030306 F membrane, respectively. Based on the result, the pure methanol flux of Solsep 030705 and Solsep 030306 F membrane decreased by 74.97 and 72.88%, respectively, after being applied for alkalinity-modified transesterification product separation.

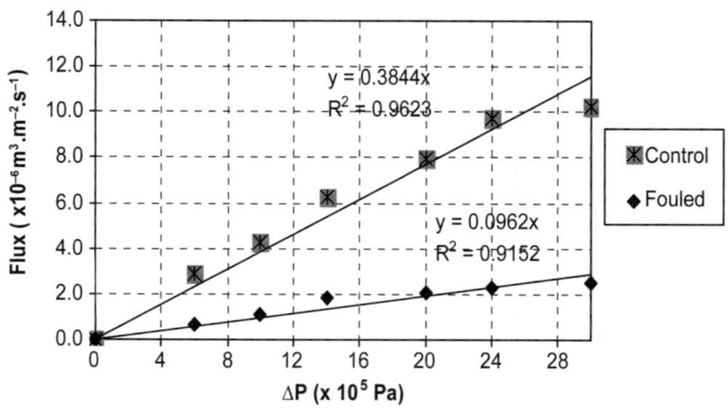

Figure 24.4. Methanol flux versus pressure before and after separation experiment (temperature, 313.15K; membrane, Solsep 030705).

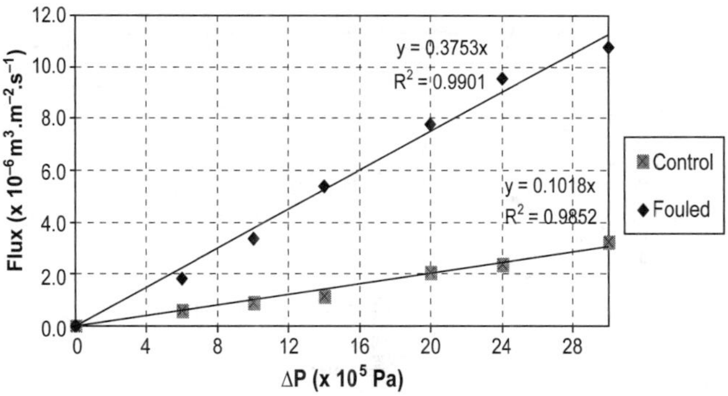

Figure 24.5. Methanol versus pressure before and after separation experiment (temperature, 313.15K; membrane, Solsep 030306 F).

Moreover, Solsep 030705 membrane gave the highest permeability flux of the alkalinity modified transesterification product compared to the other membranes. Therefore, Solsep 030705 membrane can be considered as the most suitable membrane for the biodiesel production application among the tested membranes.

24.3.4 Selectivity of Polymeric SRNF Membrane

The selectivity of polymeric SRNF membrane was identified to evaluate the capabilities of Solsep 030705 to separate the alkalinity-modified transesterification product. Figure 24.6 describes the selectivity of the tested membrane which exhibited the highest rejection for TG of about 99.8% compared to the rejection of DG was 97.16%, MG was 40.65%, and free glycerin was 75.24% at constant separation temperature of 313.15K and pressure 1000 kPa within the separation time of 30 minute. In addition, the highest percentage of methanol rejection was obtained at 30 minutes separation time with 74.98% rejection. Therefore, the rejected excess (unreacted) methanol can be considered to be recycled into the reaction medium. In addition, methyl ester also has very low rejection, which allow separation of the products from excess methanol and other impurities. The potential of SRNF membranes for this type of separation should be investigated further [18].

24.4 CONCLUSION

The modification on the transesterification product alkalinity was the major concern to enhance the performance of polymeric SRNF membranes in the biodiesel separation and purification process. The following important findings could be drawn from the study:

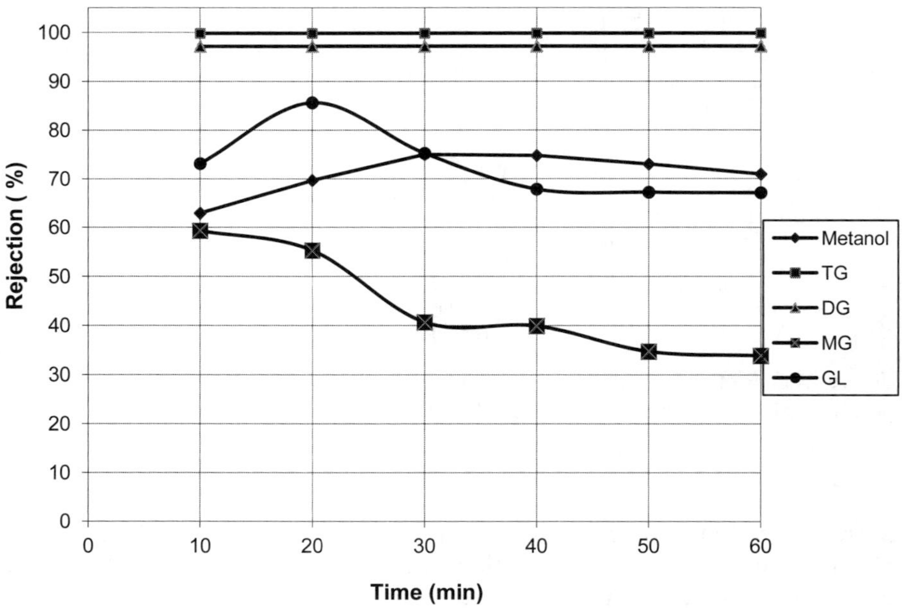

Figure 24.6. Rejection of Solsep 030705 membrane in the alkalinity modified transesterification product separation process.

1. Solsep 030307 membrane can be considered as a most suitable membrane to apply in the biodiesel separation and purification process.
2. The highest percent of excess (unreacted) methanol rejection obtained at 30 minutes separation time was 74.98% rejection and can be considered to be recycled into the reaction medium of biodiesel production system.
3. Minor differences on morphologies and functional groups of tested nanofiltration membranes were detected after use for alkalinity-modified transesterification product separation, rendering the membranes fit to be reused over a number of cycles, hence lowering the cost of material handlings.

ACKNOWLEDGMENTS

The authors gratefully acknowledge the financial support given for this work through the grants 03-01-02-SF0054 and UKM-KK-02-FRGS0002-2006.

REFERENCES

[1] Bhanushali D., S. Kloos, C. Kurth, and D. Bhattacharryya. Performance of solvent resistant membranes for non-aqueous systems: solvent permeation results and modelling. *Journal of membrane science* 189 (2001) 1–21.

[2] Dubé M. A., A. Y. Tremblay, and J. Liu. Biodiesel production using a membrane reactor. *Bioresource Technology* 98 (2006) 639–647.

[3] Cao P., A. Y. Tremblay, M. A. Dubé, and K. Morse. Effect of membrane pore size on the performance of a membrane reactor for biodiesel production. *Industrial & Engineering Chemistry Research* 46 (2007) 52–58.

[4] Cao P., M. A. Dubé, and A. Y. Tremblay. Methanol recycling in the production of biodiesel in a membrane reactor. *Fuel* 87 (2008) 825–833.

[5] Wang Y., X. Wang, Y. Liu, S. Ou, Y. Tan, and S. Tang. Refining of biodiesel by ceramic membrane separation. *Fuel Processing Technology* 90 (2008) 422–427.

[6] Lobo A., A. Cambiella, J. M. Benito, C. Pazos, and J. Coca. Ultrafiltration of oil-in-water emulsions with ceramic membranes: influence of pH and crossflow velocity. *Journal of Membrane Science* 278 (2006) 328.

[7] Zhou H., H. F. Lu, and B. Liang. Solubility of multicomponent systems in the biodiesel production by transesterification of *Jatropha curcas* L. oil with methanol *The Journal of Chemical & Engineering Data* 51 (2006) 1130.

[8] Sun Z. H. Inorganic membranes and membrane reactors. *The Korean Journal of Chemical Engineering* 18 (2001) 775–787.

[9] Bournay L. and A. Baudot. Processing for producing fatty acid alkyl esters and glycerol of high purity. U.S. Patent 7,138,536 B2, 2006.

[10] Alex H. W., D. Posarac, and N. Ellis. Assessment of four biodiesel production processes using HYSYS. Plant. *Bioresource Technology* 99 (2008) 6587–6601.

[11] Akers S. M., J. L. Conkle, S. N. Thomas, and K. B. Bider. Determination of the heat of combustion of biodiesel using bomb calorimetry. *Journal of Chemical Education* 83(2)(2006) 260–262.

[12] Gerpen J. V. Biodiesel processing and production. *Fuel Processing Technology* 86 (2005) 1097–1107.

[13] Zhang Y., M. A. Dubé, D. D. McLean, and K. Kates. Biodiesel production from waste cooking oil: 1. Process design and technological assessment. *Bioresource Technology* 89 (2003) 1–6.

[14] Yang X. J., A. G. Livingston, and L. F. Santos. Experimental observations of nanofiltration with organic solvents. *Journal of Membrane Science* 190 (2001) 45–55.

[15] ASTM-D6584. Test method for determination of free and total glycerine in B-100 biodiesel methyl esters by gas chromatography, ASTM USA, 2000.

[16] Machado D. R., D. Hasson, and R. Semiat. Effect of solvent properties on permeate flow through nanofiltration membrane: Part I. Investigation of parameters affecting solvent flux. *Journal of Membrane Science* 163 (1999) 93–102.

[17] Pavia D. L., G. M. Lampman, and G. S. Kriz (eds.), Introduction to Spectroscopy: A Guide for Students of Organic Chemistry. Brooks/Cole, Belmont, CA, 2001.

[18] Othman R., A. W. Mohammad, M. Ismail, and J. Salimon. Application of polymeric solvent resistant nanofiltration membranes for biodiesel production. *Journal of Membrane Science* 348 (2010) 287–297.

PART VI

OTHER INDUSTRIAL APPLICATIONS

25

PERVAPORATION PERFORMANCE OF METHYL TERT BUTHYL ETHER/METHANOL MIXTURES THROUGH NATURAL RUBBER/POLYSTYRENE INTERPENETRATING POLYMER NETWORK MEMBRANES

Mohd Ghazali Mohd Nawawi, Nur Azrini Ramlee,
and Fathie Ahmad Zakil

25.1 INTRODUCTION

Pervaporation is a membrane separation process in which a liquid mixture is separated by partial vaporization of the liquid and its transport through a nonporous or composite (nonporous/porous) permselective membrane, while vacuum is applied or sweep gas flows on the other side of membrane. Originally, this technique was called liquid permeation but has subsequently been termed pervaporation based on the fact that permeate undergoes a phase change from liquid feed to vapor permeate during its transport through the barrier. Consequently, energy is needed for the vaporization of permeant.

Pervaporation is attractive when it is difficult to apply distillation such as in the case of close boiling point components, fractionation of azeotropic mixtures, and isomeric mixtures [1]. Industrial application of pervaporation is

Sustainable Membrane Technology for Energy, Water, and Environment, First Edition.
Edited by Ahmad Fauzi Ismail and Takeshi Matsuura.

classified into three groups which are (1) dehydration of organic solvent; (2) removal of organic compounds from aqueous solutions; and (3) separation of organic mixtures [2].

Selection of the polymer used for pervaporation membrane is based on the molecule size of and the solubility of permeating components toward the polymeric material. Numerous research papers on the separation of methyl tert buthyl ether (MTBE) and methanol are published worldwide every year with different polymers applicable for the membrane development [3–9].

Natural rubber (NR) has been chosen as material for developing membranes in this study because of the higher fluxes through NR membranes in comparison to other nonrubber polymers [9]. NR is one type of rubber that has high molecular weights and is long and flexible, while the double bonds (C=C) provide reactive sites for the crosslinking process [10]. There are several researches in which membranes made from NR were used for pervaporation processes [9, 11].

In this study, modified membranes have been developed in order to improve pervaporation performances based on NR/polystyrene (PS) interpenetrating polymer network (IPN). This is a special polymer/polymer blend where one or both phases are cross-linked [12]. The IPNs generally can be prepared by sequential, simultaneous, or latex polymerization techniques [13]. In this study, IPNs were fabricated by the sequential technique in which the components of the second network are swollen and polymerized or cross-linked inside the previously cross-linked polymer network. The impact behavior of semi and full IPNs from NR and PS was studied by Thomas et al. [14] with special reference to blend ratio and crosslinking level of PS phase. The blend ratio and crosslinking level of PS phase will also be varied in the present study to improve membrane physical properties as well as separation of MTBE/methanol mixtures by pervaporation.

Separation of the MTBE and methanol mixtures by pervaporation was conducted using the NR/PS IPN membranes so prepared by varying the concentration of MTBE in a range from 10 wt % to 70 wt % and at different operation temperatures, 25°C and 55°C. The performance of the membranes in pervaporation was evaluated by the permeation flux (J), and separation factor (α.).

25.2 MATERIALS AND EXPERIMENTAL PROCEDURE

25.2.1 Materials

NR sample was obtained from Rubber Research Institute of Malaysia (RRIM). Styrene monomer for the IPN synthesis was supplied by Merck (Darmstadt, Germany). Divinylbenzene (DVB) was purchased from Sigma-Aldrich Chemie (GmbH, Germany), while dicumyl peroxide (DCP) was obtained from Aldrich Chemical Company Inc. (Milwaukee, WI).

TABLE 25.1. Series of Interpenetrating Polymer Networks

Sample Code	NR/PS Ratio	DVB (%) Content	DCP (%)
D_2N_{50}	50/50	2	1
D_4N_{50}	50/50	4	1

25.2.2 Experimental Procedure

25.2.2.1 Preparation of NR/PS IPN. NR was masticated in a two-roll mixing mill at room temperature. Purified DCP was added with 4 per hundred rubber (phr) into NR and mixed thoroughly for 15 minutes. Then, the mixture was cured at 160°C on a hydraulic press to obtain the crosslinked membrane network. A series of IPN membranes were prepared using different amounts of initiator. The cured NR membranes were weighed and kept immersed in inhibitor-free styrene monomer containing initiator and 2 and 4% of DVB as the cross-linker for PS phase. The membranes were swollen for different time intervals to obtain different weight percentages of polystyrene. The swollen samples are kept at 0°C for a few hours to achieve equilibrium distribution of styrene monomer in the matrix. The swollen network sheets were heated at 80°C for 6 hours and 100°C for 2 hours, respectively, to complete the polymerization and crosslinking reaction. Table 25.1 indicates the series of IPNs.

25.2.2.2 Pervaporation Experiments. Pervaporation experiments were conducted in a continuous pervaporation unit by Solteq Company. Pervaporation cell used consisted of two detachable 316 stainless steel parts. The feed tank was made from hardened stainless steel and had a maximum solution capacity of approximately 5 L. From the feed tank, the feed mixture was circulated through the cell by a circulation pump which was controlled using a control valve. The permeate side of the membrane was connected to a cold trap immersed in liquid nitrogen, followed by a vacuum pump to supply the required driving force for pervaporation. During the separation process, the permeate pressure was maintained at 2–3 mbar, and the experiment was carried out at different temperatures between 25 and 55°C for IPN NR/PS membranes with the predetermined feed composition. After the pervaporation process reached a steady-state condition, all experimental data were collected. Performance of IPN NR/PS membranes was based on the permeation flux and separation factor. The separation factors derived from the feed and permeate concentration was defined as

$$\alpha ij = \frac{Yi/Yj}{Xi/Xj}, \tag{25.1}$$

where, X and $Y$are the weight fraction (g/g) of feed and permeate, respectively. The subscripts i and j refer to MTBE and MeOH, respectively.

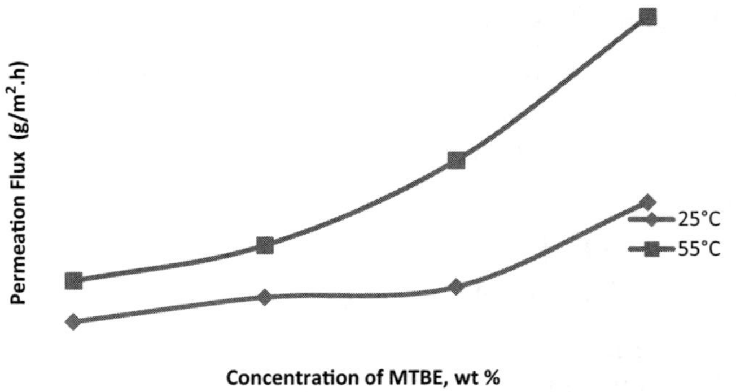

Figure 25.1. Effect of pervaporation temperature on the permeation flux of natural rubber/polystyrene membrane D_2N_{50}.

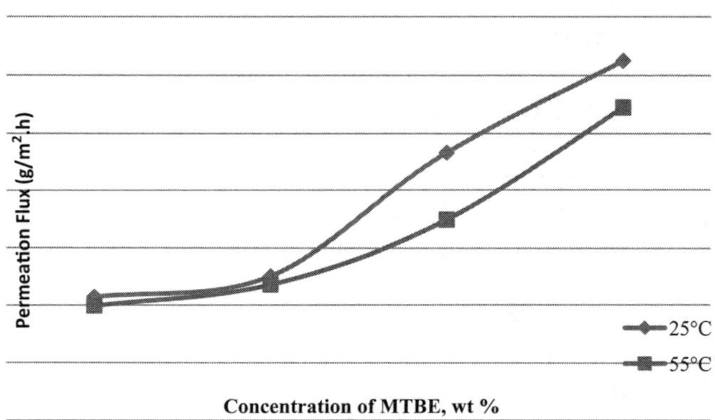

Figure 25.2. Effect of pervaporation temperature on the permeation flux of natural rubber/polystyrene membrane D_4N_{50}.

25.3 RESULTS AND DISCUSSION

25.3.1 Influence of Pervaporation Temperature on Permeation Flux

Figures 25.1 and 25.2 illustrate the effect of temperature on the permeation flux of IPN NR/PS membrane series of D_2N_{50} and D_4N_{50}. Permeation flux seems proportional to the temperature given as °C. It is worth mentioning that unlike membranes of glassy polymer [15], sorption of rubbery membranes increases with increase in feed temperature due to increased solubility of feed solution in rubber matrix containing more mobile polymer chains at higher temperature [11].

As the temperature increases, more permeation occurs through the IPN NR membranes, and this is caused by an increase in the kinetic energy of feed solution molecules. A similar observation was made by Nam and Lee [7] who reported that as the temperature increases from 25°C to 50°C, the flux increased

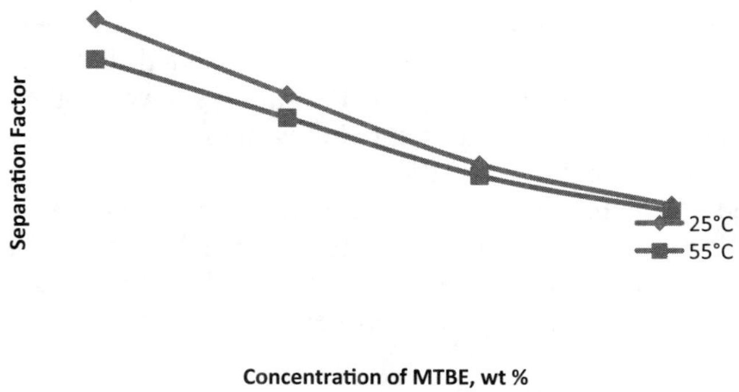

Figure 25.3. Effect of pervaporation temperature on the separation factor of IPN natural rubber/polystyrene D_2N_{50} membrane.

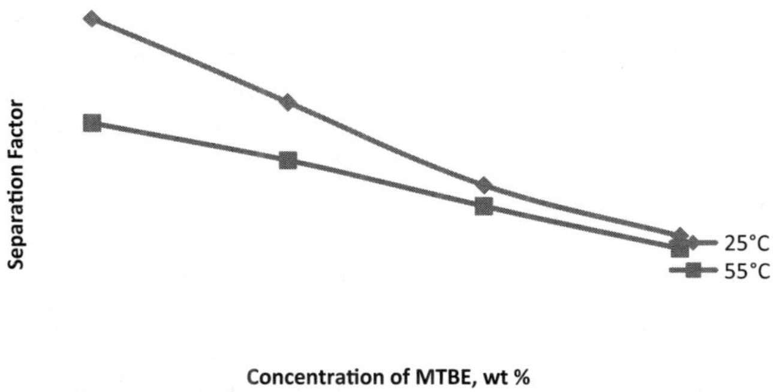

Figure 25.4. Effect of pervaporation temperature on the separation factor of IPN natural rubber/polystyrene D_4N_{50} membrane.

from 100 g/m^2·h to 300 g/m^2·h through a chitosan composite membrane modified with sulfuric acid and four surfactants.

25.3.2 Influence of Pervaporation Temperature on Separation Factor

Figures 25.3 and 25.4 show the effect of temperature on the separation factor of IPN NR/PS membrane series of D_2N_{50} and D_4N_{50}. Both figures show that the separation factor decreased with an increase in MTBE concentration. The separation factor also decreased with an increase in temperature. The decrease in separation factor at the higher concentration of MTBE in feed is accompanied by the increase in the permeation flux. With an increase in temperature and MTBE concentration, the movement of rubber chains is enhanced, allowing higher permeation of both methanol and MTBE through the membranes. Therefore, the separation factor of the IPN NR/PS membranes decreased [9].

In this research, the IPN NR/PS membranes were successfully prepared using the sequential technique to form IPN through crosslinking method. The performance of IPN NR/PS membranes for the separation of MTBE/methanol mixtures by pervaporation was investigated by measuring the permeation flux and separation factor. From the results, it is concluded that the IPN NR/PS membrane of series D_2N_{50} (with 2% DVP content) showed higher permeation fluxes both at low and high temperature but with lower separation factors. The maximum permeation flux is 132.26 g/m^2.h at 25°C and 255.22 g/m^2.h at 55°C. Membrane D_4N_{50} exhibited good characteristic compared to D_2N_{50} in terms of selectivity. The maximum separation factor is 4.5 at 25°C and 2.5 at 55°C. It is summarized that when the concentration of MTBE in feed and the temperature of pervaporation increased, the permeation flux increased and separation factor decreased.

ACKNOWLEDGMENTS

The financial support provided to this research by the Ministry of Science, Technology and Environment through IRPA project is gratefully acknowledged.

REFERENCES

[1] Wei Y. Pervaporation with latex membranes, PhD Thesis, University of Waterloo, Canada, 1993.

[2] Gozzelino G. and G. Malucelli. Permeation of methanol/methyl-t-butyl ether mixtures through poly(ethylene-co-vinyl acetate) films. *Colloids and Surfaces A: Physicochemical and Engineering Aspects* 235 (2004) 35–44.

[3] Farnand B. A. and S. H. Noh. Pervaporation as an alternative process for the separation of methanol from C4 hydrocarbons in the production of MTBE and TAME. *AIChE Symposium Series* 85(272)(1989) 89–92.

[4] Doghieri F., A. Nardella, G. C. Sarti, and C. Valentini. Pervaporation of methanol-MTBE mixtures through modified poly(phehylene oxide) membranes. *Journal of Membrane Science* 91 (1994) 283–291.

[5] Park H. C., N. E. Ramalcer, M. H. V. Mulder, and C. A. Smolders. Separation of MTBE-methanol mixture by pervaporation. *Separation Science and Technology* 30 (1995) 419–433.

[6] Sano T., M. Hasegawa, Y. Kawakami, and H. Yanagishita. Separation of methanol/methyl-t-buthyl ether mixture by pervaporation using silicalite membrane. *Journal of Membrane Science* 107 (1995) 193–196.

[7] Nam S. Y. and Y. M. Lee. Pervaporation separation of methanol/methyl t-butyl ether through chitosan composite membrane modified with surfactants. *Journal of Membrane Science* 157 (1999) 63–71.

[8] Hilmioglu N. D. and S. Tulbentci. Pervaporation of MTBE/methanol mixtures through PVA membranes. *Desalination* 160 (2004) 263–270.

[9] Ray S. K. and K. Ray. Separation of organic mixtures by pervaporation using crosslinked and filled rubber membranes. *Journal of Membrane Science* 285 (2006) 108–119.

[10] Allen P. W. *Natural Rubber and the Synthetics*. Crosby Lockwood, London, 1972.

[11] George S. C., S. Thomas, and K. N. Ninan. Pervaporation separation of chlorinated hydrocarbon and acetone mixtures with crosslinked styrene-butadiene rubber and natural rubber blends membranes. *Journal of Membrane Science* 176 (2000) 131–142.

[12] Han J. L., Y. C. Chern, K. Y. Li, and K. H. Hsieh. Interpenetrating polymer networks of bismaleimide and polyurethane crosslinked epoxy. *Journal of Polymer Science* 70 (1998) 529.

[13] Frisch H. L., G. G. Barros, and J. L. Travis, in J. E. Mark and B. Erman (eds.). *Elastomeric Polymer Networks*. Prentice Hall, New York, 1992.

[14] Thomas S., A. P. Mathew, S. Packirisamy, and H. J. Radusch. Effect of initiating system, blend ratio and crosslink density on the mechanical properties and failure topography of nano-structured full-interpenetrating polymer networks from natural rubber and polystyrene. *European Polymer Journal* 37 (2001) 1921–1934.

[15] Samit K. R., S. B. Sawant, J. B. Joshi, and V. G. Pangarkar. Separation of methyl-tert-buthyl alcohol (MTBE)-methanol by pervaporation. *Journal of Applied Polymer Science* 74 (1999) 2645.

26

P-XYLENE SEPARATION FROM TERNARY XYLENE MIXTURE OVER SILICALITE-1 MEMBRANE: PROCESS OPTIMIZATION

Yin Fong Yeong, Ahmad Zuhairi Abdullah,
Abdul Latif Ahmad, and Subhash Bhatia

26.1 INTRODUCTION

Separation of para-xylene (*p*-xylene) from its isomers is an important operation in the petrochemical industry [1]. *p*-Xylene is the feedstock for the pure terephthalic acid (PTA) production, which in turn is used in the production of polyester resin and fibers. Therefore, it has the largest commercial market value among the three isomers of xylene.

In recent years, zeolite membranes have been in focus because of their characteristics such as well-defined micro pore structure, good thermal and structural stability. The characteristics of zeolite membrane could be utilized in gas, vapor, and liquid separation based on their properties adsorption, preferential diffusion, or molecular sieving (size exclusion) [2–4]. Zeolite membrane offers significant advantages compared to existing technologies in the separation of close-boiling point hydrocarbons such as xylene isomers, which are difficult to separate using complex energy-intensive processes [5, 6]. Among the zeolite membranes, MFI-type zeolite membranes (ZSM-5 or Silicalite-1) appear as potential candidates for xylene separation because of their pore

Sustainable Membrane Technology for Energy, Water, and Environment, First Edition.
Edited by Ahmad Fauzi Ismail and Takeshi Matsuura.

structure near to the kinetic diameters of the isomers (0.58 nm for *p*-xylene and 0.68 nm for both *m*-xylene and *o*-xylene).

The separation of xylene isomers using MFI zeolite membranes has been reported by various researchers. Keizer et al. [7] obtained separation factors of *p*-xylene <1.0 at 25°C and >200 at 102–142°C, for 0.31 kPa *p*-xylene, and 0.26 kPa *o*-xylene binary mixture. The results showed that the separation factors were significantly dependent on the operating temperature. Xomeritakis et al. [8] investigated the separation of xylene isomer vapors with oriented MFI membrane in the temperature range of 22–275°C and feed partial pressure range of 0.7–0.9 kPa. They found that the separation performance of these membranes was directly related to the synthesis conditions and the resulting membrane microstructure. Tsapatsis and coworkers [3, 5] reported the best result of MFI-type zeolite membranes for vapor permeation separation of xylene isomers. The b-oriented MFI zeolite membrane gave *p*-xylene permeance about 2×10^{-7} mol/m^2.s.Pa with *p*-/*o*-xylene separation factor up to 500 at 200°C. Tarditi et al. [1] found that ZSM-5 membrane was superior to the silicalite-1 in xylene separation because of its higher permeance flux of 1.16×10^{-5} mol/s m^2, separation factors of 4.4, and thermal stability up to 400°C.

In the present research, process optimization of important operating variables for separation of *p*-xylene from its isomers, *m*-xylene and *o*-xylene over silicalite-1 membrane is reported. A silicalite-1 membrane was synthesized on an α-alumina support and characterized by X-ray diffraction and scanning electron microscopy. The response surface methodology (RSM) coupled with central composite design (CCD) was employed to study the interactions between the two separation process variables: temperature and *p*-xylene feed partial pressure that influence the three responses: *p*-xylene flux, *p*-/*o*-xylene separation factor, and *p*/*m*–xylene separation factor. Three models based on ANOVA analysis correlating the process variables to the responses are proposed. The optimum value of the variables was determined using minimization/maximization of response functions.

26.2 METHODS/THEORY

Direct in situ hydrothermal crystallization was used to synthesize silicalite-1 membrane [9]. The synthesis mixture was prepared according to the molar composition as 5TEOS: 1TPAOH:1000 DDI H_2O described in Reference [10]. After hydrothermal crystallization for 1 day at 175°C, the membrane was washed with deionized water and dried at 100°C overnight. Finally, the silicalite-1 membrane was calcined at 420°C for 15 hours with a heating and cooling rate of 0.5°C/min. The crystalline structure of the membrane was determined by XRD analysis using X-ray diffractometer (Philips PANanalytical X-Pert PRO) with CuKα radiation ($\lambda = 1.5406$ Å) in the 2θ range of 5° to 50°. The morphology and thickness of the membrane were determined using Zeiss Supra 35 VP scanning electron microscopy (SEM).

The vapor permeation data for xylene separation were obtained using the vapor permeation test rig [11]. The feed partial pressure of ternary xylene mixture in the feed stream entering the permeation cell was adjusted by regulating the carrier flow through the saturator. The composition ratio of the ternary xylene mixture in the feed was similar to that of the industrial reactor: 65% *m*-xylene, 14.5% *p*-xylene, and 20.5% *o*-xylene. The total pressure in either side of the membrane was maintained as atmospheric pressure and the volumetric flow of the sweep stream was set at 20 cm^3/min. A Hewlett–Packard, Model 5890 series II gas chromatograph was used to analyze the feed, permeate, and retentate streams. The permeation flux of xylene was calculated from its molar fraction (determined by gas chromatography [GC]), the volumetric flow of the permeate stream, and the membrane area available for permeation [11].

Design of experiments (DoE) was used to obtain the optimum value of process variables in the ternary xylene mixture separation studies. The independent variables selected were (1) A, temperature and (2) B, *p*-xylene feed partial pressure. 2^2 full factorial CCD for the two variables was employed. The range of process variables studied is shown in Table 26.1. The total numbers of 13 experiments were conducted based on CCD, as shown in Table 26.2. Three responses namely *p*-xylene flux, *p*-/*o*-xylene separation factor, and *p*-/*m*-xylene separation factor were observed. Each response was used to develop statically analyzed-based model which correlated the responses to the two xylene separation process variables.

26.3 RESULTS AND DISCUSSION

Figure 26.1a shows the XRD pattern for silicalite-1 membrane. All the major peaks consistent with the XRD pattern of silicalite-1 reported in the literature were present [9]. This confirms that the zeolite layer on the support consists of crystalline silicalite-1 membrane. Figure 26.1b presents the SEM micrograph of the surface morphology of the silicalite-1 membrane. Silicalite-1 crystals grown on the silica layer exhibited a complete crystalline phase and coffin-like crystal shape, where the morphology was similar to the conventional MFI-type zeolites [9].

TABLE 26.1. Experimental Independent Variables

		Level and Range (Coded)		
Parameter	Factor Code	–1	0	+1
Temperature, °C	A	150	200	250
p-Xylene feed partial pressure, kPa	B	0.10	0.18	0.26

TABLE 26.2. Experiment Matrix of 2^2 Center Composite Design

	Levels (Coded Values)		Factors		Responses		
Run	(A)	(B)	Temperature, °C (A)	*p*-Xylene Feed Partial Pressure, kPa (B)	*p*-Xylene Flux, $\times 10^6$ mol/m^2.s (Y1)	*p*-/*o*-Xylene Separation Factor (Y2)	*p*-/*m*-Xylene Separation Factor (Y3)
1	−1	−1	150	0.10	3.50	4.5	5.9
2	+1	−1	250	0.10	2.10	5.3	6.7
3	−1	+1	150	0.26	6.23	9.6	10.2
4	+1	+1	250	0.26	5.40	11.2	12.2
5	−1	0	150	0.18	6.29	5.8	7.2
6	+1	0	250	0.18	4.39	6.0	7.5
7	0	−1	200	0.10	2.82	10.7	11.3
8	0	+1	200	0.26	5.99	14.4	15.5
Repeated runs							
9	0	0	200	0.18	5.29	19.8	20.1
10	0	0	200	0.18	5.43	18.6	21.8
11	0	0	200	0.18	5.68	18.2	20.9
12	0	0	200	0.18	5.57	20.1	19.7
13	0	0	200	0.18	5.26	19.5	21.4
Mean					5.45	19.2	21.0
Mean error					±0.23	±1.04	±1.08
Standard deviation					0.18	0.81	0.88

Table 26.2 shows the experimental data of the three responses. The *p*-xylene flux (Y1) was in the range between 2.10×10^{-6} mol/m^2.s and 6.29×10^{-6} mol/m^2.s. The *p*-/*o*-xylene separation factor (Y3) varied from 4.5 to 20.1, whereas the *p*-/*o*-xylene separation factor (Y4) varied from 5.9 to 21.8. All three response values obtained were within an experimental error of ±5%. The models were selected for three responses based on the highest polynomial where the additional terms were significant and were not aliased. The final models in terms of coded factors for Y1 to Y3 are presented in Equations 26.1–26.3, respectively.

p-xylene Flux (mol/m^2.s),

$$\mathrm{Y1} = 5.337\times10^{-6} - 9.503\times10^{-7}\mathrm{A} + 1.583\times10^{-6}\mathrm{B} - 3.815\times10^{-8}\mathrm{A}^2 - 9.711\times10^{-7}\mathrm{B}^2 + 1.440\times10^{-7}\mathrm{AB} - 7.627\times10^{-8}\mathrm{A}^2\mathrm{B} + 3.932\times10^{-7}\mathrm{AB}^2 \tag{26.1}$$

p-/*o*-xylene Separation Factor,

$$\mathrm{Y2} = 18.41 + 0.42\mathrm{A} + 2.45\mathrm{B} - 9.46\mathrm{A}^2 - 2.84\mathrm{B}^2 + 0.22\mathrm{AB} \tag{26.2}$$

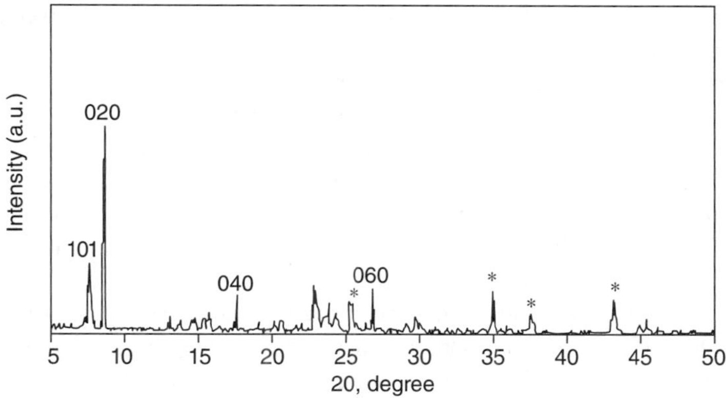

(a)

(b)

Figure 26.1. XRD pattern (a) SEM micrographs (b) for silicalite-1 membrane coated on α-alumina support. (*marks the peak coming from α-alumina support.)

$$\begin{aligned} &p\text{-}/m\text{-xylene Separation Factor,} \\ &Y3 = 20.21 + 0.42A + 2.42B - 9.62A^2 - 3.59B^2 + 0.44AB \end{aligned} \tag{26.3}$$

Equations 26.1–26.3 are subjected to temperature (*A*) range within 150–250°C and *p*-xylene feed partial pressure range (B) from 0.10 kPa to 0.26 kPa. All three models were significant at 95% confidence level with "Prob > F" less than 0.05. For a model to be reliable, the response should be predicted with a reasonable accuracy by the model when compared with the experimental data. The value of correlation coefficient, R^2 for Y1 is 0.9984, Y2 is 0.9546, and Y3 is 0.9689, respectively. Since these values are higher than 80%, it indicates that there is a good agreement between experimental data and predicted value. Therefore, the suggested models are valid for the present study, and it could be used to predict the three responses under temperature and *p*-xylene feed partial pressure.

The interaction effects of the temperature (A) and *p*-xylene feed partial pressure (B) for the three responses are presented by the three-dimensional response surface plot shown in Figure 26.2a–c. The permeation fluxes of *p*-xylene through silicalite-1 membrane did not change much. The temperature dependence of the flux for *p*-xylene was consistent with those separation results reported in the literature [10, 12]. The *p*-xylene flux is typical for molecular permeation in zeolite membrane controlled by an adsorption-diffusion mechanism. Under this mechanism, the increase in temperature reduces the surface coverage which lowers the driving force for mass transfer but is counterbalanced by higher values of the diffusion coefficient which increase the diffusivity of the adsorbed molecules [12]. The membrane is selective to *p*-xylene in the temperature range of 150–250°C with higher flux (one magnitude order) compared with the other two bulkier isomers (not shown). Increasing *p*-xylene feed partial pressure increased the permeation driving force and thus, increased the *p*-xylene flux (Fig. 26.2a). This observation was in agreement with the results reported by Gu et al. [12] and Xomeritakis et al. [13], because the low pressures result in low *p*-xylene flux.

The three-dimensional response surface plots for *p*-/*o*-xylene separation factor and *p*-/*m*-xylene separation factor are shown in Figure 26.2b,c, respectively. Due to identical kinetic diameters of *m*-xylene and *o*-xylene (6.8 Å), the interaction effects between temperature and *p*-xylene feed partial pressure on *p*-/*o*-xylene separation factor was similar as that on the *p*-/*m*-xylene separation factor. This observation shows that the transport behavior of *m*-xylene and *o*-xylene are similar. An optimum *p*-/*o*-xylene separation factor and *p*-/*m*-xylene separation factor was observed at 200°C and thereafter decreased as the temperature was further increased. This could be due to the drop in *p*-xylene flux with the increase in temperature (Fig. 26.2a). The temperature dependence of *p*-xylene separation was consistent with reported results [12].

As shown in Figure 26.2b,c, the *p*-/*m*-xylene separation and *p*-/*m*-xylene separation factors increased with the increase in *p*-xylene feed partial pressure. The decrease in separation factors was observed when the *p*-xylene feed partial pressure reached 0.18 kPa. This was due to the hardly distorted membrane pore structure that maintains the size exclusion effect for large molecules (*m*-xylene and *o*-xylene) at low *p*-xylene feed partial pressure. This result shows that the *p*-xylene loading level greatly influenced the shape-selectivity of *p*-xylene over *m*-xylene and *o*-xylene. *p*-Xylene sorption on the membrane caused distortion of the zeolite pores as the *p*-xylene feed partial pressure is further increased (Fig. 26.2a). This caused the increase in the rate of *m*-xylene and *o*-xylene penetration through the pores. Therefore, within the *p*-xylene feed partial pressure range of 0.18–0.26 kPa, *m*-xylene and *o*-xylene fluxes increase, and the separation factors decrease (Fig. 26.2b,c).

The optimum conditions preset at high and low level ranges for two process variables, temperature, and *p*-xylene feed partial pressure for separation

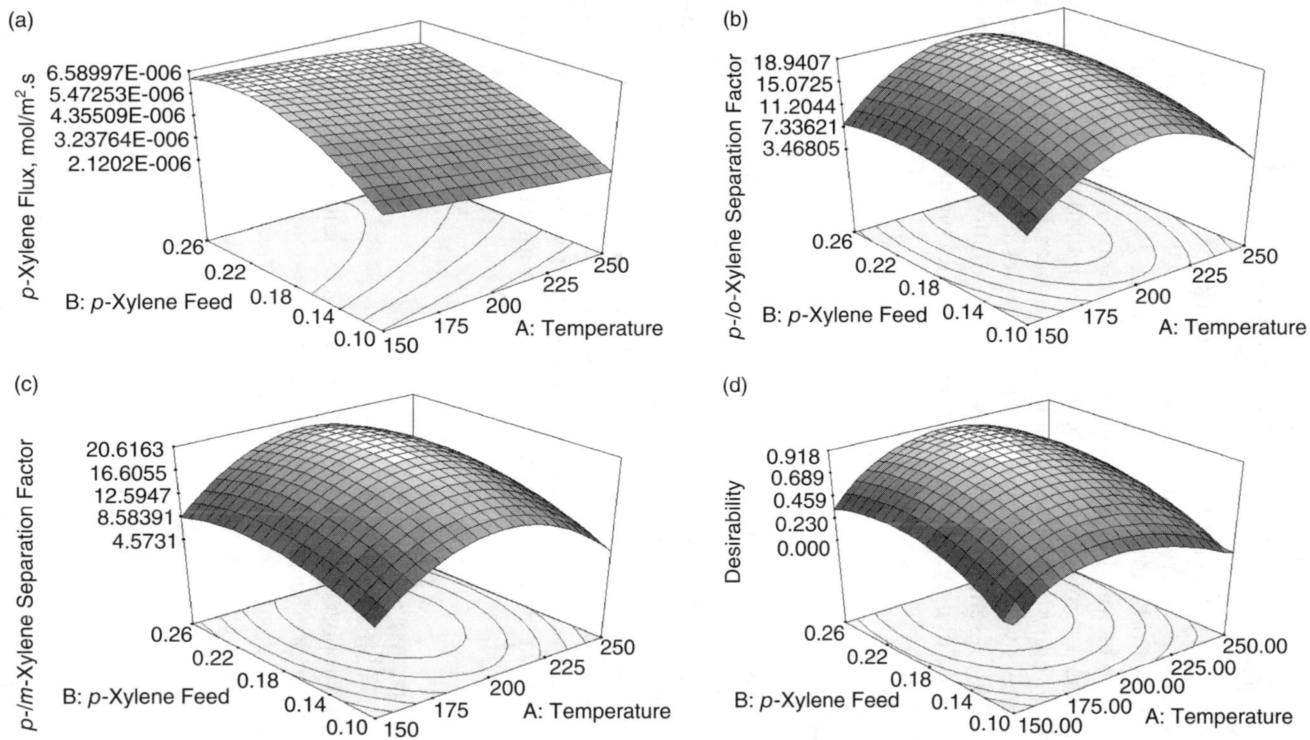

Figure 26.2. Effect of temperature and *p*-xylene feed partial pressure on the (a) *p*-xylene flux, (b) *p*-/*o*-xylene separation factor, and (c) *p*-/*m*–xylene separation factor, (d) desirability of optimum condition.

studies of ternary xylene mixture over silicalite-1 membrane were obtained using numerical optimization feature of Design Expert 6.0.6 Software. The main goal was set to optimize the *p*-xylene flux, *p*-/*o*-xylene separation factor, and *p*-/*m*-xylene separation factor, respectively. The optimum solution for the three responses was determined at temperature of 198°C and *p*-xylene feed partial pressure of 0.22 kPa, with *p*-xylene flux of 5.94×10^{-6} mol/m^2.s, *p*-/*o*-xylene separation factor of 19, and *p*-/*m*-xylene separation factor of 20, respectively. Figure 26.2d shows the desirability for the optimum solution with the combined desirability of 0.92. The result suggests near ideal operation of the separation process within the range of conditions tested. The DoE approach reported in the present work provides better flexibility for the membrane process separation of *p*-xylene in comparison with the conventional separation process study.

26.4 CONCLUSIONS

The optimum condition for the separation of *p*-xylene from its isomers over silicalite-1 membrane was obtained by DoE. The interactions between temperature and *p*-xylene feed partial pressure were investigated by CCD coupled with RSM. The optimum value of three responses was at temperature of 198°C and *p*-xylene feed partial pressure of 0.22 kPa, with *p*-xylene flux of 5.94×10^{-6} mol/m^2.s, *p*-/*o*-xylene separation factor of 19, and *p*-/*m*-xylene separation factor of 20, respectively. The simulated data from the proposed models were in agreement with the experimental data with an average value of ±2.90%.

ACKNOWLEDGMENT

The financial support provided by Ministry of Science, Technology and Environment under e-Science Fund Grant (Account No: 6013319), Universiti Sains Malaysia under Research University Grant (Account No: 811043) and Ministry of Higher Education under FRGS (Account No: 6070021) are duly acknowledged.

REFERENCES

[1] Tarditi A. M., S. Irusta, and E. A. Lombardo. Xylene isomerization in a membrane reactor: Part I: the synthesis of MFI membranes for the p-xylene separation. *Chem. Eng. J.* 122 (2006) 167–174.

[2] Nair S., Z. Lai, V. Nikolakis, G. Xomeristakis, G. Bonilla, and M. Tsapatsis. Separation of close-boiling hydrocarbon mixtures by MFI and FAU membranes made by secondary growth. *Micropor. Mesopor. Mater.* 48 (2001) 219–228.

[3] Lai Z. and M. Tsapatsis. Gas and organic vapor permeation through b-oriented MFI membranes. *Ind. Eng. Chem. Res.* 43 (2004) 3000–3007.

[4] Gump C. J., V. A. Tuan, R. D. Noble, and J. L. Falconer. Aromatic permeation through crystalline molecular sieve membranes. *Ind. Eng. Chem. Res.* 40 (2001) 565–577.

[5] Lai Z., G. Bonilla, I. Diaz, J. G. Nery, K. Sujaoti, M. A. Amat, E. Kokkoli, O. Terasaki, R. W. Thompson, M. Tsapatsis, and D. G. Vlachos. Micro structural optimization of a zeolite membrane for organic vapor separation. *Science* 300 (2003) 456–460.

[6] Nikolakis V., G. Xomeritakis, A. Abibi, M. Dickson, M. Tsapatsis, and D. G. Vlachos. Growth of a faujasite-type zeolite membrane and its application in the separation of saturated/unsaturated hydrocarbon mixtures. *J. Memb. Sci.* 184 (2001) 209–219.

[7] Keizer K., A. J. Burggraaf, Z. A. E. P. Vroon, and H. Verweij. Two component permeation through thin zeolite MFI membranes. *J. Memb. Sci.* 147 (1998) 159–172.

[8] Xomeritakis G., Z. Lai, and M. Tsapatsis. Separation of xylene isomer vapors with oriented MFI membranes made by seeded growth. *Ind. Eng. Chem. Res.* 40 (2001) 544–552.

[9] Wang Z. and Y. Yan. Controlling crystal orientation in zeolite MFI thin films by direct in situ crystallization. *Chem. Mater.* 13 (2001) 1101–1107.

[10] Lai Z., M. Tsapatsis, and J. P. Nicolich. Siliceous ZSM-5 membranes by secondary growth of b-oriented seed layers. *Adv. Funct. Mater.* 14 (2004) 716–729.

[11] Yeong Y. F., A. Z. Abdullah, A. L. Ahmad, and S. Bhatia. Process optimization studies of p-xylene separation from binary xylene mixture over silicalite-1 membrane using response surface methodology. *J. Memb. Sci.* 341 (2009) 96–108.

[12] Gu X., J. Dong, T. M. Nenoff, and D. E. Ozokwelu. Separation of p-xylene from multicomponent vapor mixtures using tubular MFI zeolite membranes. *J. Memb. Sci.* 280 (2006) 624–633.

[13] Xomeritakis G. and M. Tsapatsis. Permeation of aromatic isomer vapors through oriented MFI-type membranes made by secondary growth. *Chem. Mater.* 11 (1999) 875–878.

27

AMMONIA REMOVAL FROM SALINE WATER BY DIRECT CONTACT MEMBRANE DISTILLATION

ROSALAM SARBATLY AND CHEL-KEN CHIAM

27.1 INTRODUCTION

Ammonia is the most toxic compound found in aquaculture systems. Fish excrete unionized ammonia directly from their gills into the water, in an amount that depends on the species, body weight, water salinity, water temperature, feeding, and food rations [1–5]. Additionally, the fish excrete fecal solids as waste products of protein metabolism, which may undergo leaching to form finer solids and a dissolved ammonia by-product. The larger solids that reside and accumulate in a recirculating aquaculture system (RAS) will be mineralized faster than in a nonrecirculating system because of the pumping effect [6] and water turbulent forces [7] in this type of system. Therefore, the amount of ammonia in an RAS increases over time.

Most biological membranes are more permeable to unionized ammonia (NH_3) than to ionized ammonia (NH_4^+) [8]. Seawater species have been reported to be more sensitive to ammonia toxicity than freshwater species; the mean acute toxicity values for 32 freshwater species are 2.79 mg NH_3/L and 1.86 mg NH_3/L for 17 seawater species [9]. The concentration of ammonia in

Sustainable Membrane Technology for Energy, Water, and Environment, First Edition.
Edited by Ahmad Fauzi Ismail and Takeshi Matsuura.

the blood plasma of the fish increases when the ambient ammonia level in the water increases [10], leading to an increase in the mortality rate of the fish [11] and inhibition of juvenile growth [12]. Generally, biological treatments are primarily used to remove the ammonia from aquacultural water [13, 14]. However, a significant reduction in both nitrification and denitrification has been reported when the water salinity increased from 2% to 10% or was at a level higher than 20 g NaCl/L [15–19]. Even when salt-tolerant microbes are utilized, aerobic and anaerobic treatments are inhibited at salt levels greater than 3% [20, 21].

Ammonia is readily dissolved in water, where it exists as unionized ammonia and ionized ammonia. Dissolved NH_3 vaporizes from the water into the air at high temperatures, where it may have a direct impact on the environment as a greenhouse gas; at lower temperatures, the vaporized ammonia condenses and dissolves into the water. Even when salt-tolerant microbes are utilized, gas-liquid membrane separation is considered an alternative technique to remove ammonia from saline water. Ammonia is a volatile component, and only the volatile components can pass through the membrane; nonvolatile components, such as the salts, will be retained in the feed solution. Additionally, the irreversibility of the membrane separation does not allow the ammonia gas to redissolve into the water after separation has occurred. In this study, the effect of a sodium chloride additive during the fabrication of a polyvinylidene fluoride (PVDF) membrane and the effect of the resulting membrane porosity on ammonia removal from synthesized saline water by direct contact membrane distillation (DCMD) were investigated.

27.2 METHODS/THEORY

27.2.1 Membrane Fabrication

PVDF (typical M_r = 530,000), supplied by Aldrich, was used to fabricate the flat sheet membranes. The solvent, *N*,*N*-dimethylacetamide (DMAc), was supplied by Acros. Sodium chloride (NaCl), supplied by Lab-Scan, and distilled water were used as the dope solution additives, as shown in Table 27.1. The solution was mixed in a bottle and stirred on a stirring hot plate (IKA WERKE) at 50°C for about 6 hours to obtain a homogeneous solution. The dope solution

TABLE 27.1. Membrane Formulations and Pore Size

Membrane	PVDF/DMAc/ Water/NaCl (wt%)	$d_{avg} \pm \Delta d_{avg}$ (μm)	Median (μm)
PVDF0	10/90/–/–	–	–
PVDF1	10/88/2.00/0.00	0.101 ± 0.04	0.090
PVDF2	10/88/1.78/0.22	0.090 ± 0.03	0.082
PVDF3	10/88/1.67/0.33	0.103 ± 0.05	0.089

was then left for about 3 days to allow any bubbles to escape. Next, the dope solution was poured onto a smooth glass plate (15 × 15 cm) at room temperature (about 20–25°C), where it was cast using a casting knife. The casting thickness was controlled at 315 μm with adhesive tape placed at the sides of the glass plate. Immediately after the glass plate with the casting film was immersed in a coagulation bath containing 300 mL distilled water at 9–11°C, phase inversion occurred. The membranes were thoroughly washed with water and stored in a water bath. There were 10 membranes from each group of PVDF0, PVDF1, PVDF2, and PVDF3 selected and then air-dried at 55°C for about 2 hours. These membranes were characterized by porosity and pore size.

27.2.2 Membrane Porosity Measurement

The gravimetric method shown in Equation 27.1 was used to determine the membrane porosity, ε:

$$\varepsilon(\%) = \frac{100(m_w - m_d)/\rho_w}{(m_w - m_d)/\rho_w + m_d/\rho_p}, \tag{27.1}$$

where m is the mass, and ρ is the density. The subscripts d, w, and p indicate the dried membrane, wet membrane, and polymer, respectively. The value of ρ_p for PVDF is 1.74 g/cm^3.

27.2.3 Membrane Pore Size Measurement

A direct method using scanning electron microscopy (SEM) was used to measure the membranes' average pore size [22]. The membrane samples were immersed in methanol for 10 minutes. The sample was then coated with platinum before it was imaged to observe the pore sizes. The mean membrane pore size, d_{avg}, and the standard deviation (SD), Δd_{avg}, were statistically determined using Equation 27.2:

$$d_{avg} = \frac{\sum_{i=1}^{n} d_i f_i}{n} \quad \text{and} \quad \Delta d_{avg} = \sqrt{\frac{\sum_{i=1}^{n} f_i (d_i - d_{avg})^2}{n}}, \tag{27.2}$$

where n is the total number of pores of which the sizes have been measured, d_i is a midpoint of a pore size class occurring with a frequency, f_i.

27.2.4 Preparation of Feed Solution

Concentrated ammonia solution, GR (25%) Merck, the NaCl, and distilled water were used to prepare the feed solution. The NaCl was used without further treatment and purification to prepare the saline water. Initially, 35 g/L of NaCl was dissolved in the distilled water to produce the saline water which

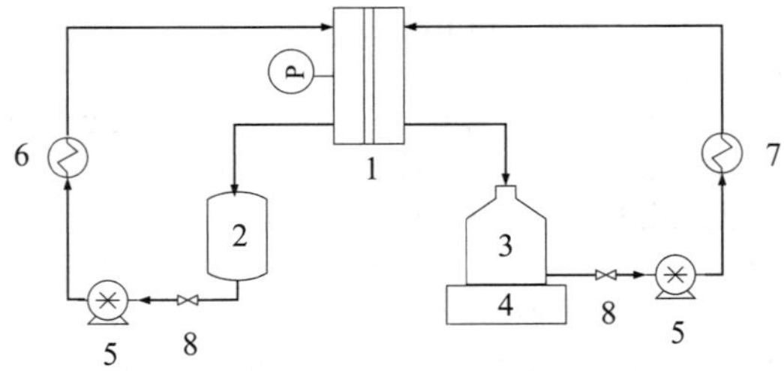

Figure 27.1. Schematic of the DCMD setup (1: DCMD module; 2: feed tank; 3: fluid condensing tank; 4: electronic balance; 5: gear pumps; 6: heater; 7: chiller; 8: valves).

mimicked the salinity of the seawater. A pipette was used to measure 0.2 mL of the ammonia solution which was dropped into the saline water. Then the solution was mixed in a stopped conical flask and stirred on the stirring plate at room temperature for 5 minutes. The salicylate method was used to measure the ammonia concentration in the solution which indicated 23 mg/L NH_3.

27.2.5 DCMD Experiment

The DCMD equipment is shown schematically in Figure 27.1. A membrane with an effective area of 56.40 cm^2 was placed into the DCMD module. The feed solution was pumped by a gear pump (driver model 75211-35, Cole-Parmer Instrument Co., Vernon Hills, IL) to pass through a heater (Ecoline 003/E100, Lauda, Germany) prior to entering the DCMD module, where it was in direct contact with the surface of the membrane. The feed solution was recirculated into the feed tank. In this study, distilled water used as the condensing fluid was pumped by a gear pump (driver model 75211-15, Cole-Parmer Instrument Co.) through another side of the membrane. The temperature of the chilled condensing fluid was controlled by a digital temperature controller (model 9101, PolyScience, Niles, IL) prior to entering the module in a cocurrent configuration; it was recirculated into the condensing fluid tank. An electronic balance was placed at the bottom of the condensing fluid tank.

During the DCMD experiment, the feed solution was heated to 50°C while the condensing fluid was cooled to 15°C. The cross-flow velocity of both fluids in the module was 0.210 cm/s. The experiment was run for about 2 hours. At time intervals of 20 minutes, 1 mL samples were taken from the feed tank, and the ammonia content was measured using the salicylate method. The efficiency of ammonia removal, E_a, was calculated by Equation 27.3 [23]:

$$E_a(\%) = \frac{100(c_{a,0} - c_{a,t})}{c_{a,0}}, \qquad (27.3)$$

where $c_{a,0}$ and $c_{a,t}$ are the equilibrium concentrations of NH_3 initially and at time t in the feed solution, respectively.

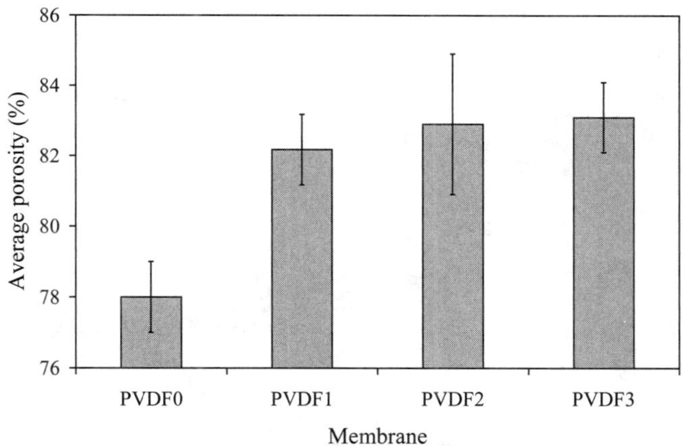

Figure 27.2. The effect of additives and sodium chloride concentration on the average membrane porosity.

27.3 RESULTS AND DISCUSSION

27.3.1 Effect of Sodium Chloride Additive

Variation of the sodium chloride additive concentration in the formulation induced changes in the membrane porosity; the porosity of PVDF0, PVDF1, PVDF2, and PVDF3 were 78.0, 82.2, 82.9, and 83.8%, respectively, as illustrated in Figure 27.2. The average porosity of membranes fabricated with additives, either distilled water or sodium chloride, increased from 78% when they were fabricated without additives to about 82–84%.

Figure 27.3 shows the effect of membranes with different formulations, that is, sodium chloride concentration on the distribution of pore sizes in the membranes. In the membranes fabricated with 2 wt % distilled water (PVDF1), about 22% of the pores, which was the greatest number of pores, have the sizes between 0.074 and 0.089 μm in the membranes. The addition of sodium chloride, either 0.22 wt % (PVDF2) or 0.33 wt % (PVDF3) has significantly reduced the frequency of pores to have the sizes between 0.074 and 0.134 μm. However, variation in the sodium chloride concentration affected the pore size distribution differently. The greatest frequency of the pores have the sizes between 0.059 and 0.074 μm which accounts for about 27 and 20% of the pores in the PVDF2 and PVDF3 membranes, respectively. The number of pores with sizes between 0.059 and 0.104 μm in the PVDF2 membrane was higher than that in the PVDF3 membrane, whereas the number of pores in the PVDF3 membrane is relatively higher than that in the PVDF1 and PVDF2 membranes for the pore sizes between 0.119 and 0.194 μm.

The mean membrane pore size and SD were statistically determined using Equation 27.2, and the values are shown in Table 27.1. The mean pore sizes for the PVDF1, PVDF2, and PVDF3 membranes were 0.101, 0.090, and

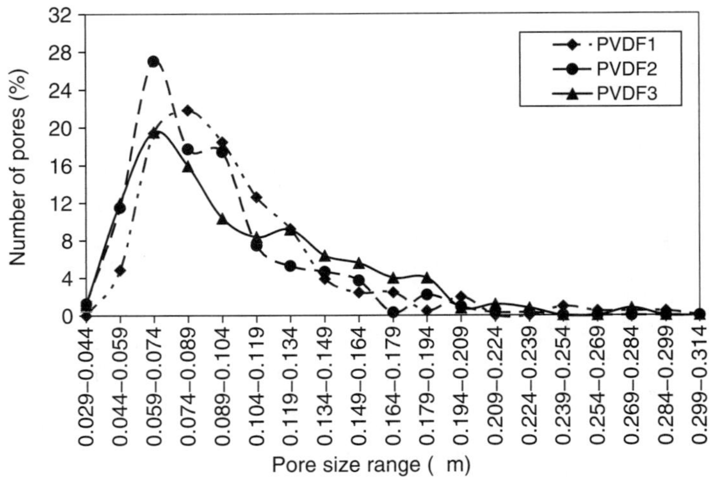

Figure 27.3. The effect of variation of NaCl concentration on membrane pore size distribution.

0.103 μm, respectively, with broad size distributions due to the outliers on the pore sizes data that are the pore sizes between 0.194 and 0.314 μm, resulting in large SD values of 0.04, 0.03, and 0.05, respectively. Therefore, the median would be better than the mean to describe the pore sizes data where the median pore sizes of the PVDF1, PVDF2, and PVDF3 membranes were 0.090, 0.082, and 0.089 μm, correspondingly.

27.3.2 Separation of Ammonia by DCMD

The pore sizes of the porous membranes that were fabricated in this study ranged from 0.029 to 0.314 μm. Molecular collisions are an important transport mechanism of a volatile component through a porous membrane. The molecules of the volatile component lose energy after every collision, resulting in an increase in the conductive energy loss and, as a consequence, a reduction in permeate flux through the membrane. Therefore, the use of a high porosity membrane is favorable to reduce heat loss by conduction because the thermal conductivity of gas/air is smaller than that of the membrane materials [24]. Thus, PVDF1, PVDF2, and PVDF3 membranes with high porosity were selected in this study. A membrane from each group of PVDF1, PVDF2, and PVDF3 membranes were selected where their porosities were approximately 82, 83, and 84%. Figure 27.4 illustrates the ammonia removal efficiency by the PVDF membranes during 2 hours of operation by DCMD. The efficiencies of the PVDF1, PVDF2, and PVDF3 membranes after 2 hours were approximately 14, 17, and 20%, respectively.

The ammonia removal efficiency in this study is quite low compared to values in the literature [23, 25]. We attribute this to the use of distilled water

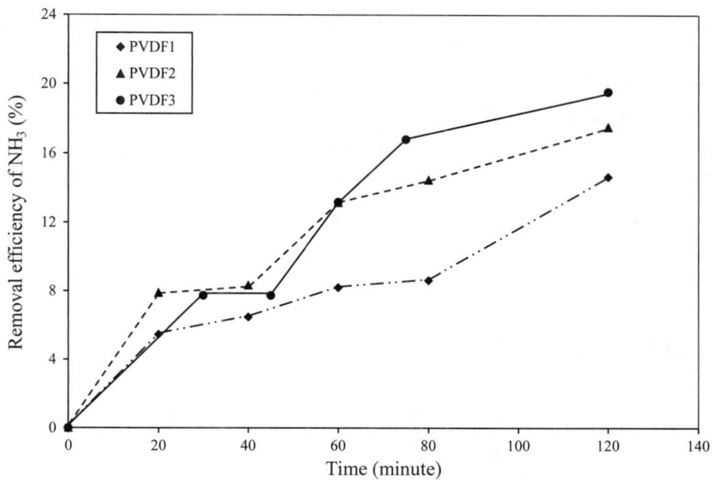

Figure 27.4. The effect of porosity on the removal efficiency of NH_3. (— · · —) PVDF1, porosity 82%; (– – – – -) PVDF2, porosity 83%; (————) PVDF3, porosity 84%.

for the condensing fluid instead of acidic solution, since ammonia/water affinity is lower than ammonia/acidic solution for the following reason. The chemistry of conversion between unionized ammonia and ionized ammonia in water is a reversible process:

$$NH_3 + H^+ \leftrightarrow NH_4^+$$

The amount of hydrogen ions present in an acidic solution is greater than in water. According to Le Châtelier's principle, the equilibrium of ammonia–ammonium shifts to right when there is an increase in the amount of hydrogen ions in the water. Thus, the required amount of NH_3 increases to reduce the number of hydrogen ions. Because of this phenomenon, the amount of ammonia gas absorbed into an acidic solution is greater than that absorbed into water.

27.4 CONCLUSIONS

A porous membrane has been successfully fabricated from PVDF with a sodium chloride additive. The additive increased the membrane porosity from 78% to 82–84%. For the membranes without the NaCl additive (PVDF1 membranes), the greatest frequency of the pores size between 0.074 and 0.089 μm occupied about 22% of the total pores, where the midpoint of the pore sizes was 0.082 μm. About 27 and 20% of the pores in the membranes with the NaCl additive, PVDF2 and PVDF3 membranes, respectively, have the sizes between 0.059 and 0.074 μm, which was the greatest frequency for the membranes. The midpoints of the pore sizes for PVDF2 and PVDF3 membranes were 0.067 μm. Clearly, the addition of additive in the membrane formulations has reduced the pore size. For the ammonia removal by DCMD experiment, the greater membrane porosity results in better ammonia removal efficiency. The efficiencies of the membranes which were having porosity of about 82, 83, and 84%, after 2 hours operation were approximately 14, 17, and 20%, respectively.

ACKNOWLEDGMENT

Financial support from Ministry of Science and Technology Innovation Malaysia (MOSTI) (SCF 0055-AGR-2008) is greatly appreciated.

REFERENCES

[1] Dosdat A., F. Servais, R. Métailler, C. Huelvan, and E. Desbruyères. Comparison of nitrogenous losses in five teleost fish species. *Aquaculture* 141 (1996) 107–127.

[2] Leung K. M. Y., J. C. W. Chu, and R. S. S. Wu. Effects of body weight, water temperature and ration size on ammonia excretion by areolated grouper (*Epinephelus areolatus*) and mangrove snapper (*Lutjanus argentimaculatus*). *Aquaculture* 170 (1999) 215–227.

[3] Crear B. J. and G. N. R. Forteath. Feeding has the largest effect on the ammonia excretion rate of the southern rock lobster, *Jasus edwardsii*, and the western rock lobster, *Panulirus cygnus*. *Aquacult. Eng.* 26 (2002) 239–250.

[4] Wu B. and S. Sun. Ammonia and urea excretion of the nemertean *Procephalothrix simulus* iwata: effects of salinity, temperature, body weight and amputation. *J. Exp. Mar. Biol. Ecol.* 337 (2006) 13–18.

[5] Merino G. E., R. H. Piedrahita, and D. E. Conklin. Ammonia and urea excretion rates of California halibut (*Paralichthys californicus*, Ayres) under farm-like conditions. *Aquaculture* 271 (2007) 227–243.

[6] McMillan J. D., F. W. Wheaton, J. N. Hochheimer, and J. Soares. Pumping effect on particle sizes in a recirculating aquaculture system. *Aquacult. Eng.* 27 (2003) 53–59.

[7] Brinker A. and R. Rösch. Factors determining the size of suspended solids in a flow-through fish farm. *Aquacult. Eng.* 33 (2005) 1–19.

[8] Eddy F. B. Review paper—ammonia in estuaries and effects on fish. *J. Fish Biol.* 67 (2005) 1495–1513.

[9] Randall D. J. and T. K. N. Tsui. Ammonia toxicity in fish. *Mar. Pollut. Bull.* 45 (2002) 17–23.

[10] Lemarié G., A. Dosdat, D. Covès, G. Dutto, E. Gasset, and J. Person-Le Ruyet. Effect of chronic ammonia exposure on growth of European seabass (*Dicentrarchus labrax*) juveniles. *Aquaculture* 229 (2004) 479–491.

[11] Kir M., M. Kumlu, and O. T. Eroldoğan. Effects of temperature on acute toxicity of ammonia to Penaeus semisulcatus juveniles. *Aquaculture* 241 (2004) 479–489.

[12] Reddy-Lopata K., L. Auerswald, and P. Cook. Ammonia toxicity and its effect on the growth of the South African abalone *Haliotis midae* linnaeus. *Aquaculture* 261 (2006) 678–687.

[13] Van Rijn J., Y. Tal, and H. J. Schreier. Denitrification in recirculating systems: theory and applications. *Aquacult. Eng.* 34 (2006) 364–376.

[14] Davidson J., N. Helwig, and S. T. Summerfelt. Fluidized sand biofilters used to remove ammonia, biochemical oxygen demand, total coliform bacteria, and suspended solids from an intensive aquaculture effluent. *Aquacult. Eng.* 39 (2008) 6–15.

[15] Aloui F., S. Khoufi, S. Loukil, and S. Sayadi. Performances of an activated sludge process for the treatment of fish processing saline wastewater. *Desalination* 246 (2009) 389–396.

[16] Dincer A. R. and F. Kargi. Salt inhibition kinetics in nitrification of synthetic saline wastewater. *Enzyme Microb. Technol.* 28 (2001) 661–665.

[17] Dincer A. R. and F. Kargi. Salt inhibition on nitrification and denitrification in saline wastewater. *Environ. Technol.* 20 (1999) 1147–1153.

[18] Rysgaard S., P. Thastum, T. Dalsgaard, P. B. Christensen, and N. P. Sloth. Effects of salinity on NH_4^+ adsorption capacity, nitrification, and denitrification in Danish estuarine sediments. *Estuaries* 22 (1999) 21–30.

[19] McAdam E. J. and S. J. Judd. Biological treatment of ion-exchange brine regenerant for re-use: a review. *Sep. Purif. Technol.* 62 (2008) 264–272.

[20] Dincer A. R. and F. Kargi. Performance of rotating biological disc system treating saline wastewater. *Process Biochem.* 36 (2001) 901–906.

[21] Kapdan I. K. and B. Erten. Anaerobic treatment of saline wastewater by *Halanaerobium lacusrosei*. *Process Biochem.* 42 (2007) 449–453.

[22] Sakai K. Review-determination of pore size distribution: 2. Dialysis membranes. *J. Memb. Sci.* 96 (1994) 91–130.

[23] Zhu Z., Z. Hao, Z. Shen, and J. Chen. Modified modeling of the effect of pH and viscosity on the mass transfer in hydrophobic hollow fiber membrane contactors. *J. Memb. Sci.* 250 (2005) 269–276.

[24] Lawson K. W. and D. R. Lloyd. Membrane distillation. *J. Memb. Sci.* 124 (1997) 1–25.

[25] Tan X. and S. P. Tan. Polyvinylidene fluoride (PVDF) hollow fiber membranes for ammonia removal. *J. Memb. Sci.* 271 (2006) 59–68.

INDEX

Sustainable Membrane Technology for Energy, Water, and Environment, First Edition.
Edited by Ahmad Fauzi Ismail and Takeshi Matsuura.